AF862928

Progress in Medical Imaging

Vernon L. Newhouse
Editor

Progress in Medical Imaging

With 118 Illustrations, 8 in Full Color

Springer-Verlag New York Berlin Heidelberg
London Paris Tokyo

Vernon L. Newhouse
Biomedical Engineering and Science Institute
Drexel University
Philadelphia, PA 19104 USA

Library of Congress Cataloging-in-Publication Data
Progress in medical imaging.
Includes bibliographies.
1. Diagnostic imaging. I. Newhouse, Vernon L.
[DNLM: 1. Diagnostic Imaging. WN 200 P964]
RC78.7.D53P78 1988 616.07'57 88-4625

©1988 by Springer-Verlag New York Inc.
All rights reserved. This work may not be translated or copied in whole or in part without the written permission of the publisher (Springer-Verlag, 175 Fifth Avenue, New York, NY 10010, USA), except for brief excerpts in connection with reviews or scholarly analysis. Use in connection with any form of information storage and retrieval, electronic adaptation, computer software, or by similar or dissimilar methodology now known or hereafter developed is forbidden.
The use of general descriptive names, trade names, trademarks, etc. in this publication, even if the former are not especially identified, is not to be taken as a sign that such names, as understood by the Trade Marks and Merchandise Marks Act, may accordingly be used freely by anyone.

Typeset by David E. Seham Associates Inc., Metuchen, New Jersey.
Printed and bound by Arcata Graphics/Halliday, West Hanover, Massachusetts.
Printed in the United States of America.

9 8 7 6 5 4 3 2 1

ISBN 0-387-96713-3 Springer-Verlag New York Berlin Heidelberg
ISBN 3-540-96713-3 Springer-Verlag Berlin Heidelberg New York

Preface

This book is first addressed to engineers and physicists engaged in research, development, or application of medical imaging, and to students intending to enter any of these fields. It is, however, also addressed to workers in nonmedical imaging, including radar, underwater sound, nondestructive testing, robotics, machine vision, and seismology. Finally, the book is addressed to physicians and other users of medical imaging who wish to deepen their understanding of the various modalities used in their field.

The problems of keeping abreast with the veritable flood of new literature in one's own field of imaging are too obvious for comment, and the problems of keeping up with new developments in adjacent fields seem wellnigh insoluble. As a result, the worker in a specific imaging field often fails to become aware of new developments in other imaging fields that could have been useful or that might have caused him to change his approach if he had been aware of them in time.

It is with the hope of ameliorating these problems that this book has been created. This book contains a collection of interdisciplinary reviews of subtopics in medical imaging written by internationally known experts. In contrast to some overly specialized reviews, the chapters of this volume are intended to be sufficiently interdisciplinary to be comprehensible to newcomers to the subject and to workers in other fields.

Since even the subfield of imaging devoted to medicine is itself vast, a decision had to be made when planning this work as to whether or not to concentrate on one subfield of medical imaging. From the specialist's point of view, concentration on one subfield such as x-rays is advantageous because complete coverage of the area can then be attempted, although there is often considerable duplication between the different chapters. However, since the intent of this book is to encourage *interdisciplinary cross-fertilization,* it was decided to draw on writers right across the medical-imaging spectrum and thus avoid weighting the volume in favor of one particular medical-imaging modality. Hence the articles in this book are designed to acquaint the reader with important developments in several very different subfields of medical imaging and have been written in a manner simple enough to make this easy and pleasant.

The book begins with a thorough introduction and review, by Preston and Bartels, of the automatic recognition of cells and tissue types in light microscopy. This chapter may be said to represent the contributions of pattern recognition to this book. The next chapter by Herman, Trivedi, and Udupa deals with the computerized manipulation and assembly of two-dimensional scans of an organ into images of the three-dimensional organ that can be rotated in space and viewed from different directions. Their techniques are applied to x-ray tomograms and are shown to produce images of remarkable "plasticity" that are useful not only in medical diagnosis but also in planning surgery. The third chapter, by Sorenson and Niklason, describes techniques for reducing the image degradation produced by scattered radiation in chest radiography and demonstrates remarkable improvements in image quality. This chapter, though also concerned with x-rays, differs from the preceding chapter in that hardware rather than software techniques are used for image enhancement. The fourth chapter of the book, by Karp and Muehllehner, reviews the basic principles of positron-emission tomography and describes the recent advances in instrumentation, including the development of detector coding schemes in high-resolution scanners and the relative benefits of using time-of-flight information. This modality is of particular importance in research, since it is able, using substances that mimic natural metabolites such as deoxyglucose, to provide otherwise unobtainable information on organ function. Chapters 5 through 9 by the editor and his collaborators, Cathignol and Chapelon, describe the advantages of pseudo-random codes as transmitted signals for ultrasonic flow measurement, imaging, and medium characterization. These chapters are both hardware- and communication-theory-oriented and are believed to be the first published review of these topics.

In conclusion, it is a pleasure to thank Eric Saunders for his editorial assistance in the preparation of this book.

Drexel University, May 1988 V.L. NEWHOUSE

Contents

Contributors

PETER H. BARTELS
Optical Sciences Center, University of Arizona, Tucson, Arizona 85721, USA

D. CATHIGNOL
Inserm, Unité 281, 151 Cours Albert Thomas, 69003 Lyon, France

J.Y. CHAPELON
Inserm, Unité 281, 151 Cours Albert Thomas, 69003 Lyon, France

GABOR T. HERMAN
Medical Image Processing Group, Department of Radiology, University of Pennsylvania, Philadelphia, Pennsylvania 19104, USA

JOEL S. KARP
Department of Radiology, University of Pennsylvania, Philadelphia, Pennsylvania 19104, USA

GERD MUEHLLEHNER
Department of Radiology, University of Pennsylvania, Philadelphia, Pennsylvania 19104, USA

VERNON L. NEWHOUSE
Biomedical Engineering and Science Institute, Drexel University, Philadelphia, Pennsylvania 19104, USA

LOREN T. NIKLASON
Department of Radiology, University of Alabama at Birmingham, Birmingham, Alabama 35294, USA
Department of Radiology, University of Michigan Hospital, Ann Arbor. Michigan 48109, USA

KENDALL PRESTON, JR.
Department of Electrical and Computer Engineering, Carnegie-Mellon University, Pittsburgh, Pennsylvania 15213, USA

JAMES A. SORENSON
Lunar Radiation Corporation, 313 West Beltline Highway, Madison, Wisconsin 53713, USA

SUSHMA S. TRIVEDI
Resonex Incorporated, 720 Palomar Avenue, Sunnyvale, California 94086, USA

JAYARAM K. UDUPA
Medical Image Processing Group, Department of Radiology, University of Pennsylvania, Philadelphia, Pennsylvania 19104, USA

1
Automated Image Processing for Cells and Tissue

KENDALL PRESTON, JR., AND PETER H. BARTELS

Introduction

After four decades of research, automation of certain aspects of the visual inspection of both cells and tissue has become a reality. This chapter not only reviews the current status of the automation of image analysis in cytology and histology, but also presents a historical account of this field extending from its inception.

Levels of Automation

Automation exists on several levels. There is *full automation* where operator interaction with the computer vision system is unnecessary except, perhaps, for loading and unloading the specimen. There is *partial automation* wherein the operator is alerted by the computer vision system to assist when objects are encountered that the system cannot classify. There is *task-specific automation* with continuous operator interaction in defining regions of interest, while the computer vision system automatically conducts preprogrammed feature extraction and/or classification tasks.

Typical of full automation was the early red blood cell counter by Casella (Great Britain) as well as modern bacterial colony counters. The robot microscopes for use in the analysis of images of human white blood cells, red blood cells, and platelets are typical of partial automation. Here, interaction with the technologist is necessary only when abnormal cells are encountered by the machine.

Task-specific automation is exemplified by the early Quantimet machine (Great Britain) wherein the user first converted a gray level image to a bilevel image by means of thresholding and then used the machine to automatically generate a list of object measures (area, perimeter, major axes, etc) either for the entire field of view or for specific regions of interest. In other machines, now made by many companies, a light pen or stylus/tablet combination is employed by the user to conduct "image editing" prior to measurement extraction by the machine.

This chapter excludes any discussion of what might be called "image processing" where the user employs relatively unsophisticated electronic and mechanical aids. Examples range from object outlining using such tools as the "camera lucida," point photometric measurements of optical density carried out by microspectrophotometry, and the use of point and intercept counting employing various grids and reticles designed for that purpose. This is not to imply that these aids are not important to research in the fields of cytology and histology, but merely that the methods employed are considered by the authors of this chapter to fall far outside the realm of true automation.

Today's Working Environment

The field of diagnostic cytology and histology, ie, the visual analysis of human cells and tissues, still relies on a vast number of individuals who perform many hundreds of millions of visual examinations per year without the assistance of automation (the total U.S. work force in this field is estimated to be of the order of 100,000 pathologists and technologists). These examinations are a vital part of national health screening, diagnosis, and treatment. Gross expenditures associated with this activity is on the order of $5 to $10 billion per year, ie, several percent of the United States annual health budget. The effort expended in this work far exceeds that expended in any other visual-image-analysis task, such as the inspection of photographs taken by civilian Earth resources and weather satellites as well as by military reconnaissance systems.

The hematopathologist or technologist who performs these examinations mentally catalogs the shape, size, texture, and color of the cells and tissues observed for the purpose of categorizing the types of structures found in the specimen and, of course, searching for departure, if any, from normality. For example, the average worker performing blood cell analysis observes thousands of cells per day and during the visual examination, keys data into either simple mechanical counters or into the computer. In some institutions manual transcription of the counts onto the patient's record is performed whereas in others a machine printed record is produced by a robot microscope. The visual blood cell examination is tedious. It therefore is often performed poorly. As reported by Sencer [1967], 40% of the laboratories examined by the National Communicable Disease Center (now the Center for Disease Control) did unsatisfactory visual blood-smear examinations in the 1960s. Similarly, the American College of Obstetricians and Gynecologists states that in visual examinations of the Papanicolaou (PAP)-stained cervical smear, 20% to 40% of abnormal smears are falsely classified as normal. Yet, the outcome of such examinations are used by physicians as an indicator of the patient's state of health and this information thus influences the overall treatment plan. Clearly, there is a need for automation. Tissue examinations lead either to the manual transcription of notes by the pathologist or to the typewritten transcription of dictated comments. So far, no robot microscopes are available to add data to the pathology transcript.

Automated Inspection Methods

The automation of blood-cell-image analysis produces more quantitative results than visual examination by replacing the eye-brain system with the combination of a television scanner and computer and usually leads to improved diagnosis and treatment. The human eye-brain system is still almost universally used to analyze other types of cells and tissue and performs feats that no computer program has yet attained or, some believe, is ever likely to attain. On the other hand, research over the past 20 years has provided evidence that computer analysis of digitized imagery of cells and tissues can produce clear-cut diagnostic discrimination on the basis of barely visible or not even perceived image properties, as discussed below. At present, image-analyzing engineers are working to design systems that form a partnership between the technologist or the pathologist and the automated microscope.

Much of the success of this work depends critically on the engineer's understanding of the clinical problem and of the full diagnostic objectives. There must be appropriate application of the image-analytic procedures, of the techniques of pattern recognition, and, most importantly for the future, of the proper "understanding" of the information offered by diagnostic imagery to a machine-vision system. There must be recognition of the value of expert system design for the analysis of imagery of cells and tissues.

Sample Preparation

Because of the minute size of the structures being examined, high-resolution microscopy is required for the examinations of cells and tissue. The automated microscope must maintain the specimen in a single focal plane for accurate observation. Cell-containing body fluids must be spread as a uniform thin film on a glass substrate. Tissue samples must be rendered as a thin slice mounted similarly. Furthermore, a significant part of the preparation of the specimen is the addition of chromatic labels by means of staining with various organic chemicals. The way in which the specimen is mechanically and chemically processed prior to examination by the automated microscope is of the utmost importance in achieving repeatable and reliable results through automation.

Historical Perspective

The uninitiated reader may feel that the application of computer vision to the microscopic examination of cells and tissues is a new and only recently developed field. This is far from the truth. An important aspect of this chapter is to provide the reader with a survey of the field so that the many facets of its gradual evolution may be understood.

The late 1940s and early 1950s saw the evolution of important work in the utilization of the television microscope for quantitative microscopy. Television microscopes were built and their output connected to relatively simple pulse-height and pulse-length analyzers. At this time—more than three decades ago—simple structures, such as red blood cells smeared on microscope slides, could be counted and sized at full video rates. This was initially reported in 1953 by Causley and Young [1953].

In the 1960s transistorization led to the same dramatic improvements in computer systems for automated microscopy as it did in general-purpose computation. Research systems were built in several laboratories whose performance, for some tasks, approached or exceeded that of the human technologist.

The 1970s saw a new era that began with the announcement of the first commercial computerized microscopes that automated the clinical examination of human blood cells. Others were offered by numerous companies as general-purpose systems for use in research. This trend has continued into the 1980s and, owing to the dramatic decreases in cost made possible by integrated circuitry, task-specific automation has spread into hundreds of research laboratories. Full and partial automation, however, have not prospered, owing to the difficulty of automating clinical visual tasks in a cost-effective manner.

For the person who wishes to avoid reading the thousands of articles published to date, several valuable surveys are recommended. For an interesting historical perspective on the early stages of television microscopy, the general surveys by Parpart [1951], Flory [1959], and Berkley [1959] are pertinent. Other important compendia were published in the late 1960s, namely, those edited by Montgomery [1962], Ramsey [1968], and Tolles [1969]. These covered work in image analysis for chromosomes, bacterial colonies, and human white blood cells, and described systems for monitoring the airborne bacteria used in bacterial warfare. Besides these compendia, certain individual authors reviewed the field. They are Bushor [1961], Shackel [1963], Eaves [1967], Neurath [1971], and Ledley and Ruddle [1966].

In the 1970s Preston and Rutovitz [1971] published a survey as part of the proceedings of the Conference on Multiphasic Health Testing held in Davos, Switzerland, by the Engineering Foundation of New York. Mellecker [1972] edited a collection of papers extracted from the 1973 Engineering Foundation Research Conference on Automatic Cytology and further compendia from subsequent Engineering Foundation conferences were edited by Mayall [1974, 1976, 1977, 1979]. A survey of digital picture analysis citing 410 references was written by Preston [1976]. Books were published in both Europe and the United States by Evans [1970], Wied et al [1970a], Simon et al [1975], Koss and Coleman [1979], Preston and Onoe [1976], Chermant [1978], and Pressman and Wied [1979].

The 1980s have seen further books by Onoe et al [1980] and Baak [1983], as well as a new review by Preston [1985], and a "symposium by correspondence" that began publication with the March 1985 issue of *Analytical and Quantitative Cytology*.

The sections that follow expand on the previous surveys. Progress is described decade by decade. Note, however, that the year 1979, which was the founding year for the first journal in the field, *Analytic and Quantitative Cytology*, is placed with the 1980s.

THE 1950S

The television scanner initiated research on electronic image analysis for both cells and tissues. Initially, the cathode ray tube (CRT) was used to produce "flying-spot" television microscope scanners where the bright spot that was formed on the face of the CRT by the electron beam was imaged onto the microscope slide. This image was raster-scanned over the slide, transmitted light was detected by a photomultiplier, and, in early video-processing circuits, descriptors were extracted from the video signal often using vacuum-tube electronics. In these early years, mechanical Nipkow-disk television scanners were also employed to generate a raster scan as in the primitive television systems of the 1930s. Young and Roberts [1952] produced the first flying-spot scanner capable of size analysis (see also Walton, 1952; Hawksley et al, 1954; and Causley and Young, 1953). The first high-precision photometric scanning was perfected by Caspersson et al [1955], which led to the development of the Universal Microspectrophotometer, model 1 (UMSP1) by Zeiss [1964].

DNA/RNA Photometry

Measurement of optical absorption by cells in the ultraviolet made possible the quantitative measurement of the major nuclear constituents, namely, deoxyribonucleic acid (DNA) and ribonucleic acid (RNA). Today's work in cytophotometry followed the scientific groundwork laid by Caspersson [1950] in these early years. Land et al [1949] and Zworykin and Hatke [1957] produced "color-translating" microscopes generating the world's first "pseudocolor" images by using color television monitors to display information at ultraviolet wavelengths invisible to the human eye. Other research on photometry is described by Ramberg [1958], Pollister and Moses [1949], Moses [1952], Skeggs [1957], Barer [1957], Williams [1959], Montgomery and Bonner [1959], and Barer and Wardley [1961].

Video-Signal Analysis

When objects of interest appear on the microscope slide at high contrast, boundaries are well defined. The video signal appears as a series of pulses as the flying spot crosses from light to dark regions. If the objects on the microscope slide are small, there are many short pulses; if large, many long pulses, etc. Thus early work on video-signal analysis used pulse-length analyzers to quantitate this information. By adding pulse-height analysis "gray level" information was also quantitated. Although such measurements are relatively crude, the special-purpose electronic circuits required are simple and video-rate-image analysis was feasible. Great Britain was the center of much of this early work as described by Coates [1956] and Taylor [1959].

Berkley [1959] first described an ingenious circuit for pulse analysis called "intercept association," which combined data from two scanning lines so as to use information on connectivity. Pulses generated on one video scan were stored in a "delay line" and recalled in synchronization with pulses produced during the next line scan. Berkley's method permitted the association of sequential intercepts of the same object. By comparing the time of occurrence of intercepts from pairs of scan lines, it could be determined whether intercepts were of separate objects or from the same object. Using intercept association, Mansberg [1957] was able to produce video-rate circuitry for bacterial colony counting. One should note that such circuitry generates errors when objects have complex (convoluted) shapes. Sometimes multiple counts are generated for a single object. Still, intercept association circuits formed the basis for commercial image-analyzing microscopes produced in the 1960s. Furthermore, by accumulating pulse-height information on the individual objects defined by intercept association, it is possible to measure both object area and optical density. This can also be carried out as a function of the wavelength of illumination.

The Cytoanalyzer Project

In the 1950s Mellors et al [1952] reported that cancer cells in PAP-stained cervical smears could apparently be distinguished from normal cells by measurement of nuclear size and optical density. This discovery led Tolles [1955] of the Airborne Instruments Laboratory to launch a multimillion dollar program to develop a fully automated microscope system for screening cervical smears. This so-called "Cytoanalyzer" system incorporated automatic slide feed and focus and circuitry, which generated a two-dimensional histogram of nuclear size versus nuclear optical density, as described by Horvath et al [1956] and Tolles and Bostrom [1956]. A manufacturing prototype of the Cytoanalyzer as well as production versions of some early commercial automated microscopes are shown in Figure 1.1. The Cytoanalyzer program proved to be overambitious and was destined to fail in the 1960s 6 years after its inception. The reasons for its demise lay in the failure of the engineers involved to recognize problems inherent in sample preparation and difficulties involved in conducting complex pattern-recognition tasks by special-purpose computer electronics.

Color Plate I

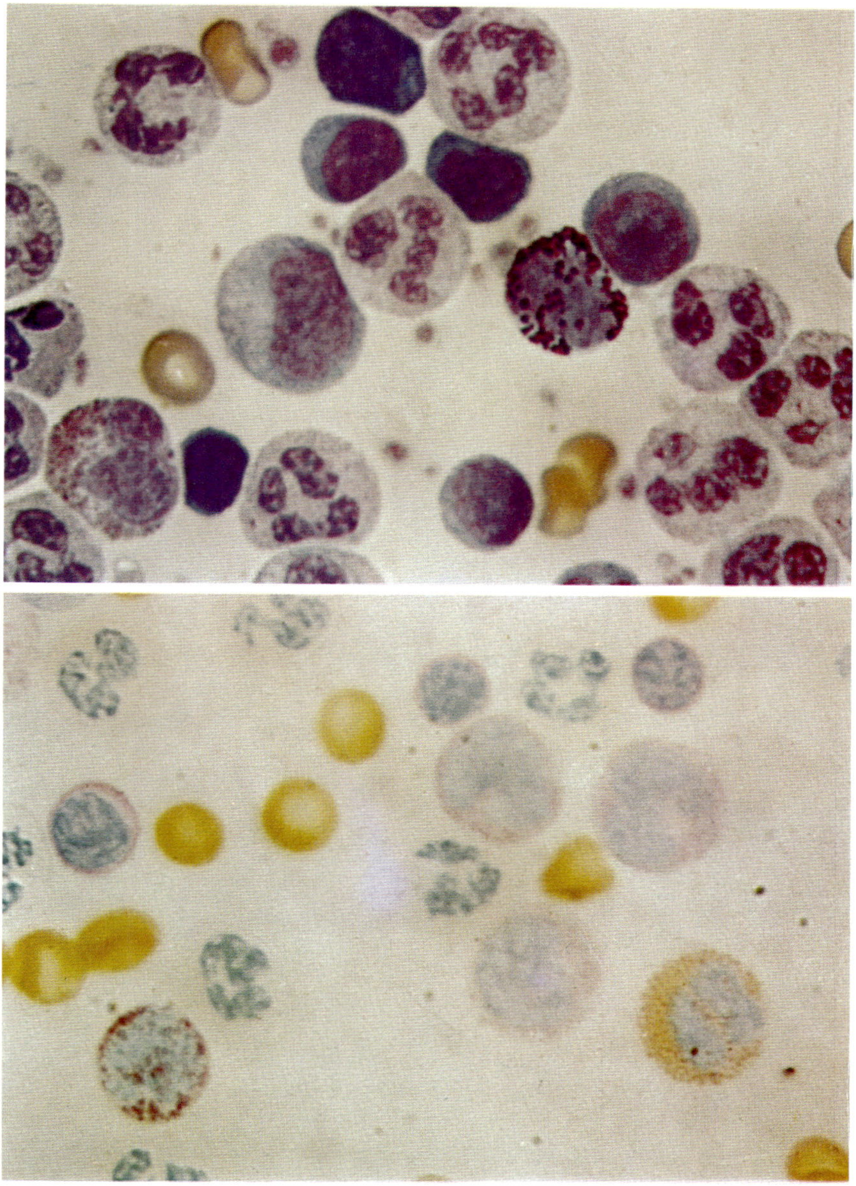

FIGURE 1.5. Stained-buffy-coat smears of human blood showing (*top*) the standard Wright-Giemsa stain and (*bottom*) the trichrome stain developed specifically for machine reading of human blood smears by Ornstein and Ansley [1974]. For further discussion, see page 26.

FIGURE 1.6. The "brown revolution" occurred when immunoperoxidase stains were introduced in the 1970s. The images above are of human breast and colon tissue showing both negative and positive interactions using an immunoperoxidase stain specific to estrogen receptor. (Images produced by the research diff3/50 at Carnegie-Mellon University from samples furnished by the Eastern Oncology Group at the University of Wisconsin.) For further discussion, see page 27.

FIGURE 1.8. (*Top*) Traditional television images generated by CELLSCAN/GLOPR mirror scanner showing electronic images produced in **(a)** yellow and **(b)** green and their associated optical transmission histograms. Note the trimodality in **(c)** where the modes correspond to nucleus, cytoplasm, red cells and background plasma, respectively. The cell under study is a hematoxylin-eosin-stained neutrophil in a human blood smear produced by spinning. (*Bottom*) Full-color mosaic by multiresolution imagery produced by the research diff3/50 showing scenes from a needle biopsy of human liver tissue varying from a field size of 5,000 × 20,000 μm (left) to 170 × 170 μm (upper right). For further discussion, see page 30.

Color Plate III

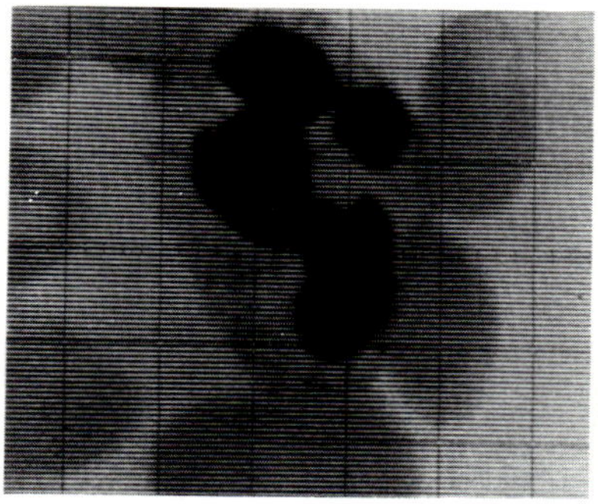

(a) Yellow Image

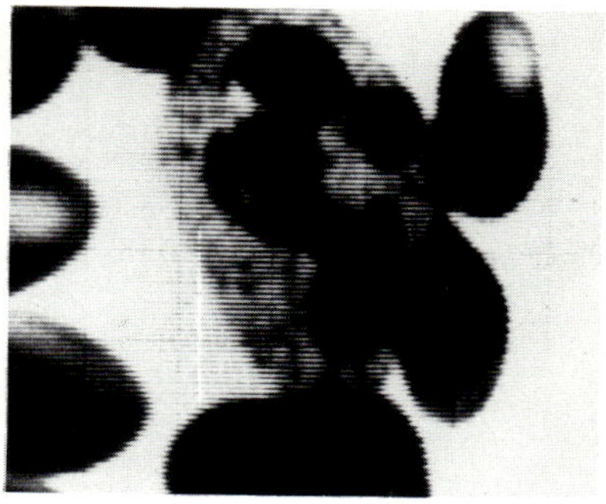

(b) Green Image

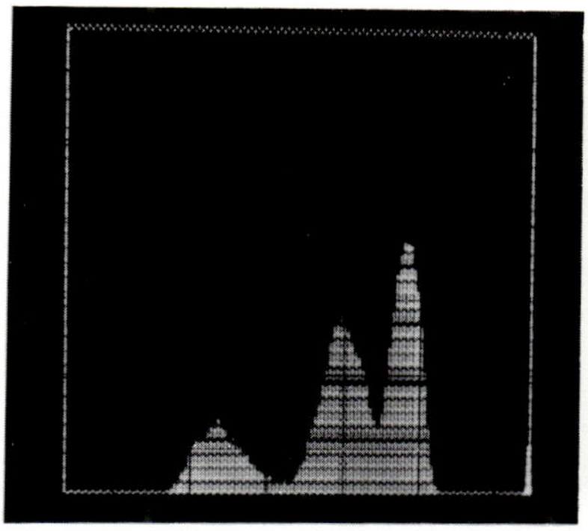

(c) Yellow Histogram

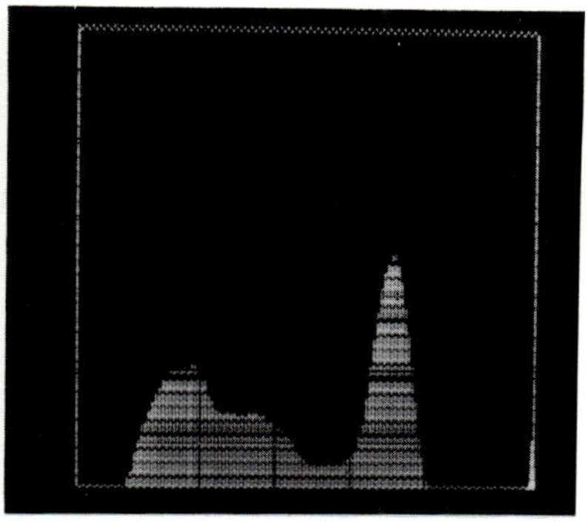

(d) Green Histogram

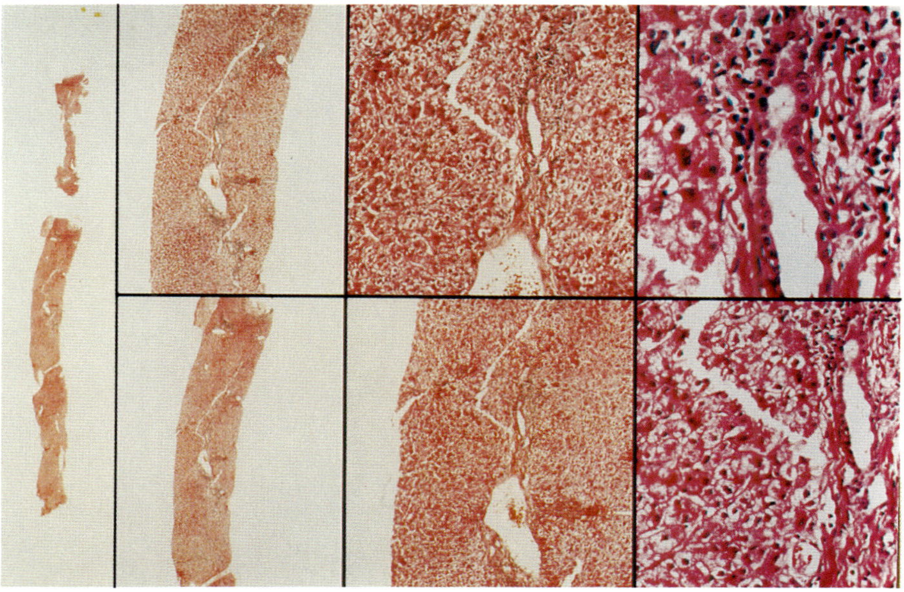

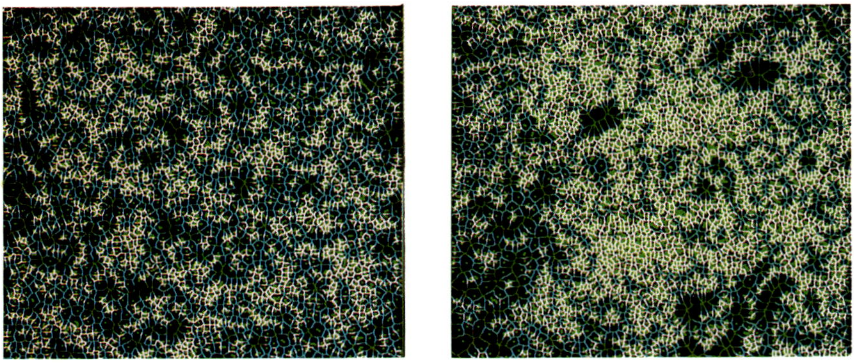

FIGURE 1.13. Exoskeletons generated using the (median − 1) cellular logic filter for both normal (*left*) and abnormal (*right*) 512 × 512 images of human kidney tissue. White indicates the portions of the arcs of the exoskeletons formed in the initial stages of the iterative algorithm; green, in the intermediate stages; blue, in the final stages. The exoskeleton arc length histogram readily identifies the normal or quantitates the degree of abnormality of the tissue architecture. For further discussion, see page 41.

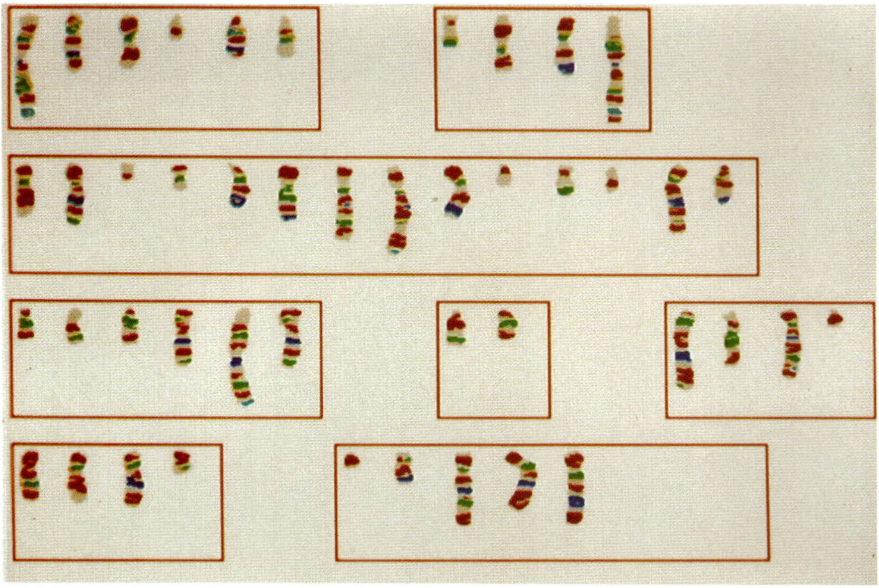

FIGURE 1.17. Chromosomes color-coded according to the size, location, and optical density of each band [TK Ten Kate, 1985]. For further discussion, see page 56.

Color Plate V

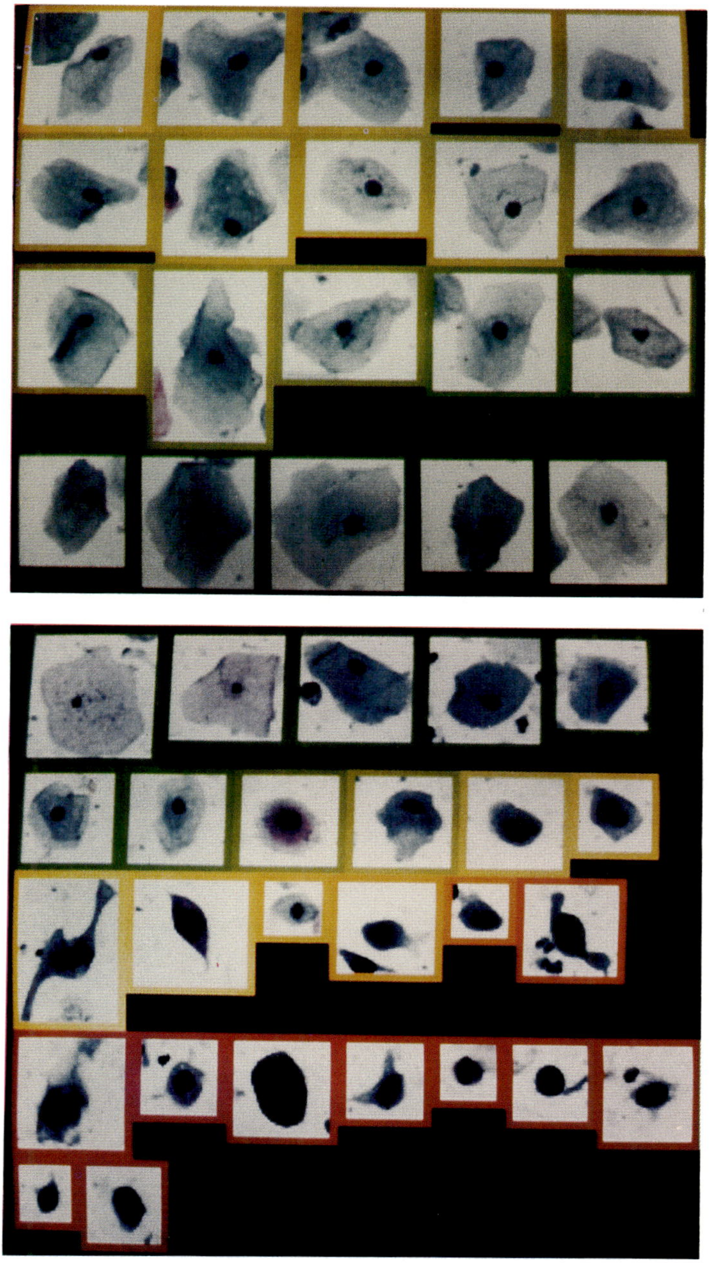

FIGURE 1.20. By ordering cell images on a color-graphics computer display, it is possible to illustrate for the observer the subtle colorimetric, morphologic, and textural properties that take place over the range of normal to malignant. *Top* shows normal-appearing intermediate cells; *bottom*, abnormal cells from the ectocervical epithelium. For further discussion, see page 71.

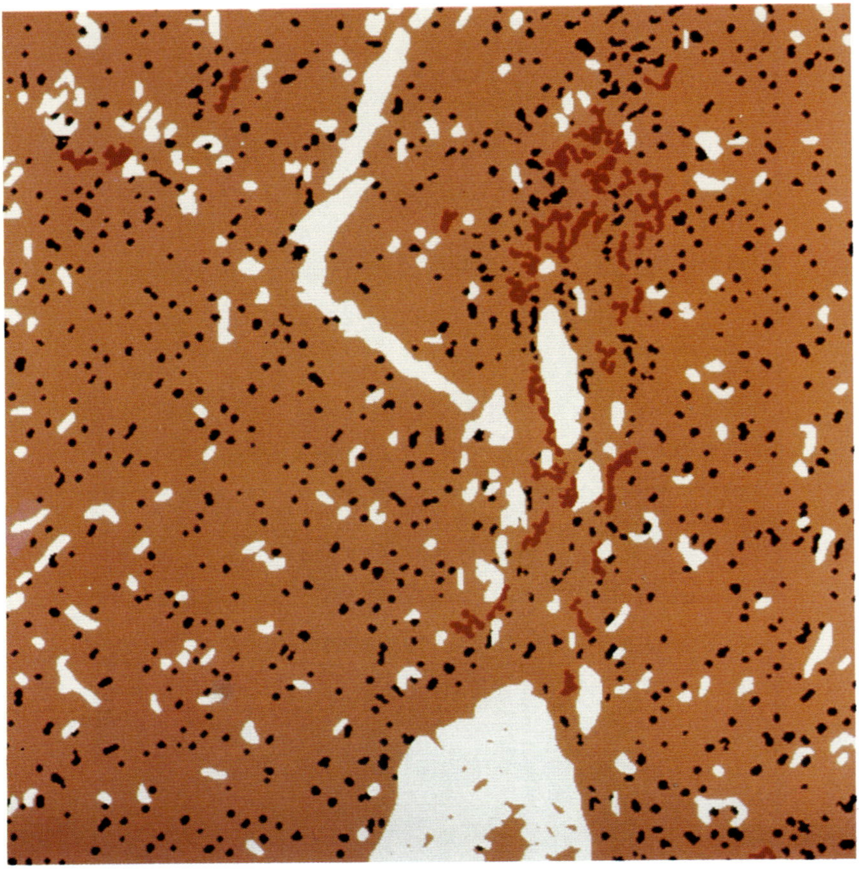

FIGURE 1.21. Computer analysis of the middle right image from Figure 1.8 (*bottom*), which extracts all normal cell nuclei (black), the nuclei of inflammatory cells (red), as well as the cytoplasmic/sinusoid and the cytoplasmic/vessel boundaries. For further discussion, see page 73.

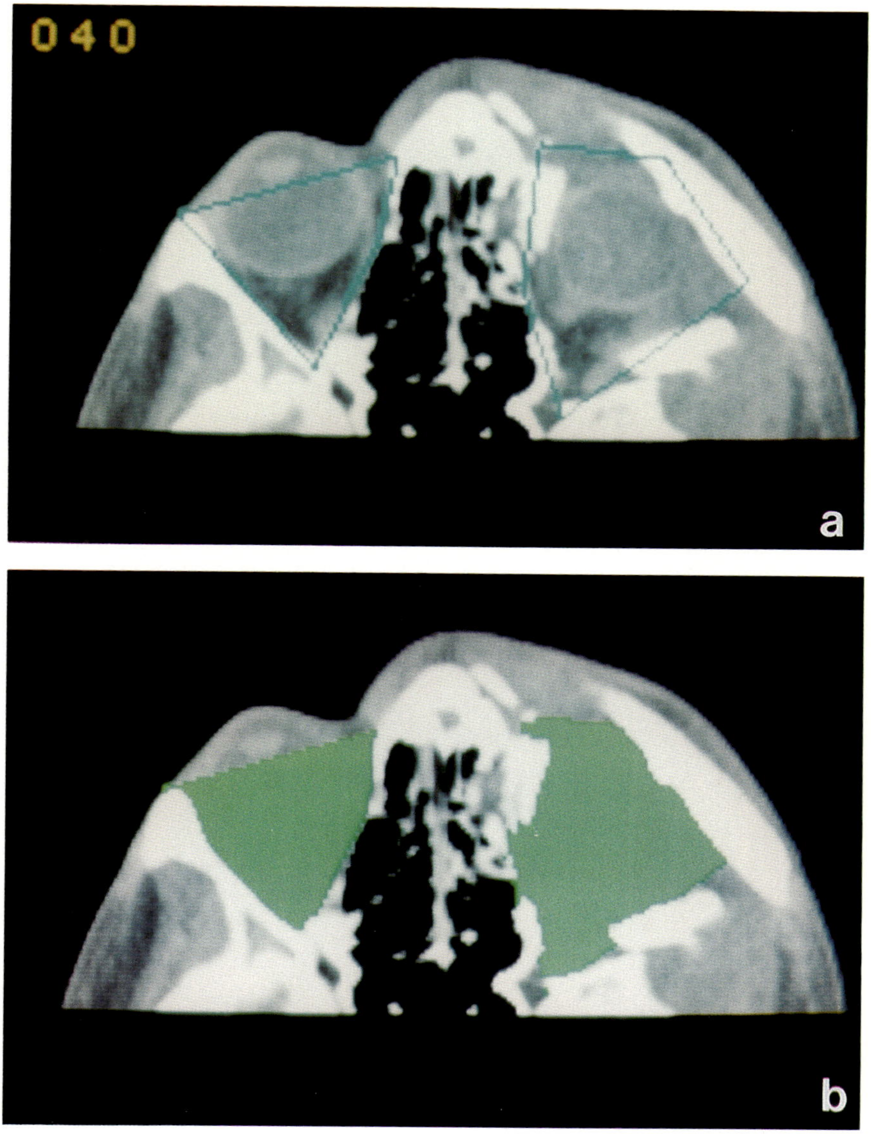

FIGURE 2.17. Computation of the orbital volume using 2D interactive manipulation. (**a**) The user roughly outlines the orbital region on individual slices. (**b**) The interactive software computes the actual orbital region (shown in uniform grey) by restricting thresholding to the user-specified region. The volume of the orbit is output as a by-product when the surface detection algorithm described in Subsection 2d is applied to the resulting binary array (Udupa, 1981). For further discussion, see page 150. (Reproduced with permission from National Computer Graphics Association from J.K. Udupa, "Display and Analysis of 3D Medical Images Using Directed Contours," *NCGA '85 Technical Sessions Proceedings*, Dallas, Texas, v. III, pp. 145–155, 1985.)

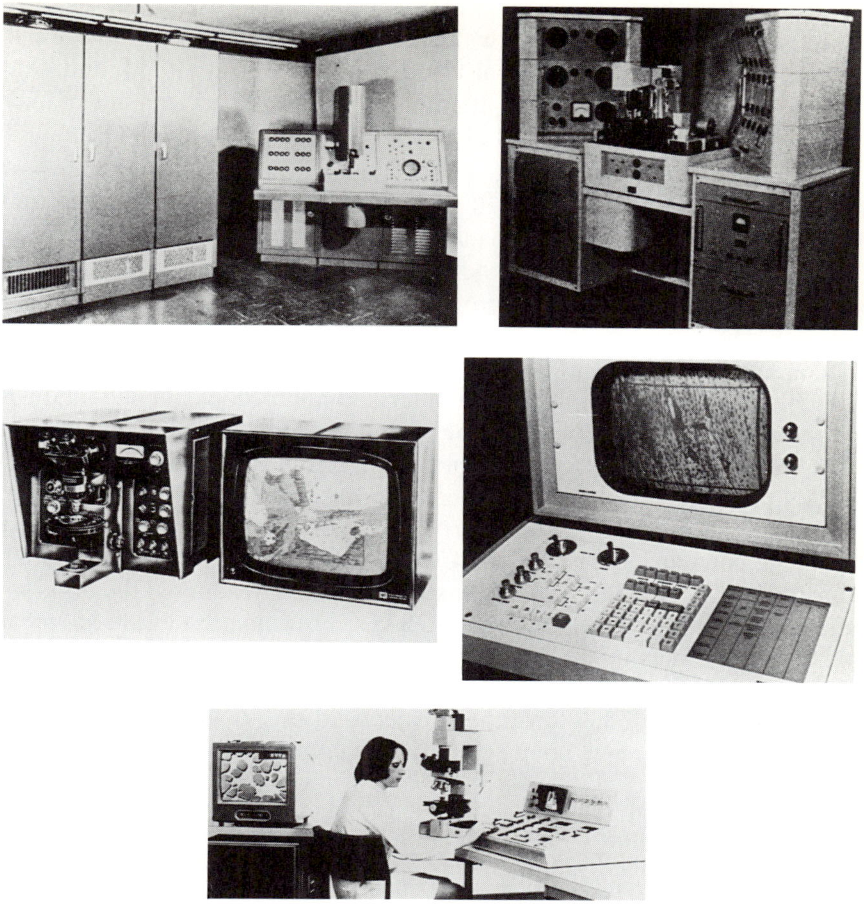

FIGURE 1.1. Early image analyzers for automated microscopy were (*left to right, top to bottom*) the Cytoanalyzer, the red blood cell counter of Cassella, the Quantimet of IMANCO, the Omnicon of Bausch and Lomb, and the Videoplan of Zeiss.

THE 1960S

By the early 1960s, although the Cytoanalyzer was able to measure nuclear size and optical density at high speed (several square centimeters per minute), artifacts in the sample preparation were a monumental problem. These consisted of clumps of blood cells, strands of tissue and mucus, and overlapping epithelial cells that caused an unacceptable level of false positives. It was still hoped, however, that, despite this high false-positive rate, the Cytoanalyzer could be used to screen cervical smears at rates faster than that accomplished by human technologists. But, despite extensive work in improving sample preparation, tolerable false-positive rates could only be achieved with an intolerable number of false negatives. Any serious researcher should review the literature on attempts to salvage this project in papers by Pruitt et al [1959], Bostrom et al [1959], and Diacumakos et al [1962a–c]. Other attempts to automate cervical-smear analysis are given by Rosenberg et al [1965] and Spriggs et al [1968].

White Blood Cell Analysis (CELLSCAN)

Despite the failure of the Cytoanalyzer, the 1960s saw the continuation of further efforts in applying the television microscope to a variety of applications. Many of these were presented in the proceedings of a conference sponsored in 1962 by the New York Academy of Science on "Scanning Techniques in Biology and Medicine" edited by Montgomery [1962]. Interest grew in using computerized microscopy for the analysis of human white blood cells. The U.S. Atomic Energy Commission (USAEC) initiated the first step in this direction by funding the CELLSCAN program at the Perkin-Elmer Corporation, Norwalk, Connecticut. By 1961 a slow-scan vidicon microscope and special-purpose computer had been constructed by Izzo and Coles [1962] working for Preston [1961]. During 1964 to 1966 the National Institutes of Health adopted the program after it was dropped in 1962 by the USAEC. Automation of blood-smear analysis proved to be far easier than that of cervical-smear analysis and CELLSCAN showed immediate signs of success, as described by Ingram et al [1968]. For details on other projects in this field see Prewitt and Mendelsohn [1966] and Young [1969]; for a comprehensive tabulation see Table 1.1.

CYDAC, TICAS, and Other Projects

The team of scientists at Airborne Instruments Laboratory who had developed the Cytoanalyzer designed a new image-logging microscope called CYDAC (Cytologic Digital Analysis System), which was installed at the University of Pennsylvania [Nadel, 1965]. It was used for DNA photometry and for chromosome analysis [Mendelsohn et al, 1964a–b, 1968, 1969a–b]. Other programs in chromosome analysis commenced at the University of Pittsburgh [Wald and Preston, 1968], in Great Britain [Rutovitz, 1967, 1968a–b; Gilbert, 1966], the National Biomedical Research Foundation [Ledley, 1966], the New England Medical Center [Neurath et al, 1966, 1969], and in Italy [Gallus et al, 1967].

Simultaneously, a general-purpose system called TICAS (Taxonomic IntraCellular Analysis System) for the digital analysis of microscope images was built at the University of Chicago by Wied et al [1968a–c]. Sandritter et al [1967] applied a Zeiss digital-scanning microscope to the analysis of chromatin in cell nuclei. The National Institutes of Health, in a cooperative program with Perkin-Elmer, established Spectre II, a system for general-purpose cell-image analysis [Stein et al, 1969].

Commercial Systems

The first commercially available system, called "Quantimet," appeared and went into manufacture at Image Analyzing Computers (IMANCO) in Great Britain. It is described by Beadle [1971] and Fisher [1971]. Other commercial systems were announced and marketed by Zeiss, Leitz, Wild Herbrugg, and Bausch and Lomb. They are described in publications by Zimmer [1969], Weibel et al [1971], Morton [1972], and Mueller [1974], respectively.

Other Instruments

Military automated microscopes were under development by the Department of Defense for use in the detection of bacterial warfare attacks as described by Nelson et al [1968]; Mansberg and Segarra [1962] developed an instrument for scanning brain tissue for neuron counting; Salomon and Musgrave [1969] continued work on bacterial colony counting; Norgren [1969] reported on an instrument for high-speed grain counting in "autoradiographs;" Soffen [1969] built an automated television microscope for detecting life on Mars; Oberst et al [1969] and Mawdesley-Thomas and Healey [1969a–c] used the automated microscope for measuring goblet cells in sections of lung tissue.

THE 1970S

The 1970s saw the culmination of 10 years of research on the automation of the human white blood cell differential count with the first public announcement (at the 1970 Engineering Foundation Research Conference on Multiphasic Health Testing) of the success of the Perkin-Elmer CELLSCAN/GLOPR (Golay Logic Processor) system built under sponsorship of the Department of Defense Internal Research and Development Program for automating white blood cell analysis. CELLSCAN was hailed as a "landmark which points the way" by Neurath [1971]. A full report is given by Ingram and Preston [1970]. Other applications are reviewed by Bahr and Wied [1970].

Commercial Clinical Instruments

Work at the University of Pennsylvania by Miller [1972, 1973] led to the establishment of the Geometric Data Corporation (now a subsidiary of SmithKline Beckman) and the design, manufacture, and marketing of the Hematrak®, the world's first commercial robot microscope. Research at Rush-Presbyterian Hospital in Chicago by Bacus [1970] and Bacus and Gose [1972] led to the development of another instrument—the LARC® (Leukocyte Analyzing and Recognition Computer) by Corning Glass Works. Reports on LARC are contained in publications by Megla [1972, 1973]. By the mid-1970s about 50 LARCs and Hematraks were in clinical use. At that time Perkin-Elmer announced their commercial white blood

TABLE 1.1. Characteristics of certain research systems used in computerized microscopy as well as two early commercial systems.

System Researcher Affiliation	Cytoanalyzer I Tolles, WE (1956) Airborne Instr Lab	Cytoanalyzer II Bostrum, RC (1963) Airborne Instr Lab	CYDAC Mendelsohn, ML (1968) University of Pa Lawrence Livermore Lab	CELLSCAN Preston, K, Jr (1962) Perkin-Elmer Corp	Cellscan/GLOPR Preston, K, Jr (1972) Perkin-Elmer Corp
Light source	Tungsten	Tungsten	CRT	Tungsten	Xenon arc
Wavelength			503 nm	Green	540 and 570 nm
Wave band					± 50 nm
Flux/pixel	15-μ lumen				10 nW
Optics	45× objective		100× oil objective	100× oil objective	100× oil objective
Focal length	4.0 mm				
Numerical aperture			1.25	1.3	1.3
Correction	Achromat		Planachromat	Achromat	Achromat
Spread function	5-μm circular aperture	2-μm circular aperture			
Modulation transfer function			50% at 1,150 cc/mm	50% at 1,800 cc/mm	50% at 1,800 cc/mm
Search scanner	1,800 rpm, 30-hole Nipkow disk with 6467 PMT	1,800 rpm, 48-hole Nipkow disk with PMT	Each cell manually located	Each cell manually located	20-Hz mirror scanner
Search rate	50 mm²/min	25 mm²/min			0.06 mm²/min
Focus method	Reference surface	Reference surface			Minimize high video freq
Focus response	NA	NA			1–2 seconds
Stage increment (X, Y, Z)	3 μm/line	1 μm/line			Continuous at 1.2 mm/min
Increment time (X, Y, Z)					NA
Stage travel (X, Y, Z)	10 × 20 mm				10 × 10 mm
Acquisition time	NA				30–60 s
Bandwidth	0.25 MHz				18 kHz

Image scanner	(Same as search)	(Same as search)	CRT with PMT detector	DAGE slow-scan vidicon	Same as search scanner
Pixel spacing	Continuous	Continuous	0.25 μm	0.1 μm	0.2 μm
Pixels/s	900 lines/s	1,240 lines/s	6,000	18,000	10,000
Bits/pixel	Analog	Analog	8 bits	1 bit	8 bits
Pixel array format	250 μm/line		192 × 192	300 × 300	128 × 128
Frame time			6.4 s	5 s	6 s
Bandwidth			2.3 kHz	10 kHz	18 kHz
Processor	Special purpose	Special purpose	Off-line, gen'l. purpose	Special purpose	Special purpose (GLOPR) Gen'l. purpose (Varian 6201)
Image-memory capacity				90,000 pixels	4,096 words
Measurements made	4	5	72 colorimetric 8 morphologic	15	35
Measurement time				10 min	40 s
Recognition time					
Processing rate	2,000 cells/min	1,000 cells/min		4 cells/h	20–40 cells/h
Output	Abnormal and total cell count	Abnormal and total cell count			Data sheet (1 page)
Software system			Cells		GLOL
Data bank			200 cell images		Measurements on 8,000 cells 500 cell images

TABLE 1.1. *Continued*

System Researcher Affiliation	BIOSCAN Wald, N (1968) Univ of Pittsburgh	NICHHD Castleman, KR Jet Propulsion Lab	SCANIT Neurath, PW/ Dew, B (1974) Tufts Med Cntr/MIT	ALMS Castleman, KR Jet Propulsion Lab	LARC Megla, GK (1973) Corning Glass Works	ACS 1000* Powell, R. Honeywell Corp
Light source	Litton CRT		Xenon arc	Tungsten		
Wavelength	P 36	Green	8 Wavebands used	Variable by filter	Orange, blue/green	
Wave band						
Flux/pixel						
Optics	97× oil objective	100× oil objective	100× oil objective	Selectable by turret		
Focal length	1.23	1.25	1.40			
Numerical aperture						
Correction						
Spread function		0.35 μm		10-, 20-, and 50-μm apertures (in plane of PMT)	5.0-μm aperture (search)	4.0-μm aperture (search) 0.5 μm aperture (focus)
Modulation transfer function		28% at 1000 cc/mm	30% at 500 cc/mm			
Search scanner	Hughes 3076 mW laser	Vidicon, 63× Obj	8-μm aperture silicon diode	Not used in search mode		Mirror scanner with 2 (search/focus) photodiodes
Search rate	300 mm²/min	6 mm²/min	9 mm²/min			38 mm²/min
Focus method	Focus insensitive	Max high video freq	Max high video freq			
Focus response	NA	1 s	2 s		0.2 s	0.1 s
Stage increment (*X*,*Y*,*Z*)				5 μm, 5 μm, 0.05 μm		
Increment time (*X*,*Y*,*Z*)				5 ms, 5 ms, 1 ms (max)		
Stage travel (*X*,*Y*,*Z*)						

Acquisition time	0.4 s	10–20 s	0.33 s		0.2 s	0.15 s
Bandwidth						
Image scanner	Litton CRT and 4249 B Centronic PMT	(Same as search)	Plumbicon	Plumbicon/mirror scanner and RCA 8645 PMT	Plumbicon	Standard color display for operator
Pixel spacing	0.1 μm	0.2 μm	0.2 μm	0.1 μm to 5 μm		
Pixels/s	200	15,750	1,000,000	3,300		NA
Bits/pixel	8 bits	8 bits	6 bits	8 bits		NA
Pixel array format	512 × 512	480 × 525	256 × 256	Up to 1,024 × 1,024		NA
Frame time	4 min	17 s	0.07 s	Up to 5 min	0.13 s	NA
Bandwidth				2 kHz		4 MHz
Processor	Dig Equip Corp KAIO + PDP 7	PDP 11/40	Nova 840	IBM 1130	PDP minicomputer	Intel Microprocessor
Image-memory capacity	2 images (524 K bytes)	15 · 10⁶ bytes	112 · 10⁶ bytes	512 K disk		NA
Measurements made	12	6 (homogeneous)/ 22 (banded)	100	NA	8	NA
Measurement time	3 min	58 s	"Several minutes"	NA	0.3 s	NA
Recognition time	5 min	21 s (for 46 chromosomes)		NA	0.05 s	NA
Processing rate	2 karyograms/h	8 karyograms/h	25 cells/h in "data collection" mode	NA	2 cells/s (max) 2000 cells/h (avg)	NA
Output	karyogram (calcomp)	Patient report (2 pages)		NA	Printed lab ticket	Printed lab ticket
Software system		Mini-VICAR		VICAR (available off-Line)		
Data bank	45 cell images	600 cell images	"Several thousand"			

CRT = cathode ray tube; PMT = photomultiplier tube; GLOL = Golay logic language; NA = not applicable; PDP = an abbreviation used by Digital Equipment Corp. for their computers, e.g., PDP 11, PDP 10 . . . :

*The ACS-1000 was a white blood cell finding machine wherein the technologist did visual classification from a color TV display.

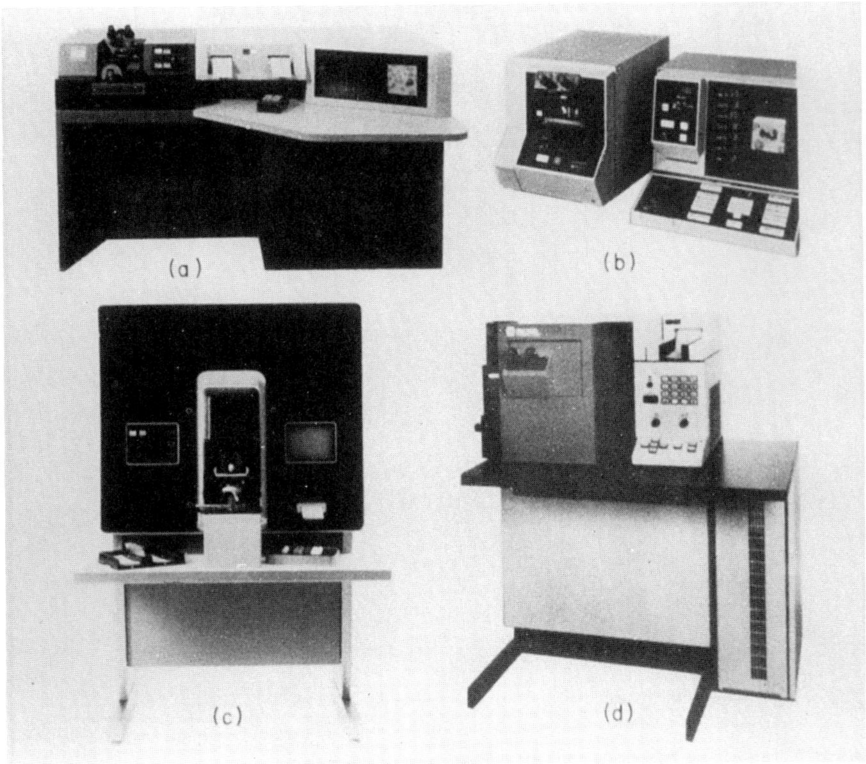

FIGURE 1.2. During the 1970s, four major companies produced robot microscopes for use in automated blood-smear analysis. These are the **(a)** diff3/50® of Coulter Biomedical Research Corporation, **(b)** the LARC® of Corning Glass Works, Inc, **(c)** the SmithKline-Beckman Hematrak®, and **(d)** the Abbott Laboratories, Inc ADC-500.

cell differential counter, the diff3®, which was subsequently sold to Coulter Biomedical Research Corp. (See Figure 1.2, Tables 1.1 and 1.2.)

The Hematrak and the diff3 employed special-purpose computers for image analysis—the former being nonprogrammable; the latter, programmable. The LARC employed a general-purpose computer. When a second-generation LARC was required to meet the high-speed image-analyzing capability of the Hematrak and diff3, Corning decided that the development cost was too great and stopped manufacture. Abbott Laboratories announced the ADC-500 based on work patented by Green and Neurath [1974], and Green [1976a–b] but quantity manufacture never commenced. Smithline [1967] and Brain [1974] patented other similar systems.

Initially the technologist using these robot microscopes was required to manually load and unload each microscope slide. With the advent of the diff3 and the Abbot Laboratories ADC-500, magazine-loaders became available (Green, 1979a–b, 1980). The Hematrak also adopted this procedure. By the end of the 1970s both the Hematrak and the diff3 had been produced at a level of about 400 total units. The LARC and ADC-500 were abandoned.

U.S. Research Efforts

Research at the Tufts New England Medical Center continued in the 1970s [Neurath and Brenner, 1972; Brenner et al, 1974, 1977]. Neurath's group cooperated with Gallus in Italy [Gallus and Neurath, 1970; Neurath et al, 1970], with Rutovitz in Great Britain [Green and Neurath, 1974], and with the Massachusetts Institute of Technology (MIT) [Eden et al, 1973; Dew et al, 1974]. Other work at MIT is described by Young [1970, 1972]. Research at Perkin-Elmer ended with the move of Preston to Carnegie-Mellon University and the sale of the diff3 product line to Coulter. The CYDAC project left the University of Pennsylvania and was installed at the Lawrence Livermore Laboratory. This new facility is described by Mayall [1974b].

The TICAS project at the University of Chicago was significantly expanded [Wied et al, 1968a, 1970a]. A comprehensive analytical program package offering scene segmentation, information extraction, supervised and unsupervised learning algorithms, multivariate statistical analysis, and data-base management was assembled [Wied et al, 1977a]. Information requirements for the reliable recognition of ectocervical cells were systematically determined [Bartels et al, 1981]. Material from other body sites was examined, and the technology was transferred to the following: Koss et al [1975] at Montefiore Hospital in New York for the study of bladder cancer; Kimzey et al [1980] at the cell-image analysis laboratory at NASA, Houston, for the analysis of lymphocytes; Bartels and Olson [1980] at the University of Arizona for the analysis of lymphocytes; Greenberg at Baylor University, Houston [see Baky et al, 1980] for the analysis of cells from the respiratory tract. In Washington the National Biomedical Research Foundation became allied with Georgetown University and produced an automated microscope called MACDAC [Golab, 1970]. The National Aeronautics and Space Administration established a biomedical image-analysis program at Jet Propulsion Laboratories. Work at this location on chromosome analysis is reported by Castleman and Wall [1973] and on muscle tissue by Castleman et al [1978].

At Washington University work on counting grains in autoradiographs was reported by Wann et al [1973]. At the National Institutes of Health the laboratory research produced using Spectre was documented by Lemkin et al [1974] and Carman et al [1974], and Prewitt and Wu [1978] studied the cellular structure of bladder tumor tissue from micrographs digitized at Jet Propulsion Laboratory. The Mayo Foundation built a laboratory for computerized microscopy [Robb, 1973]. Fuster et al [1977] used this facility to study the distribution of myocardial fiber diameters

TABLE 1.2. Tabulation of the major characteristics of five commercial systems (the Coulter differential was never manufactured) developed during the 1970s.

System Source Affiliation	ACS 1000 R. M. Powell Honeywell	Coulter Differential L. Smith Cognos Corp	diff3 G. T. Paul Perkin-Elmer Corp	Hematrak® M. N. Miller Geometric Data	LARC® C. H. Rogers Corning Glass
Optics					
Objective					
Magnification	100	50	40	40	100
Numerical aperture	1.25	1.4	1.0	1.0	1.3
Type	Planapochromat	Planapochromat	Planachromat	Planachromat	Planachromat
Immersion medium (index)	Oil (1.5)	Oil (1.518)	Oil	Oil (1.518)	Oil (1.515)
Ocular					
Magnification	10	20	12.5	10	10
Type	B & L	Zeiss	KPL	KPL	(Adjustable focus)
Illumination					
Light source	100 W Tungsten	150 W Xenon	60 W Tungsten	CRT	80 W Tungsten
Wavelength(s) (nm)	570	—	510, 580	420, 520, 650	Orange, Blue-Green
Bandwidth(s) (nm)	40	—	30, 20	—	—
Modulation transfer function	10% at 2,000 cc/mm	40% at 2,000 cc/mm	—	—	—
Scanners					
Mechanical					
Slide magazine capacity	1	20	14	1	1
Stage increments (*X*, *Y*, *Z*) (μm)	4, 4, 0.08	5, 5, 0.1	3, 3, 0.1	350, 350, 0.25	5, 5, 0.1
Increment time (*X*, *Y*, *Z*) (ms)	1	2.3, 2.3, 1.2	1.2, 2	70 (*X*, *Y*)	2, 2, 3
Total travel (*X*, *Y*, *Z*) (mm)	20, 75, 0.5	—	20, 10, 0.5	—	60, 20, 0.25
Search mode					
Type of scanner	Mirror	Plumbicon	Linear diode array	CRT	Mirror
Scan lines/s	1,000	15,000	1,000	20,000	500
Spot velocity (μ/ms)	288	—	—	—	150
Line length (μm)	144	250	200	300	≈ 300

Line width (μm)	4	—	3	300	≈ 7
Line spacing (μm)	4	1	3	—	≈ 5
Coverage (mm^2/min)	34	—	36	—	45
Focus method	Max high video freq	Move objective	Not used	Max high video freq	Move objective
Focus response (s)	0.15	—	NA	—	Real time
Cell acquisition time (s)	0.35 (avg)	—	0.15	0.05	0.10
Video bandwidth	40 kHz	—	75 kHz	—	—
Imaging mode					
Type of scanner	Color TV display	Plumbicon	Plumbicon	CRT	Plumbicon
Pixel spacing (μm)	0.5	0.2	0.4	0.25	0.4
Acquisition rate (pixels/s)	NA	—	$2.5 \cdot 10^6$	—	—
Acquisition time (s)	NA	—	0.02	—	—
Format (x pixels, y pixels)	NA	256, 256	128, 128	112, 80	48, 48
Video bandwidth	4 MHz	—	1 MHz	—	—
Bits/pixel	NA	—	6	—	6
Image processor					
Type of computer	Intel 8008	Minicomputer	Mini and special	HP2,105	PDP 8M
Image memory (words)	NA	—	16–20 k	—	16 k
Operating system (words)	6 k	—	32 k	8 k	—
Measurements made	NA	—	20 (red cell)/50 (white cell)	96	8
Measurement time (ms)	NA	—	500	Real time	—
Recognition time (ms)	NA	—	7	30	
Processing rate (cells/h)	Ltd. by Human Recog.	—	3,500–4,000 (Av. white) 5,000–6,000 (Av. red)	12,000 (peak)	5,500 (av) 8,000 (peak)
Output	Lab ticket; computer entry	Printed tape	Printed lab ticket; computer link	Printed lab ticket	Printed lab ticket; computer interface

NA = not available; B & L = Bausch and Lomb; KPL = an abbreviation used by Leitz Corporation for their microscope optics; CRT = cathode ray tube.

and Karnes et al [1977] used the facility to study myelinated nerve fibers. A short-lived program in cell and tissue analysis was started at Purdue University, resulting in reports on blood cells by Mui et al [1977] and Mui and Fu [1978].

Research Efforts in Europe and Japan

During the 1970s European cell- and tissue-analysis facilities continued their earlier work and new laboratories were established. A program in the analysis of blood cells was initiated at CERN by Gelsema and Powell [1972] as described in Gelsema et al [1973]. Contributions continued from the Karolinska Institute and other groups in Sweden, notably the comprehensive work by Stenkvist et al [1978] and that of Holmquist [1978]. In Great Britain the research group in Edinburgh continued with the work of Rutovitz [1968, 1970]. Tucker and Gresham [1971] and Husain et al [1970] continued research in cervical prescreening automation and Bentley and Lewis [1970] in red cell morphology.

In West Germany the TUDAB project started for the automation of prescreening for cervical cancer [Burger et al, 1977, 1980]. Also supported in West Germany were the laboratories of Soost [see Otto et al, 1979a] in Munich and at the GSF in Munich under the direction of Burger [Burger et al, 1977, 1980]. In Stuttgart the FAZYTAN project was initiated under the direction of Bloss [see Reinhardt et al, 1976, 1979]. In Hannover, Liedtke and Aeikens [1980] and Liedtke and Kappei [1983] began bladder cancer cell studies. Research teams in the Netherlands, under the direction of Ploem [Ploem et al, 1974] and Vooijs [see Zahniser et al, 1979, 1980], maintained an active research program. In East Germany Voss [1974] worked on methods for the counting and sizing of overlapping red blood cells, and Simon [1975] initiated studies in histopathology, especially for the assessment of liver sections. Kunze et al [1980] in Dresden worked on the analysis of lymphocytes. HeLa cells were enumerated by Aus et al [1974]. Aus et al [1974] and Volker ter Meulen et al [1972] also worked on image-analysis programs in virology. Christov et al [1974] and Moore [1974] performed automated analyses of cells extracted from thyroid tissue. Grimelius [1978] did work on the computer analysis of endocrine tissue. Computerized analysis of liver-tissue images was reported by Simon et al [1975].

In Japan the late 1970s also saw the systematic development of the CYBEST cervical cancer prescreening system at Toshiba [Tanaka et al, 1976, 1977a–d, 1979].

Advances in General-Purpose Instrumentation

New instruments that came into use during the 1970s are the Leitz Videomat and TAS (Texture Analysis System), Bausch and Lomb Omnicon, Joyce-Loebl Magiscan, and the Imanco 720 [Cole and Bond, 1972; Robertson et al, 1978; Fisher and Bond, 1971; and Rosenquist, 1976]. Use of the Leitz TAS system in microscopy is described by Kline and Serra [1972], Serra [1972], and their co-worker, Delfiner [1972], as well as by Moore [1972]. Langston and Thurlbeck [1978] discuss the use of such image analyzers for lung morphometric measurements. Use of the Bausch and Lomb Omnicon for quantitation of nuclear-cytoplasmic changes in bladder epithelium is described by Stinson et al [1977] and for measurements of Golgi-impregnated neurons by Wann et al [1973]. Application of the Zeiss Videomat to the study of nuclear diameters in lymph nodes is described by Crocker and Curran [1979]. Other surveys appear in Weibel et al [1971] and Bradbury [1979]. The latter discusses general problems and approaches in the analysis of microscopical images.

Immunological Methods

The revolution in sample preparation techniques due to the increasing availability of immunological stains had an impact on automated microscopy in the 1970s. An introduction to this field is contained in Sternberger et al [1970] and in a book by the same author [1979]. Quantitation of the immunological reaction is given by Streefkerk et al [1973] and Frasch et al [1978].

The 1980s

The year 1979 (which is treated here with the 1980s) might be called a "watershed" for work in the computer analysis of cells and tissues. Meeting in Israel in 1979 under the leadership of Pavlidis of Princeton University and sponsored by the Dahlem Foundation, Fu, Poeppl, Preston, Prewitt, Schneider, and Zajicek planned the Dahlem Workshop on Biomedical Pattern Recognition and Image Processing [1979], which was held in Berlin in May of that year. The proceedings of this workshop became a development plan for new laboratories throughout Europe.

New Literature

In the interval 1979 to 1985 several books appeared. Pressmann and Wied [1979] prepared a compendium on automation in cancer cytology, as did Koss and Coleman [1979]. Catsimpoolas [1980] edited a book on cell analysis. Eisert and Mendelsohn [1982] prepared one concerning the use of cell analysis for biological dosimetry, ie, the recognition and quantitation of environmental threats. Other important books were those of Weibel [1979] on stereological methods, that of Aherne and Dunnill [1982] and Baak and Oort [1983] on morphometry, and that of Adelh and Signor [1981] on shape recognition. Onoe et al [1980] reported on the Japan conference on real-time medical-image processing. Fu [1979] edited a special issue of the IEEE Proceedings on the subject of pattern recognition and image processing. The International Academy of Cytology began the publication of the new journal, *Analytical and Quantitative Cytology*, and, shortly thereafter, the Society for Analytical Cytology was formed and publication of the journal *Cytometry* commenced. As of this writing, these journals have published some 5,000 pages of new literature encompassing approximately 500 new journal articles. Of these, approximately 100 are referenced in the following sections. Thus, the transition interval between the 1970s and 1980s was one of enormous productivity and significance.

Commercial Image Processors

Use of commercial image processors for task-specific automation proliferated in the 1980s. Although use of these machines in manual mode (light pen, stylus/tablet) is strictly outside of the purview of this chapter, some results are mentioned so as to indicate the breadth of studies being carried out using this popular but painstaking technology. Work by Slavin et al [1980] traced boundaries of both epithelium and mucosa in jejunal biopsies by the light pen. Boon et al [1980, 1981] used the graphics tablet to trace cytoplasmic and nuclear area in cells from follicular carcinomas of the thyroid, smears of gastric mucosas, and bladder tumors. Crocker and Curran [1979] used the Zeiss Videomat to study nuclear diameters in lymphomas. Gross and Rothfeld [1985] used the Quantimet 720 to perform quantitative immunocytochemical studies of the brain, hypothalamus, and pituitary. Similar studies were conducted by Benno et al [1982] in animals while quantitative DNA studies were reported by Pak et al [1982] and by Teplitz et al [1983] with particular emphasis on pulmonary cytology. Baak et al [1981a–b] used the Zeiss Videoplan to study cutaneous malignant melanomas as well as endometrial carcinomas. Other workers used simple graphics tablets and microcomputers in studies of tissue in a variety of diseases. The Leitz TAS was employed in morphometric studies of myeloblasts specifically for the purpose of calculating relative DNA content. Pasteur et al [1984] used the Quantimet 700 for cell analysis in bronchial carcinomas. A graphic tablet and minicomputer (Hewlett Packard 2647A) was used by Gamel and McLean [1982–1984] for the purpose of sizing nucleoli in melanomas of the eye. Excellent work using the more traditional mechanisms of scanning microspectrophotometers continued to be employed by Casperson's group at the Karolinska Institute. Sample publications are those of Auer et al [1980] where DNA quantitation was performed for cells in human mammary carcinomas. Other similar studies of DNA quantitation are those of Atkin et al [1979] and of Sprenger et al [1979]. Useful general surveys of the entire field of semiautomated microscopical image analysis were written by Bradbury [1979].

Automation of Cervical Cancer Prescreening

Tucker [1983] continued with development of the CERVISCAN machine and with the development of an array processor for studies in DNA quantitation [Tucker, 1976]. The AEG Telefunken Research Institute constructed FAZYTAN as a high-resolution prescreening system for gynecological specimens with reports provided by Bloss et al [1979] and by Ott et al [1981]. Similarly, the BioPEPR cervical smear image-analyzing system was developed in the Netherlands as reported by Oud et al [1981a]. Also in the Netherlands Al and Ploem [1979] used the LEYTAS television analysis system for studies of cervical cytology. In Japan work continued on CYBEST [Tanaka et al, 1981; Mukawa et al, 1983]. A major study of cervical cells digitally recorded at Jet Propulsion Laboratory was done in cooperation with the University of California at Los Angeles [Rosenthal et al, 1983; Stern et al, 1982]. At Rush-Presbyterian St. Lukes Hospital, work continued on cervical smear analysis by Bacus et al [1982]. Use of TICAS for identification of ectocervical cells is reported by Wied et al [1981a]. Bartels et al [1986] built a high-speed special-purpose multicomputer system for cervical cell image analysis.

White Blood Cell Analyzers

In the commercial white blood cell analyzing instrument field, Coulter Electronics announced the closing of its facility for manufacturing the diff line of machines in late 1983. SmithKline-Beckman followed suit 2 years later. The reasons for Coulter's drastic decision are given by Beard [1984], who quotes Wallace Coulter as saying, "We lost several million a year making, selling, and maintaining the microscopes. The problems turned out to be bigger, and the market smaller, than all of us in the business thought. We believe that, since the microscope is such a universal tool, an automated microscope would find a place. It hasn't yet." At the time there were approximately 300 of the diff3-line instruments in place and SmithKline-Beckman had sold a total of about 700 Hematrack systems. Detailed documentation on the demise of the U.S. white blood cell analyzers is provided in a retrospective by Preston [1987].

In Japan, Hitachi and Omron commenced production of white-blood-cell differential machines, which were sold primarily to the Japanese and Asian communities. In the United States new research systems were developed for automated cell and tissue analysis by Perceptive Systems Inc. and Cell Analysis Systems. (See Figure 1.3.)

There is no question that the hematological community has benefited immensely. Sample preparation is now automated and blood cell specimens are under far better quality control than when manual methods were employed. The machines deployed can now record statistically significant data banks and conduct epidemiologic studies on an enormous volume of information. The future of these machines is still problematic. Flow systems operating at thousands of cells per second offer serious competition.

During this period of time, owing to the appearance of the automated white blood cell differential counting systems, it was difficult to obtain research support in this field on a national level. Except for the group at Tufts New England Medical Center [Lester et al, 1978], efforts wound down in a number of countries. Doctoral theses were done at the Uni-

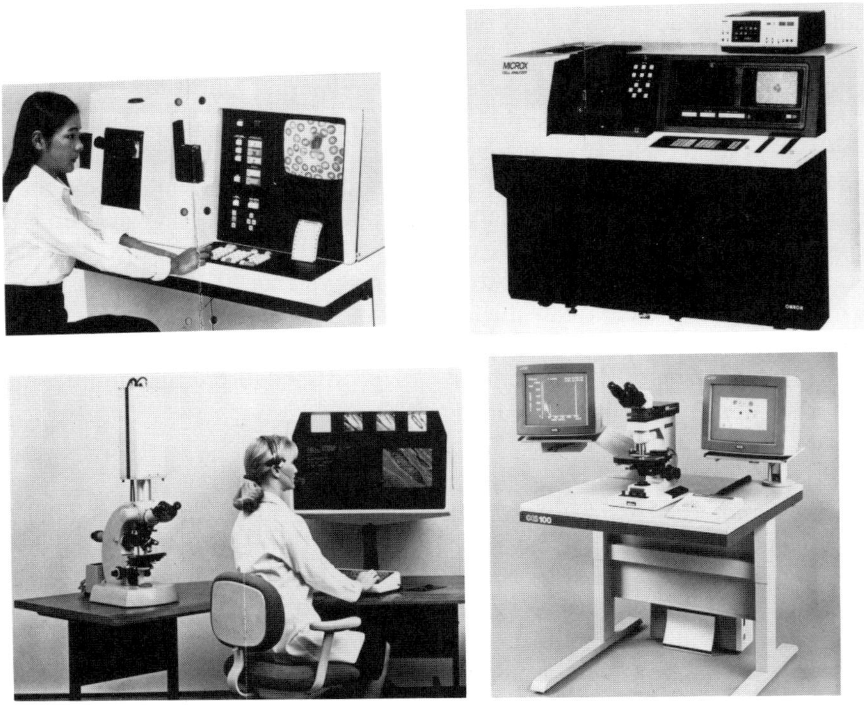

FIGURE 1.3. Automated microscopes for the late 1980s are (*left to right, top to bottom*) the white blood cell differential machines of Hitachi and Omron, the PSI-COM of Perceptive Systems Inc., and the CAS 100 of Cell Analysis Systems.

versity of Groningen by Bins [Bins et al, 1981] and at the University of Amsterdam by Landeweerd [1981]. [See also Landeweerd and Gelsema, 1978.] Zajicek et al [1983a–b] in Israel pursued studies on nucleated red blood cells.

Bladder Cancer Studies

Another major area of effort is in the study of bladder washings for cancerous cells. The Laboratory for Analytical Cytology at Montefiore Hospital (Bronx, New York) performed major studies in this area [Sherman et al, 1981, 1983; Koss et al, 1978a–b, 1980a–e, 1984a–b; Koss et al, 1980d]. An important new instrument for urinalysis was introduced by International Remote Imaging Systems. This remarkable machine executed urinalysis by capturing images *in vivo* with the capability of sorting and displaying these images as shown in Figure 1.4.

Breast, Pulmonary, and Bone Cancer Studies

Another important area of study is that of mammary carcinoma. Single-cell image analysis was compared with flow DNA analysis by Auer et al [1980b] and by Cornelisse et al [1983] as well as by Van Driel-Kulker et

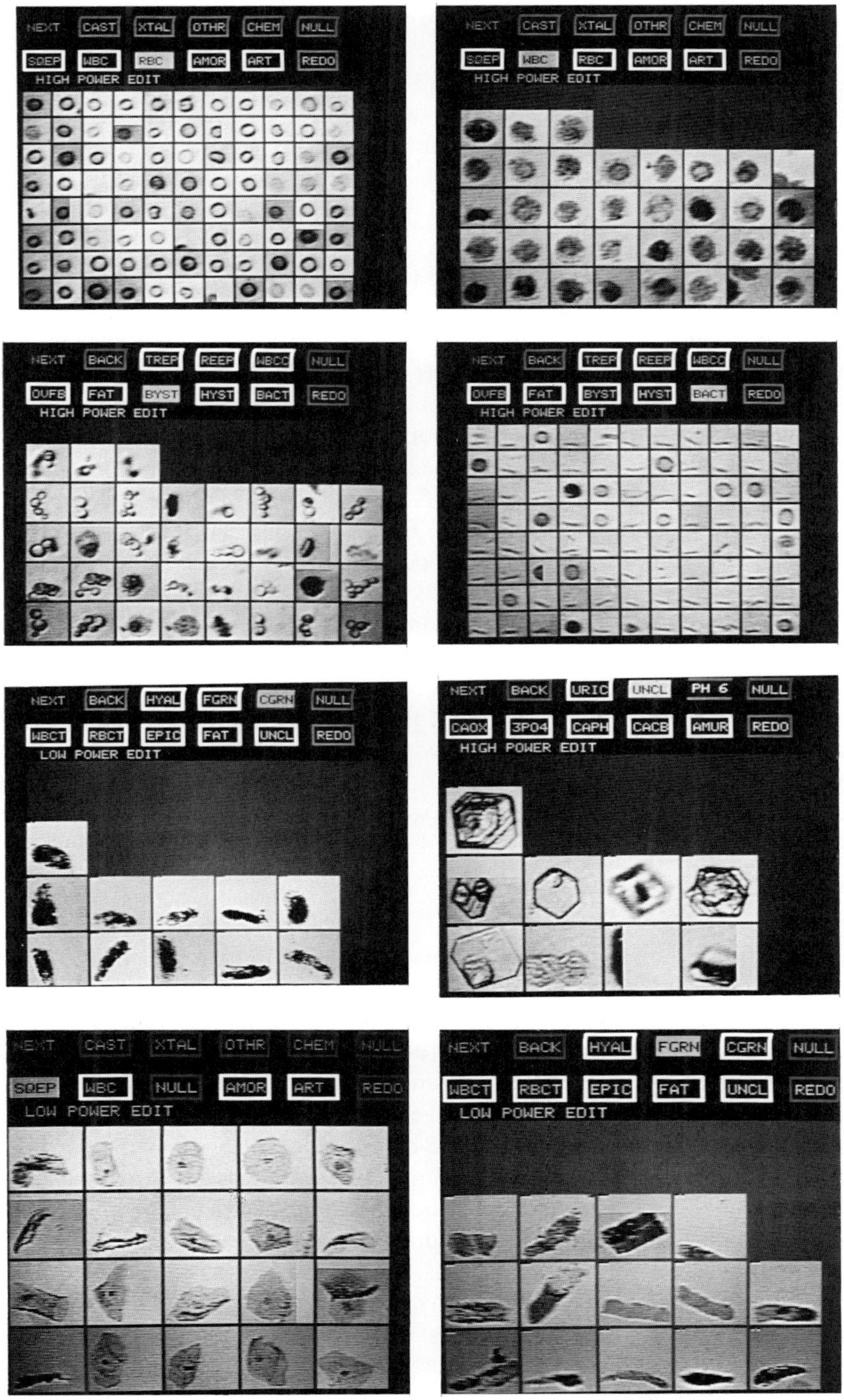

FIGURE 1.4. Displays generated by the Yellow Iris® urinalysis workstation showing (*left to right, top to bottom*) *in vivo* images of red blood cells, white blood cells, budding yeast cells, bacteria, coarse granular casts, cystine crystals, squamous epithelial cells, and fine granular casts.

al [1982], who employed the LEYTAS (Leyden Texture Analysis System) system. Pulmonary cancer cytology was studied by Winkler et al [1981] and that of the bone of Kreicbergs et al [1980a–b]. Several of these latter studies compared flow analysis with high-resolution automated microscopy.

General Studies

The TICAS program, as a whole, generated considerable literature in many areas of application. The hierarchical classification methods of TICAS were explained in a paper by Bartels et al [1984]. Wied et al [1980] discussed applications in uterine cancer cytology as did Bibbo et al [1981], and Wied et al [1984] studied further applications in cervical cytology. Many studies on lymphocytes were carried out. An example is the report by Bartels and Olson [1980, 1984]. The latter paper of these two reports is on environmental-induced effects. Olson and Bartels [1981] conducted work on the differentiation of splenocytes and lymphocytes in animals. Other laboratories reporting are the Central Research and Development Department of Dupont and the Quantitative Microscopic Group at the Medical and Scientific University of Grenoble [Gauvain et al, 1984], who are developing SAMBA. Zajicek and Meinzer [1982] in Israel studied cells in colon crypt epithelium. Finally, it is worth mentioning the cell-cycling studies of Abmayr et al [1982a] as well as the study on automated cytophotometry by Rueter et al [1980].

Methodology

This section discusses the "tools" used in the automation of image processing for cells and tissue starting with sample preparation. Following this, scanning- and image-digitizing systems are reviewed. Finally, scene segmentation and cell recognition are covered.

The simplest instrument for cell and tissue analysis is exemplified by the Zeiss MOP, which uses an image of the specimen projected onto a tablet. The tablet, either by acoustic or magnetostrictive means, permits the user to trace cellular or tissue components with a stylus. An associated computer makes measurements on the contours traced. This task-specific automation is slow and tedious and does not represent the state of the art in automated cell and tissue analysis. As is stated by Wann et al [1973], their system of this kind for measuring cell and nuclear size by manual means permits the analysis of only 200 to 400 cells per hour. This compares with 10,000 cells measured per hour with commercial automated white blood cell differential counting systems, and from 30,000 to 60,000 objects per hour for the tissue-analyzing system described by Castillo et al [1982], and approximately 50,000 objects per minute (or 3,000,000 per hour) that can be scanned by the ultrafast laser scanner microscope at the University of Arizona. Of course, some laboratories combine both manual with the semiautomatic methods. A particularly useful review giving comparisons is the book by Weibel [1979].

Sample Preparation

The automation of sample preparation, ie, the physical layering of cells or tissue sections on the microscope slide and staining thereof, holds the key to success in the automated image analysis for cells and tissue. It was the invention of Preston and Norgren [1971] of automatic blood "spinning" as an alternative to manual smearing that made feasible automated white blood cell differential counting systems. Conversely, the failure of the Cytoanalyzer is attributed, to a large extent, to its inability to distinguish cervical cells from preparation artifacts. The magnitude of this problem was not recognized until Mellors et al [1952] investigated its significance some 6 years after the project was initiated. The problem remained until the late 1970s. Electrical engineers, who pioneered in applying digital picture processing to cell and tissue analysis, have given sample preparation short shrift, owing to their lack of training in biochemistry and biomechanics.

Sections below on hematology automation, cytogenetics automation, and cytology automation treat progress in sample preparation in those areas. This section concentrates on developments of interest in other fields.

Biological Stains

Although of great importance, specific studies on the relationship of sample staining to pattern recognition success rates in automated microscopy are few. An exemplary study is that of DeCampos-Vidal et al [1973]. A quantitative comparison of results obtained using different staining procedures are given by Van Driel-Kulker et al [1980]. Other reports on the effect of staining on pattern recognition success rates are given in Gill and Jotz [1974], Ornstein and Ansley [1974], Sandritter et al [1974], Steinkamp and Crissman [1974].

DNA-Specific Stains

Interest in the quantitation of the nucleic acid content of cells can be traced to the work of Feulgen and Rossenbeck [1924]. Other reviews of protein-specific stains are given in Pollister and Ris [1947] and DiStefano [1948]. Deitch [1955] reports on naphthol yellow S and Deitch [1964] reviews work with methylene blue. Boehm and Sprenger [1968], Trujillo and Van Dilla [1972], Latt [1974], Golden and West [1974] report on advances in fluorescent staining. Davidson [1971] reviews consecutive staining methods. Reports on acridine orange are given in Adams and Kamentsky [1971], Melamed [1974], and Steinkamp et al [1973], and Green [1979c]. An example is given in Figure 1.5.

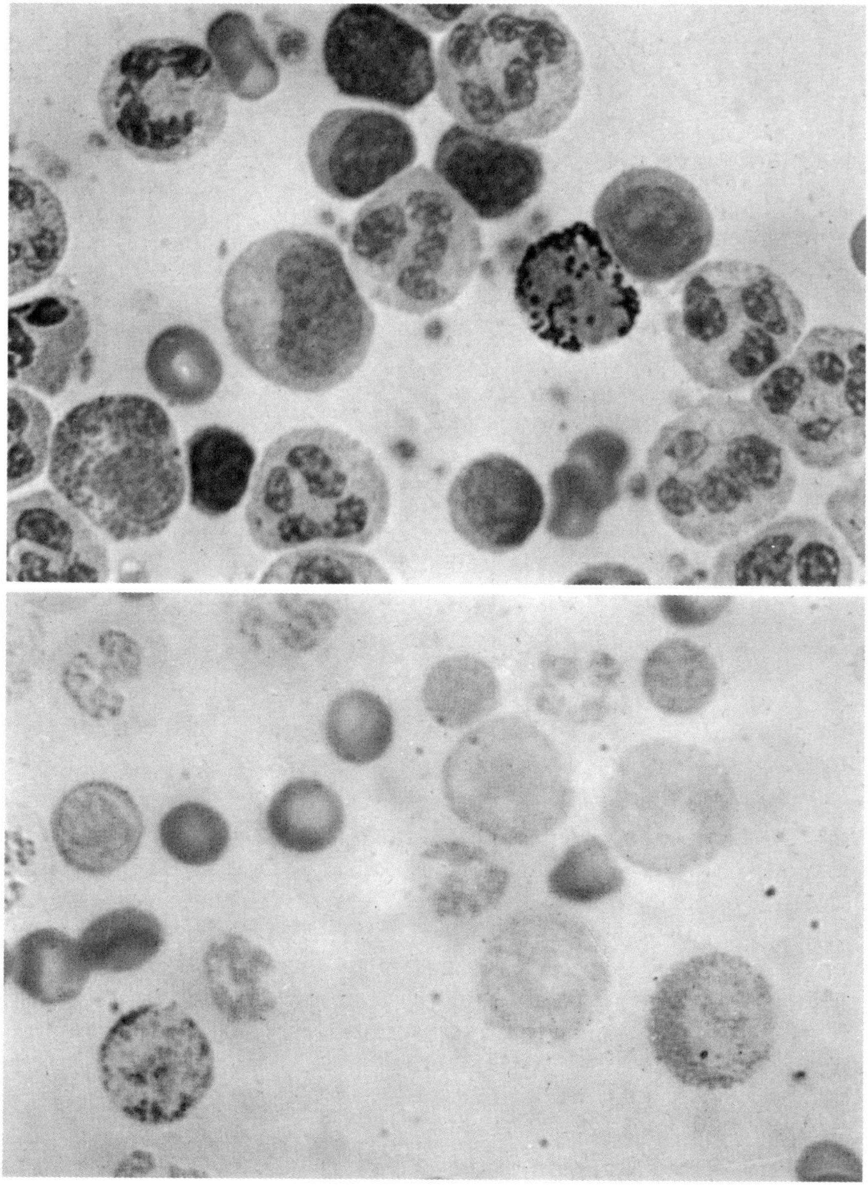

FIGURE 1.5. Stained-buffy-coat smears of human blood showing (*top*) the standard Wright-Giemsa stain and (*bottom*) the trichrome stain developed specifically for machine reading of human blood smears by Ornstein and Ansley [1974]. See Color Plate I.

Vital Stains

The important field of staining a viable cell population has also been undertaken in studies of cell kinetics by Kiefer et al [1972] and Gray [1974]. Bremeskow et al [1970] report on the DNA content of L-cells, Henry et al [1971] on stains specific to hemoglobin, Davis and Ornstein [1959] and Sweetman et al [1972] on various enzymatic stains. Stains specific to leukocyte recognition are given in Gilbert and Ornstein [1975] and also by Atamer and Groner [1973].

Immunological Stains

The revolution in immunological staining techniques is described in the book by Sternberger [1979] as well as in such papers by Mason et al [1969], Sternberger et al [1970], Rothfeld and Gross [1982], and Gross and Rothfeld [1985]. Streefkerk et al [1973] present investigations of immunohistoperoxidase procedures. An example is given in Figure 1.6. Comparisons of quantitative immunocytochemistry and radioimmunoassay are given in Briski et al [1983], and Benno et al [1982] discuss the immunocytochemistry of tyrosine hydroxylase. Frasch et al [1978] discuss quantitation of the

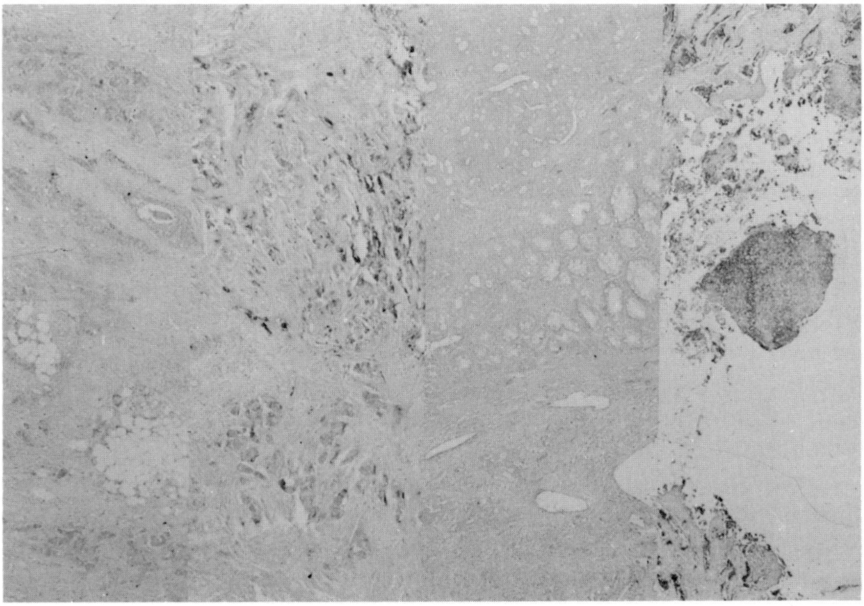

FIGURE 1.6. The "brown revolution" occurred when immunoperoxidase stains were introduced in the 1970s. The images above are of human breast and colon tissue showing both negative and positive interactions using an immunoperoxidase stain specific to estrogen receptor. (Images produced by the research diff3/50 at Carnegie-Mellon University from samples furnished by the Eastern Oncology Group at the University of Wisconsin.) See Color Plate II.

diaminobenzidine reaction for histochemical demonstration of cytochrome oxidase activity. Studies using immunocytometric methods to distinguish between human T and B lymphocytes are given in Olson et al [1974a] and in Durie et al [1978].

Stain Journals

From this extensive list of publications, it is obvious that stain biochemistry is both a complex and an active field. It is absolutely essential that the engineer entering the field of automated microscopy should be thoroughly aware of its existence. Journals such as *Stain Technology, Histochemica Acta, Histochemie,* and both the *Histochemistry Journal* and the *Journal of Histochemistry and Cytochemistry* serve as prime references.

Scanners

The scanner is the mechanism that transforms the spatial colorimetric information present in the specimen into a digital number array or matrix. The elements of this matrix provide the measured values of optical density or transmission point-by-point at a variety of wavelengths. It is almost universal practice to use incoherent illuminators to furnish light to the specimen, although the advantages of laser-scanning systems are reported by Kaufman et al [1972], Davidovits and Egger [1971], and Bartels et al [1981]. More esoteric methods of transfering spatial information from the microscope slide to the computer are under investigation by Lemons and Quate [1974], who utilize ultra-high-frequency ultrasound. Shadow-casting techniques are described by Hlinka and Sanders [1970].

Types of Scanners and Illumination Sources

The three primary types of optical scanners are shown in Figure 1.7. They are the early flying-spot scanner, the typical television image plane scanner, and the laser scanner. The image plane scanner is used most frequently with the television tube occupying a plane conjugate to the eyes of the user, which is easily instrumented by affixing the camera to the trinocular tube normally used for photomicroscopy.

Before the advent of inexpensive and reliable television tubes, mechanical scanners employing the Nipkow disk were used in such instruments as the Cytoanalyzer. Next, flying-spot scanners were employed with the CRT placed in the traditional image plane with the light generated from the face of the tube directed down the microscope toward the specimen. Such a scanner is exemplified by CYDAC as described by Bostrom and Holcomb [1963]. Early surveys of scanners of historical interest are given in Mansberg and Ohringer [1970] and in Burn [1971]. Flying-spot scanners used to excite fluorescence are described by Thieme [1966], Pearse and Rost [1969], Rost and Pearse [1971], and Ploem et al [1974].

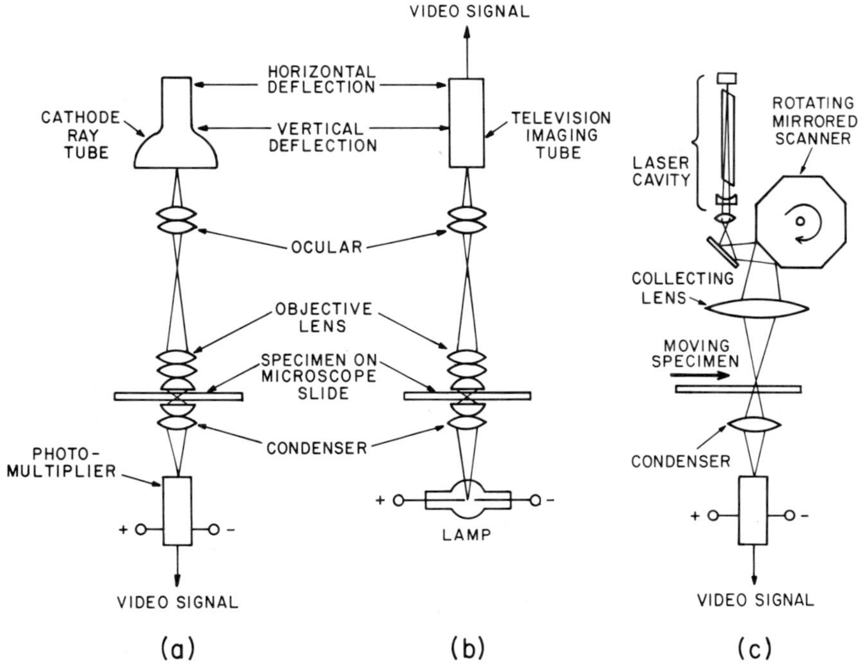

FIGURE 1.7. Schematic diagrams showing three basic scanners: (**a**) the flying-spot scanner, (**b**) the image plane scanner, and (**c**) the laser scanner.

As scanning systems became more sophisticated, both the scanner and the specimen transport mechanism were placed under computer control. In early systems, such as the Cytoanalyzer, specimen motion in the focus direction was executed by means of an open-loop servo. This servo sensed the position of the microscope slide mechanically and kept its surface at a fixed distance from the objective lens. Later systems, exemplified by CELLSCAN/GLOPR, used closed-loop servos, which continuously detected and maximized the temporal frequency content of the television signal. An alternative method of automatic focusing was applied in CYDAC based on the analysis of the photometric histogram as described in Kujoory et al [1973].

The earliest report on microscope slide scanning using a laser is that of Norgren [1969]. His system was a reduction to practice of the system patented by Preston and Norgren [1970] for the high-speed detection of metaphase cells. Other nonlaser scanning systems designed for the same purpose are described by Paton [1971], Green and Neurath [1974], and Johnson and Goforth [1974]. The most advanced laser-scanner system using two wavelengths and interrogating the microscope slide at hundreds of megapixels per second is that described by Shack et al [1982] at the University of Arizona.

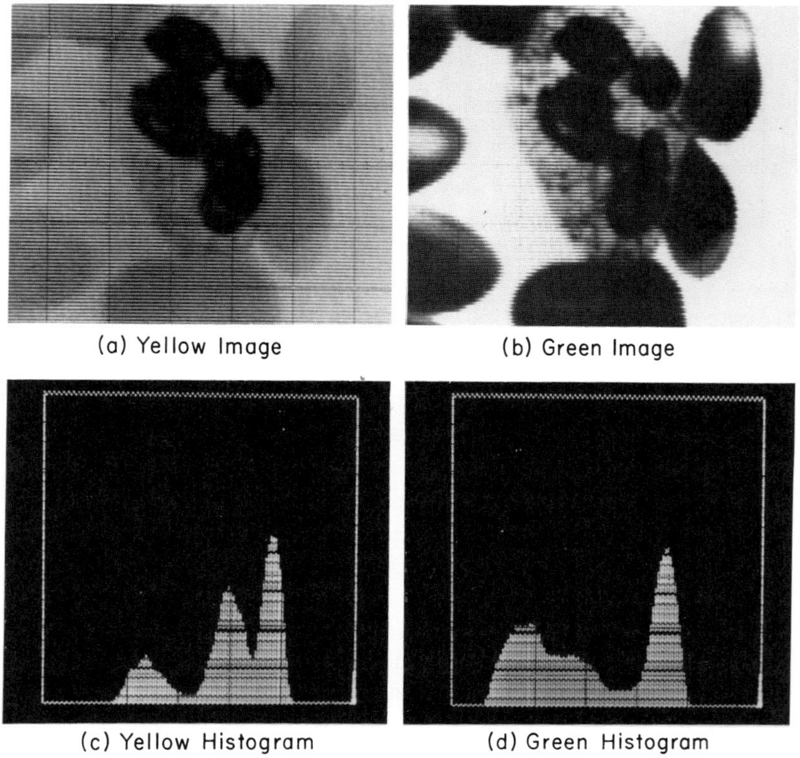
(a) Yellow Image
(b) Green Image
(c) Yellow Histogram
(d) Green Histogram

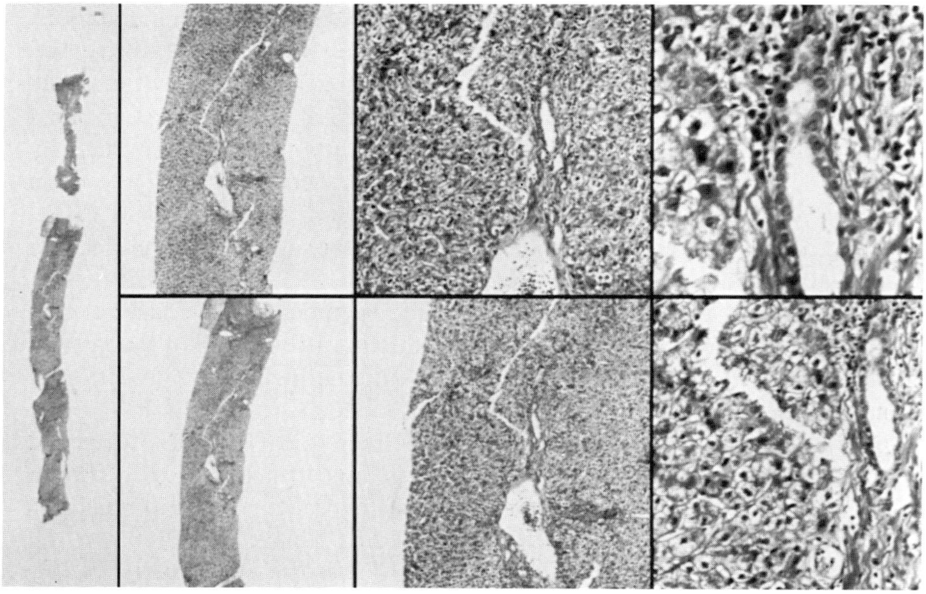

Examples of some early television images taken in the 1960s as well as some very recent work are shown in Figure 1.8. The former was taken from the monochrome scanner of CELLSCAN showing images of leukocytes scanned with a 0.1-μm line spacing over a 20 × 20-μm field. Each field contains a single cell. In contrast, Figure 1.8 also shows a full-color image generated by the Research diff3/50 system manufactured by Coulter for Carnegie-Mellon University showing a 1536 × 2048 mosaic of human liver-tissue images. The largest subimage is of a 10 × 20-mm field scanned at 22-μm line spacing, with the smallest a 180 × 180-μm field with 0.4-μm line spacing. In the case of CELLSCAN, the scanner utilized was a Dage slow-scan vidicon; in the latter case, a Plumbicon using only the central 64 × 64 picture elements from each field was employed, with the mosaic accomplished by a step-and-repeat procedure.

In quantitating the performance of the microscope scanner, the user must be aware of the modulation transfer function of both the optics and of the overall system. To preserve important detail, the frequency response of the video amplifier is often tailored by what is called "aperture correction" so as to obtain the best overall performance. Table 1.2 lists the modulation transfer function and cut-off frequencies of several automated scanners for purposes of cross-reference.

Image Digitization and Preprocessing

Typically in cell and tissue analysis, the field addressed by the video system contains from a few thousand to a few million picture points. In the analysis of individual human blood cells, the field of view is usually 25 × 25 μm, with a picture element spacing of a few tenths of a micrometer. Computer memory requirements are therefore only a few tens of kilowords. For chromosome analysis, the field of view must be enlarged to approximately 80 × 80 μm so as to encompass all of the chromosomes from the entire metaphase cell. In order to recognize banding patterns a picture point spacing of 0.1 μm is usually required, leading to storage requirements of about half a million picture points. Tissue-section analysis places an even

---

FIGURE 1.8. (*Top*) Traditional television images generated by CELLSCAN/GLOPR mirror scanner showing electronic images produced in (**a**) yellow and (**b**) green and their associated optical transmission histograms. Note the trimodality in (**c**) where the modes correspond to nucleus, cytoplasm, red cells and background plasma, respectively. The cell under study is a hematoxylin-eosin-stained neutrophil in a human blood smear produced by spinning. (*Bottom*) Full-color mosaic of multiresolution imagery produced by the research diff3/50 showing scenes from a needle biopsy of human liver tissue varying from a field size of 5,000 × 20,000 μm (*left*) to 170 × 170 μm (*upper right*). See Color Plate III.

greater burden on the computer, with the image storage requiring billions of picture points.

To combat problems in both picture point data storage and processing, many methods of data compression have been investigated. This field of study has loosely been called "preprocessing." Four basic preprocessing techniques are employed: (1) minimization of the number of bits required per picture element, (2) the packing of several picture elements per computer word, (3) use of algorithms for combining the values of connected groups of picture elements in such a fashion that fundamental information is not lost, and (4) employment of a closed-loop, multiresolution microscope, which avoids digitizing all picture elements on the microscope slide and instead searches for regions of interest at low resolution and then examines these at increasingly higher resolution.

Another important aspect of video microscopy is the understanding of the quantization noise introduced by digitization itself. Although mentioned infrequently in the literature on image processing in cell and tissue analysis, this has been thoroughly discussed in a signal processing context by Bennet [1948], Widrow [1956], and Bruce [1961]. Also frequently omitted in the literature on automated microscopy is the question of evaluating the effect of the sampling interval, ie, the distance between picture elements in the object plane, on recognition success rates. Wied et al [1970b] treats this problem. Gaven et al [1970] also discusses it in the much broader context of scanning systems for intelligence gathering.

Methods of combining picture points for the purposes of data compression are treated by Norgren et al [1967], who was able to obtain an order of magnitude reduction (from 40,000 picture points to only 4,096) by using a "white stretching" technique on each video line in addition to using a line combiner. This is illustrated in Figure 1.9. Far more sophisticated techniques for reducing the number of bits required per picture point are presented in the paper by Huang et al [1971] on "bandwidth compression." As of this writing, these techniques are infrequently employed in automated microscopy primarily because of the rapid decrease in the cost of semiconductor memory and the expense, complexity, and relatively slow speed of systems capable of bandwith compression.

Techniques for Scene Segmentation and Cell Recognition

The process of cell identification and of the analysis of tissue structures is similar in many ways to scene analysis in studies of what is called "pattern recognition." Referring to Figure 1.10 and the general survey of pattern recognition by Kolers and Eden [1968], it is found that the two major steps are: (1) generation of measurement vectors from the image data matrix and (2) classification of the image components from the measurement vectors. This section treats these two operations separately, emphasizing measurement techniques rather than classification methods.

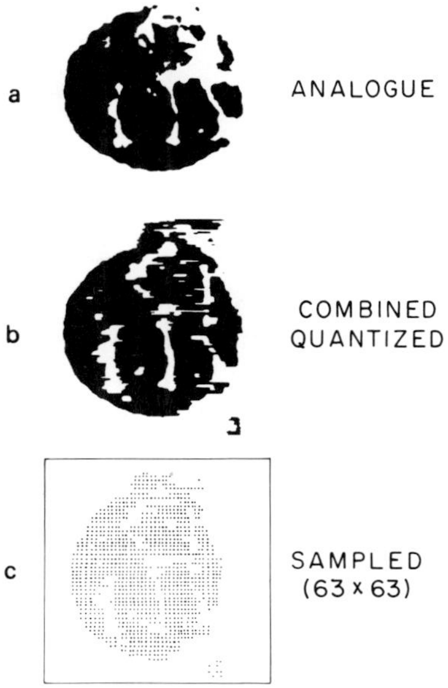

FIGURE 1.9. One of the earliest examples of data compression as applied to images of single cells is that due to Norgren et al [1967], where a 200:1 reduction was achieved of the original image **(a)** by using electronic preprocessing to "stretch" transmitting (white) region **(b)**, so as to generate a final binary image of 63×63 sample points spaced at $0.3\mu m$ intervals **(c)**. The cell under study is a peroxidase-stained granulocyte from a smear of human blood.

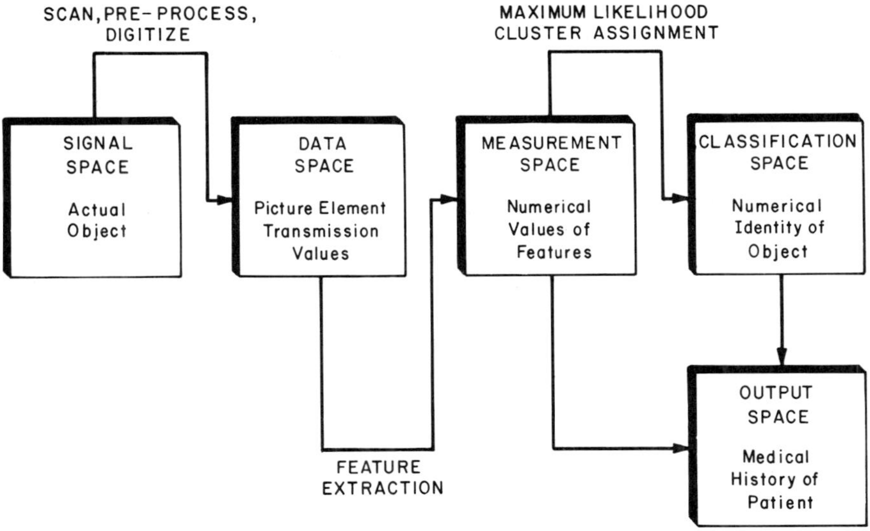

FIGURE 1.10. Block diagram of the procedure followed in automated pattern recognition whereby digital picture analysis is used to generate measurements from the data produced by scanning, preprocessing, and digitizing the cell or tissue image followed by measurement classification and storage of the final results as part of the patient's medical history.

Sample Preparation

The importance of sample preparation to the success of cell and tissue analysis cannot be overemphasized. The designer of the cell- or tissue-analysis system must make every possible effort in both mechanical and chemical preparation of the specimens to generate monolayers of cells or tissue labeled colorimetrically in such a fashion that their identity is recognizable readily not only to the human observer but also to the machine. When this is done and appropriate measurements are devised, the measurement vectors for classes of similar cells (or similar classes of tissue) should cluster and exhibit a large interclass difference with other types of cells (or tissue samples). If not, it is impossible for the classifier to determine whether the measurement vector for a new cell (or tissue) is more likely to belong to a particular class out of the many possible classes.

Classification Methods

Methods of classification for a given measurement vector are well developed. They are treated in an extensive literature, and one need only review the early paper by Fisher [1936] or the excellent text by Rao [1952] to obtain a thorough working knowledge of the field. More modern approaches to classification by hierarchical sorting should also be reviewed. See the early paper by Ornstein [1960] as well as that by Bartels et al [1984a].

Generating and Evaluating the Measurement Vector

The most difficult problem is how to find suitable analytic approaches to map the image data matrix into appropriate measurement vectors. The image itself cannot, of course, serve as the measurement vector. Typical cell and tissue images consist of from several thousand to several million picture points. There are at least 2^6 meaningful values per picture point per color. Thus, there are essentially an infinity of possible pictures. To map picture matrices directly into classification space is impossible. The crucial step in the design of the pattern-recognition system is to find the mapping that transforms the image matrix into measurement vectors with as great a reduction of dimensionality as possible. In general, since there are less than 100 cell classes to be treated in most image-analyzing situations (or, similarly, with tissue classes), the mapping into measurement vectors should be capable of reducing the information in the data array to a very small number of bits. Current systems usually perform a reduction of two or more orders of magnitude from the image matrix into measurement vectors. The final step in the process is then vector identification.

A few workers have looked into analytical methods for generating and evaluating measurements. The early work of Uhr and Vossler [1961] and Lewis [1961] is of historical interest. Recently, further analytical approaches have been taken to the formulation and evaluation of measurements. This is exemplified by the work of Yasnoff and Bacus [1984].

Types of Measurements

There are certain well-defined measurement-making techniques, and in the following subsections, this section provides a general discussion of them, namely (1) optical density histogramming, (2) stereologic analysis, (3) boundary and contour tracing, (4) counting and sizing, (5) shape analysis, (6) texture analysis, and (7) spatial frequency analysis. These seven subsections each include illustrations and are followed by a summary. Results in both cell and tissue analysis are discussed later.

Optical Density Histograms

The histogram of the frequency of occurrence of the picture point values, ie, the optical density histogram, has been found to be useful in the locating of picture points belonging to the major components of cells and tissue, namely, nuclei, cytoplasm, and vessels. Examples of optical density histograms are shown in Figure 1.8. The histogram for yellow illumination shows three modes corresponding (left to right) to nuclei, cytoplasm, and background. Nuclear and cytoplasmic boundaries may be determined easily by thresholding at the minima that occur between these modes. The bilevel masks produced by thresholding may be used at other wavelengths to obtain the color vector that describes each of the regions found. Many researchers, such as Tycko et al [1974], Prewitt and Mendelsohn [1966], Gelsema and Powell [1972], Halaby and Vance [1979], Cheng [1974], Noguchi et al [1977], and Noguchi [1982], have used image segmentation and color analysis. Other work on the statistics of these optical density histograms are presented in Kirsch [1969].

Colorimetric measurements combined with morphological measurements usually permit the final automated recognition of the structures present in the image data. Aus et al [1977] and Garbay et al [1981] furnish an example as applied to bone marrow cells. In the case where DNA-specific stains are employed, the area under the optical density histogram, the "total optical density," which is proportional to DNA content, is also useful. Direct determination of the malignancy of the cells' nuclei under study may often be made using this measure.

Stereology

The field of stereology originated more than a century ago with the work of the geologist Delesse [1846] and Glagoleff [1933]. Stereologic techniques were first applied to biology by Chalkey [1943] and extended to the field of cell and tissue analysis by Weibel et al [1966, 1971] and his co-workers Freere and Weibel [1967] and collected in the book by Weibel [1979]. The basic equations for two-dimensional stereology are

$$S_a = 2N_a$$

$$A_a = P_a \tag{1.1}$$

where A_a is the area of the component a in the image data, and S_a is its perimeter. The quantities N_a and P_a are best illustrated by referring to Figure 1.11 which shows two stylized chromosomes cut into horizontal swaths. Each swath represents the portions of chromosomes intercepted by a single scan line. The quantity N_a is the number of intercepts associated with component a, and P_a is the total number of picture points in the associated intercepts. These intercepts are sometimes called "runs," and analysis of their lengths is called the "run-length histogram," which provides some generalized information about the image data. For example, for chromosomes that are oriented vertically, there will be a large number of runs whose length is characteristic of the arm-width. If oriented horizontally, there will be a relatively smaller number of runs whose length delineates the width of the chromosome arms. From the stereologic equations, one can determine total area and average perimeter-to-area ratio. This type of analysis is useful in generating simple shape descriptors in the analysis of such simple structures as nonoverlapping bacterial colonies and blood films consisting of nonoverlapping red cells.

Counting and Sizing

Stereology is probably the most elementary method of obtaining quantitative measurements on the morphology of objects characterized by the image data matrix. Data on object area plus area-to-perimeter ratios yield estimates of size, shape, and, of course, a count of the total objects. However, if the image data contain objects of many different sizes and shapes, the stereologic method is inadequate. If, however, one could associate all intercepts of a single object with each other, it would be possible to determine the (1) area of each object, (2) total boundary points, and (3) total number of objects. Early methods of intercept association as developed by Berkley [1959] were too simplistic to handle objects of arbitrary shape. Convoluted shapes were counted differently, depending on their orientation. However, Scott and Preston [1968] solved this problem by inventing the intercept continuity computer. Whenever the scanner first intercepts an object, an intercept number is assigned; when portions of a single object merge, the assigned numbers are merged; finally, by counting the number of assigned, but nonmerged numbers, a true object count can be established independent of object orientation. This method can be implemented at video rates using relatively simple circuitry. Furthermore, by utilizing additional memory, addressed by using the intercept number itself, it is possible to determine both total area and total boundary points associated with each and every object in the field.

Boundaries and Contours

If there are clearly defined modes in the optical density histogram, it is relatively easy to select suitable threshholds and generate masks specific

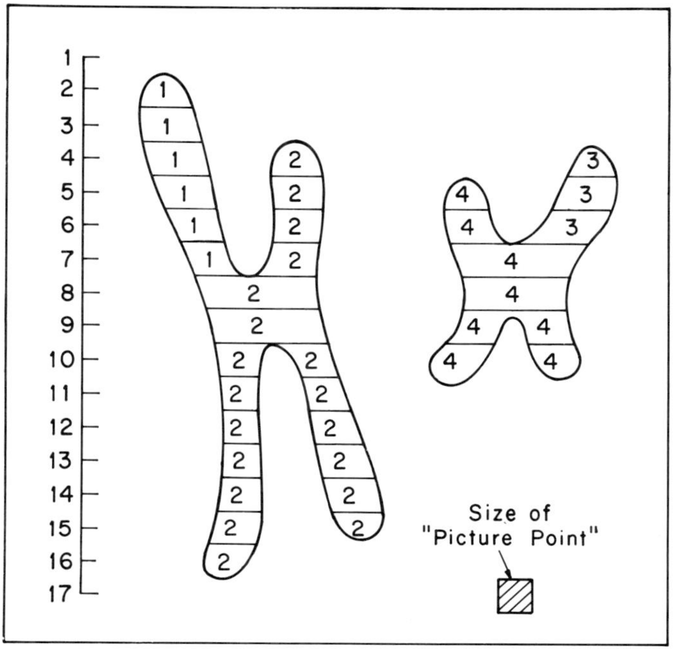

FIGURE 1.11. Illustration of "slicing" a hypothetical binary image of two chromosome-like figures into horizontal "runs." The numerical labels relate to train-continuity object counting patented by the Perkin-Elmer Corporation.

to the major components of the image. When this is not true, more complex methods of gradient tracking may be necessary to find boundaries. Boundary-following circuits were implemented in early flying-spot scanners as reported in Sprick [1956], Sprick and Ganzhorn [1959], and Broullette [1961]. Such scanners stored *x*- and *y*-deflection waveforms as boundaries were followed. From these, the boundary coordinates were determined. Later, Freeman [1961] worked on the same problem using image information in digital form. For each boundary location, he listed the vector direction of the boundary at that point. This is the digital equivalent of the earlier analog approach. Mendelsohn et al [1971] worked on boundary detection in chromosome images, and Prewitt [1972] did similar studies on boundaries in images of leukocytes. The reader should also consult parallel work in radiology where Selzer [1969] located arterial boundaries in radiographs. Finally, Hou et al [1970] and Young and Paskowitz [1974] discuss the general problem of describing biological forms. When the field of view is complex, further refinements such as those given by Preston and Dekker [1980] can be used to delineate both cell nuclei and vessels in tissue sections.

Other methods of boundary finding are given in Chien and Fu [1974], who used an "embedding metric" theory. The metric employed was that of Fischler and Elschlager [1973], which has a value at the point x given by the equation:

$$G(x) = \sum_{i=1}^{P} \sum_{j=1}^{i} g_{ij}(x_i, x_j) \tag{1.2}$$

where the kernel g_{ij} is the "global evaluation function." This function is the "cost" of embedding the ith component at the location x_i and the jth component at location x_j in the image array of P elements. Chien and Fu [1974] applied this approach to radiological images with great success.

Another elegant method for contour development has been obtained by Sklansky et al [1973], originally for application to radiographic images, but also applicable to cell and tissue analysis. Sklansky extends the figure-of-merit approach originated by Montanari [1971] and maximizes the following objective function.

$$\max_{Z_1 \ldots Z_n} \sum_{i=1}^{N} G(Z_i) + \alpha \sum_{i=1}^{N-2} q(Z_i, Z_{i+1}, Z_{i+2}) + \beta c(Z_{n-1}, Z_n, Z_1, Z_2) \tag{1.3}$$

where $G(Z_i)$ is the two-dimensional gradient modulus at the point i, $q(Z_i, Z_{i+1}, Z_{i+2})$ is the "curvature" at the point d_i, and $c(Z_{n-1}, Z_n, Z_1, Z_2)$ is the "curve length" variable. These quantities take on the following values:

$$q = \begin{cases} 1 \text{ if } Z_k, Z_{k+1}, Z_{k+2} \text{ subtend an angle of } > 135^\circ \\ 0 \text{ otherwise} \end{cases}$$

$$c = \begin{cases} 1 \text{ if } Z_n = Z_1 \text{ or } Z_{n-1}, Z_n, Z_2 \text{ subtend an angle} \geq 135^\circ \\ 0 \text{ otherwise} \end{cases} \tag{1.4}$$

where $k(i,j)$ denotes a point in the I × J image array. Still other approaches to coding edge and/or boundary data developed by workers in artificial intelligence was given by Alexander and Thaler [1971] and Saraga and Wavish [1971]. Rosenfeld and Thurston [1971] have worked on the problem of locating boundaries between regions of different texture. Bengtsson et al [1977] have applied "ratio contouring" for finding boundaries of both nucleus and cytoplasm and have also studied overlapping structures in cervical smears. Finally, Yasnoff and Bacus [1984] have used "array" measures for locating boundaries in images of similar material.

Shape Analysis

Many standard-shape measures are available once the boundary of an object has been extracted by means of scene segmentation. These standard measures are automatically generated by such machines as the IMANCO Quantimet, Bausch and Lomb OMNICON, Zeiss Videoplan, etc. They include moments, circumscribing rectangles, major and minor axes of the nearest ellipsoid, etc. These and other measures are discussed below.

Trajectory Transforms

The boundary following scanners of Sprick [1956] and Broullette [1962] directly generated the x and y trajectories of the boundary of the object being followed. For example, when traversing the boundary of a circle of radius R at a uniform velocity v, these are given by

$$x = R\cos(vt/R), \; y = R\sin(vt/R) \tag{1.5}$$

The Fourier transform of these trajectories may be used as a shape measure. The Hankel transform may be taken by transforming into r and θ coordinates. High-frequency boundary regularities may be recognized by this method, ie, both gross- and micro-boundary structure may be quantitated.

Moments and Principal Axes

The first moment of the data provided in Equation (5) provides the centroid of the object. Higher order moments about the centroid provide further shape measurements as described in Hu [1961] and Sapp [1962]. Butler et al [1966] applied these moment-measurement techniques to chromosomes. The principal axes of each object about the centroid may also be computed.

Optical Density Profiles

The image data itself or of an image component may be projected on the principal axes to form what are called "optical density profiles." Differential optical density profiles may also be employed as used in radiology by Hall et al [1973]. Statistical measures may be extracted from these profiles such as their mean, mode, standard deviation, kurtosis, skew, etc.

Circumscribing Figures

The pseudo-convex-hull of an object may be generated readily as shown in Figure 1.12 from Preston [1973]. This approach was originated by Neurath and Enslein [1968]. It is used for the location of centromeres and arm tips in chromosome analysis by Gallus et al [1968], Rutovitz [1970], and Regoliosi [1970]. As presented in Ingram and Preston [1970], a "reduced hexagonal convex hull" may be employed to generate a size histogram of indentations in the nuclear boundary of blood cells. This method is employed commercially in the Coulter diff3-50.

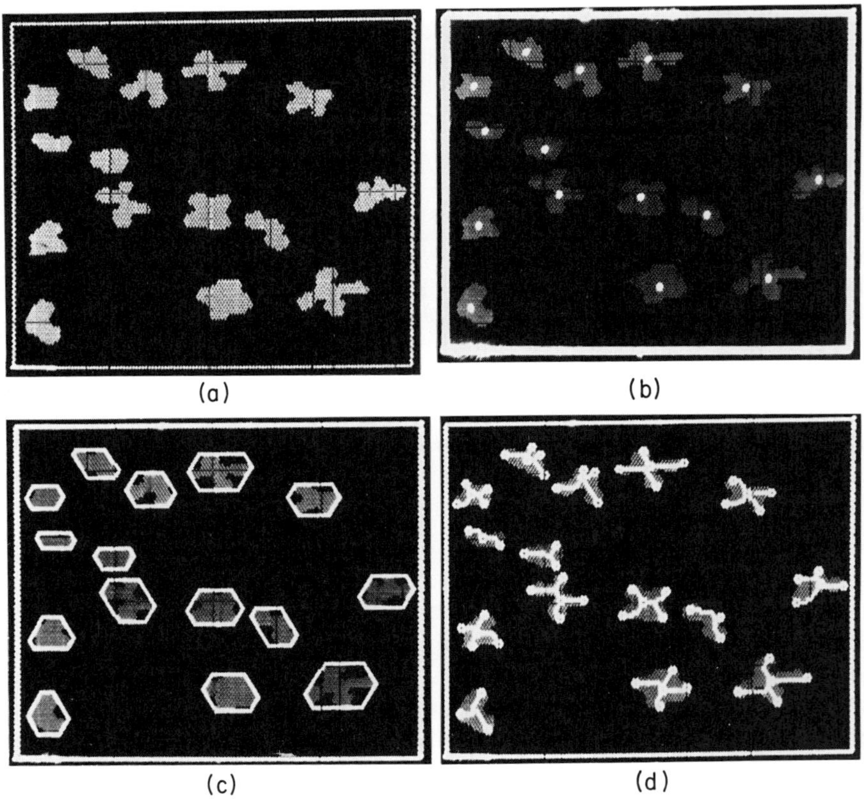

FIGURE 1.12. Illustration from Preston [1973] showing (a) a binary image of a chromosome spread, (b) centromeres located by a cellular logic reduction algorithm, (c) circumscribing hexagons generated by the cellular logic (median − 1) filter for the purpose of locating arm tips, and (d) the endoskeletons generating the centromeres and arm tips so located.

Shape Features

Heuristic work on the generation of shape features originated in the field of optical character recognition. Doyle [1959] introduced innovative methods of run-length analysis that led to quantitative measures of "cavities," "inlets," and "voids." Other work in this area is presented by Greanias et al [1957] and Kamentsky and Liu [1963]. More recent developments of shape measurements have been done by Gallus [1968], Gallus and Neurath [1970], and Gallus and Regoliosi [1974] as related to chromosome analysis. Shape recognition of nuclear boundaries is discussed by Kiefer and Sandritter [1970], Bourk and Tretiak [1973], Kiefer et al [1974], Cowan and Wann [1973], and Arcelli and Levialdi [1973]. In the last case, the authors extend the analysis into three dimensions for the purpose of examining the structures of neuronal interconnections. Further work on three-dimensional neuronal interconnection has been done by Hafford and Preston [1984].

Skeletonizing and Augreding

Augment/reduction ("augreding") methods were first studied by Dinneen [1955] and Selfridge [1955] at MIT and Kirsch et al [1957] at the National Bureau of Standards. Methods of implementing this technology were developed by Unger [1958] at Bell Telephone Laboratories, patented by Golay [1965], and reduced to practice by Preston [1961] at Perkin-Elmer. More recently, the theoretical basis of augreding transforms have been outlined by Preston [1983b] based on earlier discussions by Preston [1972c].

By using connectivity-preserving augreding, the medial axis may be constructed as devised by Blum [1964]. Other skeletonizing methods have been developed by Montanari [1968], Hilditch [1969], and Levialdi and Montanari [1970] and applied to finding the endoskeletons of chromosomes. Various analytical discussions of variations of skeletonizing algorithms are given by Levialdi [1972], Deutsch [1969], and Rosenfeld [1970]. Prewitt [1972] was the first to define the exoskeleton (endoskeleton of the background) and applied this concept to blood cell analysis. Preston [1981] extended the procedure to tissue analysis. The endoskeleton of chromosomes is shown in Figure 1.12. The exoskeleton of both normal and abnormal kidney tissue is given in Figure 1.13. The problem of skeletonizing overlapping structures is treated by Tretiak [1966], by Voss [1974]

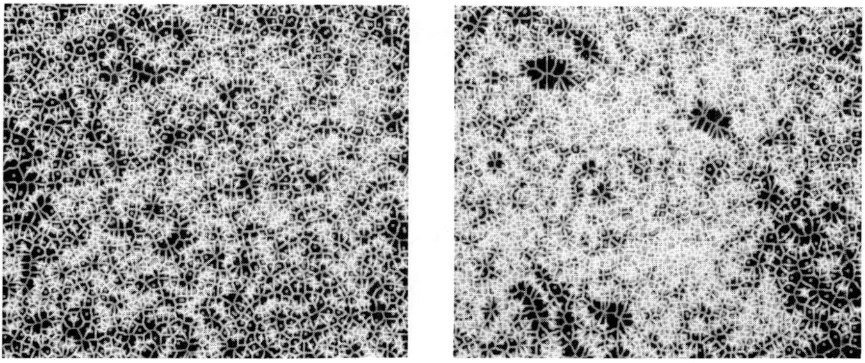

FIGURE 1.13. Exoskeletons generated using the (median − 1) cellular logic filter for both normal (*left*) and abnormal (*right*) 512 × 512 images of human kidney tissue. White indicates the portions of the arcs of the exoskeletons formed in the initial stages of the iterative algorithm; green in the intermediate stages; blue, in the final stages. The exoskeleton arc length histogram readily identifies the normal or quantitates the degree of abnormality of the tissue architecture. See Color Plate IV.

as regards overlapping red cells, and by Hilditch [in Preston and Duff, 1984] as regards chromosome clusters.

TEXTURE ANALYSIS

Measures of local variability in both optical density, color, and shape are frequently useful in determining the identity of either nucleus or cytoplasm. Following the early work of Schaphorst [1961] in the analysis of television signals using communications theory, Bartels et al [1969, 1973, 1974] used neighborhood transition probabilities within the nuclear and cytoplasmic regions as a measurement vector. Algorithms for making these measures are found in the literature on "texture analysis." For example, malignant cells often show coarse chromatin structure that serves as a valuable visual diagnostic clue.

Analysis of irregularities in the nuclear outline is given in Holmquist [1979]. Bacus [1970] devised a method of run-length analysis to quantitate variations in optical density within the nucleus. Regions exhibiting "fine" texture exhibit short runs while "coarse" textured regions, conversely, exhibit long runs. Carvalko [1972] used the Golay transform to measure nuclear texture differences between lymphocytes and monocytes. Other studies relating to texture measurement of cells and cell constituents are provided in Kirsch [1969], Jayaramamurthy [1973], Lipkin and Lipkin [1974], Harms et al [1978, 1979a, 1979b], and Vastola et al [1974]. In other areas of medical-image analysis, texture measures have been developed by Jagoe and Paton [1976] and Hall et al [1973]. Using the Fourier transform, the spatial-frequency-power spectrum (Wiener spectrum) can be measured. Fine texture contributes high spatial frequencies while coarse texture has the opposite effect. This is reviewed in the next section.

SPATIAL-FREQUENCY ANALYSIS

As discussed in texts by both Goodman [1968] and Preston [1972a], coherent light passing through the microscope slide is scattered in such a fashion that the distribution in the focal plane of a lens that collects and collimates this light is given by the Fourier transform

$$F(w_x, w_y) = \int_{-A/2}^{+A/2} \int_{-A/2}^{+A/2} f(x,y)\exp[-j(w_x x + w_y y)]\, dxdy \quad (1.6)$$

where $f(x,y)$ is the light-amplitude distribution in the illuminated field of view whose dimensions are $A \times A$. Workers using flow systems have studied this relationship extensively [Brunsting, 1972, 1974; Mullaney et al, 1969a–b, 1974; Fiel, 1970; Fiel and Scheintaub, 1973; Fiel et al, 1970; Fiel and Munson, 1970].

Norgren [1969] was the first to apply this relationship to locating metaphase spreads on microscope slides. Other workers who made analytical measurements of optical scatter are Aughey and Baum [1954], Bateman [1968], Koch and Ehrenfeld [1968], and Berkman and Wyatt [1970]. These last workers applied the technique to study the size and shape of bacteria. Bowie [1970] used the method for leukocyte identification as did Lurie et al [1969] and Rauch [1972]. Ward et al [1974] applied this approach to the analysis of reticulocytes, and Kopp et al [1974] used the Wiener spectrum to detect malignant cells in cervical smears. Bryant et al [1969] presented experimental determinations of optical scatter. Surveys of this method are given in papers by Gravatt [1971] and Wyatt [1972, 1973]. Finally, it is worth noting that an entire line of instruments for industrial inspection are based on "ring and wedge" detectors to measure the distribution of scattered light. Literature on these devices (trademarked ROSA-3) is available from Recognition Systems, Inc. (Van Nuys, California).

Measurement Summary and Comparison

The researcher confronted with the plethora of image measurements described before is sometimes confounded by which to utilize to carry out the recognition task at hand. Only in the area of leukocyte analysis is there currently some basis of comparison of measurement ability. Significant statistical data have now been gathered on both research and commercial programs in this field. It has become apparent that only a small number of carefully selected measurements are required to classify most of the types of human white blood cells. Although a thorough examination of the importance of these measures has not been published, some workers have pooled their measurements so as to make cross-comparisons possible. Table 1.3 lists measures with a priority ranking. Comparing columns in this table, it is noted that there are both consistencies and inconsistencies. In general, however, it can be seen that it is necessary to measure both nuclear and cytoplasmic shape, texture, and color. It is not clear at this writing which measurements are optimum. Until it is possible to review the algorithms utilized and to evaluate the cell-image-transfer process of the automated cell analyzers used, a definite decision on optimum measurement methods for the analysis of human white blood cells cannot be established.

In the case of chromosome analysis, an extensive cross-comparison was undertaken under the auspices of the World Health Organization between four laboratories in the United States (Jet Propulsion Laboratory, University of Colorado, Tufts New England Medical Center, and the National Biomedical Research Institute) and two in Great Britain (Medical Research Council and National Research Council). This cross-comparison was executed only for measurements of centromeric index and arm-length measurements and, as of this writing, has not been performed for the banding-pattern analysis. Results are given in Table 1.4.

Informal cross-comparison studies are conducted between laboratories in Japan, United States, and Europe in the analysis of cervical smears but no literature on this work is known to the authors.

TABLE 1.3. Measurements used for white blood cell differential count.

Researcher	Prewitt, JMS (1972)	Preston, K, Jr (1970)	Neurath/Powell*	Bacus, JW (1970)
Affiliation	Nat'l Insts of Health	Perkin-Elmer Corp	Tufts/CERN	Univ of Illinois
Material preparation	Manual smear; Gallocyanin chromalum and naphthol yellow S stain	Spun smear Wright-Giemsa stain	Manual smear Wright-Giemsa stain	Manual smear Wright-Giemsa stain
Transducer	TV microscope (CYDAC)	TV microscope (CELLSCAN)	TV photograph scanner	TV microscope
Illumination	503 nm (monochrome)	540 nm (green) 570 nm (yellow)	Wratten 44 (green) Wratten 22 (yellow)	Wratten 44 (green) Wratten 15 (yellow)
Pixel spacing	0.25 micrometer	0.20 micrometer	0.10 micrometer	0.33 micrometer
Pixel range	8 bits	8 bits	6 bits	8 bits
Total number of measurements	80	35	46	8
Types of measurements				
Optical density histogram	Mean (N, C, T) Mode (N, C, T) Intermean distance (N-C, N-T) Intermode distance (N-C, N-T) Absorbance (N, C, T) Mean ratios (N/C, N/T) Absorbance ratios (N/C, N/T)	Mode (N, C-2 colors) Mode ratios (N/C-2 colors)	Mean (N, C, T-2 colors) Absorbance (N, C, T-2 colors) Mean ratios (N/C, N/T-2 colors) Contrast† Color angle†	Mean color angle (C)† Mean ratio (N/C)

Stereologic			Average run length Run-length histogram Total pixels in runs Average run color Average run density (The above taken at high and low thresholds for N, C, T)	
Morphologic	Area (N, C, T) Perimeter (N, T) Diameter (N, T) Asphericity (N, T) (area/perim.^{2}) Area ratios (N/C, N/T) Diameter ratios (N/C, N/T) Asphericity ratios (N/C, N/T)	Area (N, C, T) Perimeter (N, T) Area dense nuclear chromatin Perimeter dense nuclear chromatin Number of lobes (N) Number of small, med., large concavities (N) Area nuclear bridges Area nuclear inlets Reduced area (N) Reduced perim. (N) Area nuclear skeleton Ends in skeleton Area nuclear voids Area cytoplasmic vacuoles	Area (N, C, T) Perimeter (N) Asphericity (N) (Perim^2 / area) Eccentricity (N) (Centroid/C.G.) Number of lobes (N) Average chain value (N) [See Freeman 1961]	Area (N) Perimeter (N) Asphericity (N) (area/perim^2)
Textural	Density moments (N, C, T) (variance, skew, kurtosis) Entropy (N, C, T)	Fine structure of dense nuclear chromatin, nuclear voids, cytoplasmic vacuoles, Number of clumps of dense nuclear chromatin	Color variance (N, C)	Variance color angle (C) Nuclear area having given edge derivative Cytoplasm area with given color angle derivative

TABLE 1.3. *Continued*

Best measurements	Mean (T), Variance (T), Absorbance (T), Absorbance ratio (N/T)	Area (T) Reduced area (N) Reduced perimeter (N) Fine structure (N) Vacuoles (C) Mode (N, C-2 colors)	Color contrast angle Mean (N, T) Area (T) Mean (T-yellow) Asphericity (N) Color variance (N) The above are best for a 6-class analysis. When a 16-class analysis was carried out, the best were: Mean (N-Yellow) Color Angle (C, T) Color contrast magnitude Mean (N, T) Mean (T-green) Contrast (N/C-green)	NA (all meas. used)

*Unpublished work.

†Color angle and contrast diagram as below:

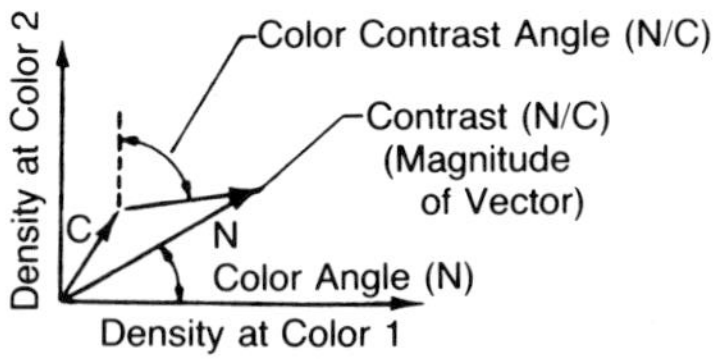

N = nucleus; C = cytoplasm; T = total cell.

TABLE 1.4. Investigators on chromosome aberration analysis.*

Laboratory	Arm length (μm)	Centromeric index (length)	Area	Centromeric index (area)
	Within laboratory overall standard deviations			
1.	0.27	0.023	0.31	0.023
2.	0.29	0.032	—	—
3.	0.27	0.032	0.33	0.029
4.	0.30	0.047	0.35	0.046
5.	—	—	0.42	0.045
6.	0.45	0.063	—	—
	Between laboratories comparative standard deviations			
3.–1.	0.14	0.024	0.22	0.022
3.–2.	0.15	0.028	—	—
3.–4.	0.18	0.039	0.28	0.037
3.–5.	—	—	0.31	0.026
3.–6.	0.30	0.052	—	—
2.–1.	0.13	0.025	—	—
2.–4.	0.19	0.039	—	—
4.–1.	0.19	0.034	0.25	0.032
4.–5.	—	—	0.25	0.033
5.–1.	—	—	0.30	0.027

*Adapted from World Health Organization Report RHL/WP/72.5, 1972.
1. Jet Propulsion Laboratory (KR Castleman); 2. University of Colorado (HA Lubs); 3. Tufts New England Medical Center (PW Neurath); 4. National Biomedical Research Institute (RS Ledley); 5. Medical Research Council (D Rutovitz); 6. National Research Council (T Kasvand).

Recognition and Classification

It is generally assumed that the generation of measurement vectors is critical to the automated analysis of cell and tissue images. If measurement data do not cluster according to the classes of cells or tissue examined, then it is better to devise new and better measures than to resort to "heroic" techniques for clustering data that are essentially unclusterable. This section, therefore, undertakes to describe the two major types of classification techniques. Both are based on the assumption that there are modes in the measurement data. These modes must first be found and then, when located, used to classify new objects. The discovery of modes is called "unsupervised classification," and the assignment of data to predefined modes is called "supervised classification."

NONSUPERVISED CLASSIFICATION

Nonsupervised classification is a problem of the most basic and fundamental nature (see Figure 1.10). Assuming the researcher has no a priori knowledge as to the inherent clustering of the measurements taken on different classes, the problem of nonsupervised classification or "learning" is to determine whether there are significant modes or "clusters" in the measurement set. Using nonsupervised classification, one can discover structures in data that are data dependent rather than dependent upon somewhat subjective and arbitrary a priori human classification. This approach to the analysis of data sets is treated extensively in the pattern-recognition literature. However, little of this literature relates directly to cell and tissue analysis. Bartels et al [1970] are the major authors who discuss nonsupervised or self-learning methods, with emphasis in cell analysis. This subject is also treated by Bhattacharya et al [1971, 1973]. Rutovitz [1968] treats applications in chromosome analysis.

One of the most frequently used general computer programs for data clustering is ISODATA, as described in Ball and Hall [1957]. Other papers of interest that utilize the human observer to assist the computer on an interactive basis are by Patrick and Fischer [1969] and Sammon [1969, 1970]. The system described by Sammon [1970] is called OLPARS and was developed for general use by the U.S. Air Force at the Rome Air Development Center. Another system operating in the same location is described by Hall et al [1969]. Marrill [1969] discusses a similar system, CYCLOPS, in use at Bolt, Beranek, and Newman. A compendium on the subject by Bullock [1971] is recommended to the dedicated reader. The above systems are general purpose, and a review of their characteristics is worthwhile for any reader.

One of the first systems specific to biomedical image analysis was that developed by Hodes [1970] at the National Institutes of Health. A more recent system developed at the M.D. Anderson Hospital is that of Johnston [undated]. The TICAS software of Wied et al [1968c] and that of the Medical Research Council as described by Hilditch [1970] provide similar facilities.

Supervised Classification

As shown in Figure 1.10 the final classification step maps the measurement vector into classification space. This mapping is "supervised" when a user has a priori knowledge about the measurement clusters. In this case the classical techniques of Hotelling [1933] and Fisher [1936] may be used to construct weighting functions, which define "hypersurfaces" isolating compartments in measurement space, which correspond to known classes of images or image components. Classification is readily performed from a set of C linear equations given by

$$c_i = \sum_{j=1}^{j=M} w_{ij} m_j \tag{1.7}$$

where the m_j are the values of the M measurements, (w_{ij}) is a matrix of weights, and c_i are the class coefficients. The maximum component of the vector c_i determines the class to which the vector m_j belongs. This classical approach to supervised data analysis is reviewed in such papers as Highleyman [1962], Karush [1963], and Sebestyen [1962]. This general field of analysis is called "decision theory." In the 1960s several researchers applied its methodology to biological samples. Other important work on using classification by linear programming is given in Mangasarian [1956], by dynamic programming by Nelson and Levy [1968], and by binary tree classifiers by Anderson and Fu [1979].

Tanimoto and Loomis [1960], Tanimoto [1961], and Ornstein [1960, 1965] investigated cases where classical decision theory was inadequate. They treated measurement data that could not be readily interpretable as residing in a Euclidean space. They proposed a metric that could be used to determine the similarity between an unknown set of measurements m_i and several standard measurement sets r_{ij} representing different types of cells (or tissues). This metric is given by

$$\sigma_j = \frac{\sum_i \min\,(m_i, r_{ij})}{\sum_i \max(max_i, r_{ij})} \tag{1.8}$$

where $\min(m_i,r_{ij})$ stands for the smaller of the numbers in the pair (m_i,r_{ij}) and $\max(m_i,r_{ij})$ for the larger.

Results

Looking back over 40 years of research and development in image analysis for cells and tissue, one sees a confusing picture. There are the early successes in the automated counting of red blood cells in Great Britain, the failure of the Cytoanalyzer program in the United States, and, subsequently, highly successful developments in the automated analysis of human white blood cells. Even in automated blood-cell-image analysis, however, economic success has eluded the manufacturers who undertook commercialization of these instruments. None of the four commercial U.S. machines developed remains in production. At the same time, despite extensive work in the development of automated sample preparation and digital-computer-image analysis of cervical smears and of chromosome preparations, there is no commercial instrument available that automates either of these tasks. This is a disappointing product for almost a half century of effort.

However, the use of some kind of computer-assisted image processing is now employed in research by many departments of cytology, hematology, and pathology. In some locations, data are gathered for clinical use and entered in the patient's history [Wied et al, 1985b]. This is particularly true for determinations of cellular DNA content in the assessment of malignancy. Also, it appears clear that use of computer quantization in immunocytochemistry may provide more useful data than immunoassay by either chemical or radiological means.

Forty years of research and development have provided few successes in total automation and few clinical applications; more decades will be required before this technology is widely utilized. Also, today the spread between the "haves" and "have nots" is increasing. An example of this anomaly are statements of the difficulty of measuring 400 to 500 cells in a histologic section while Preston [1985], using the Carnegie-Mellon Coulter Electronics Research diff3-30 at Presbyterian Hospital in Pittsburgh, electronically "harvests" the images of 5,000 cell nuclei from histologic sections of the lymph node and sizes them for the purpose of small-cell/large-cell lymphoma differentiation in a period of 10 minutes.

The following sections provide examples of today's state of the art for automation of cell analysis for blood smears, cervical smears, and chromosome preparations. Since automation is yet to be achieved in tissue analysis, this subject is omitted.

Automation of Blood-Smear Analysis

Commercial automation has occurred in this field based on the early work of Ingram et al [1968] and Prewitt and Mendelsohn [1966]. In these early investigations, digitized images of only a few hundred cells were machine analyzed. Results, although encouraging, were statistically inclusive. Full experimental automation, but only for a sample of about 2,000 white blood cell images, was next reported by Ingram and Preston [1970] (see Figure 1.14). Their system (CELLSCAN/GLOPR), although fully automated, ie, with provision for focusing, searching, locating, measuring, and classifying white blood cells without human intervention, operated at the rate of only about one cell image per minute. Experimental confirmation of machine results was made by comparison with visual counts conducted by a few observers on samples taken from only seven patients. This work plus that of Miller [1973], Bacus [1970], and Green [1974] led to commercial developments by the Coulter Electronics Corporation, the Geometric Data Division of SmithKline-Beckman, Corning Glass Works, and Abbott Laboratories. As reported by Miller [1973] of Geometric Data, for a sample of some 20,000 cell images, an overall false-positive rate of 3% was achieved with some misclassification problems occurring in the rarer cell types such as myelocytes, promyelocytes, nucleated red cells, blasts, and plasmacytes. However, Corning Glass and Abbott Laboratories stopped production in the 1970s and Coulter and SmithKline in the 1980s. Omron and Hitachi still manufacture similar machines in Japan, with about 300 in place.

The major reasons for success in automating the white blood cell differential count were (1) ease of automating sample preparation; (2) the work was performed by semiprofessional staff so that there was no legal barrier to automation; and (3) the task itself was well-defined and documented.

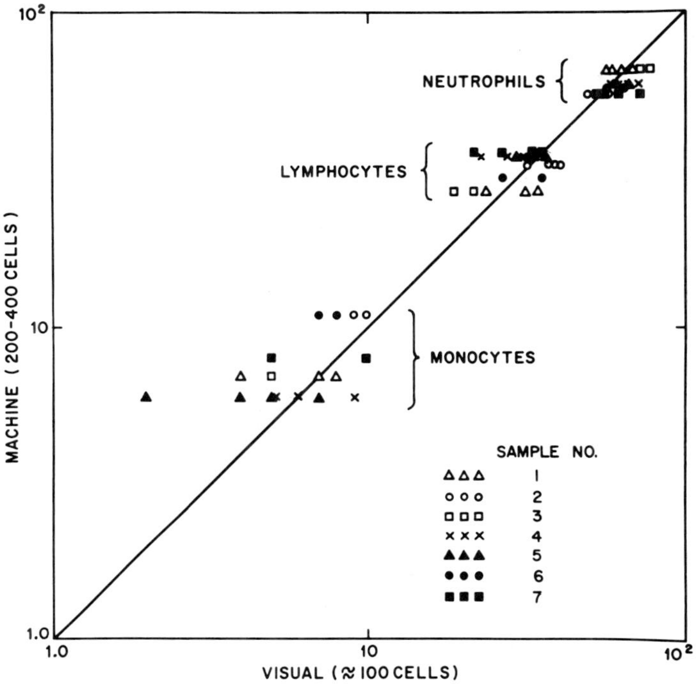

FIGURE 1.14. Results of the first reduction to practice of a prototype robot microscope for recognizing the major types of human white blood cells. Seven blood smears (from seven different subjects) were automatically scanned by the Perkin-Elmer CELLSCAN/GLOPR as well as being visually examined by four separate clinical technologists. Points lying on the diagonal line indicate exact agreement between machine and human observations. In general, the disagreement in the counts shown is as great or greater among the participating technologists as between the machine count and the average result of the technologists taken together.

One of the simplest sample-preparation systems is that used in hematology. Blood is drawn from a vein or by puncturing the finger. It is applied to the microscope slide, with the only chemical treatment being that for inhibiting coagulation. Traditionally, a few drops of blood are placed at one end of the microscope slide, and a smear is produced by the "squeegie" action of a second microscope slide or coverslip that is used as a smearing tool. As shown in Figure 1.15, the result is a layer of blood that is thick at one end and thin at the other. After smearing, the blood is fixed by air drying and is stained biochemically using "Wright's stain." Many classical texts, pamphlets, and manuals describe the details of this process. The reader is referred to Diggs et al [1954], Wintrobe [1956], Harper [1968], Miale [1967], and the U.S. Army Manual [1963]. In the thick portion of the smear, cells are so clumped as to be unsatisfactory for visual analysis and at the opposite end are too widely separated and/or disrupted. Since,

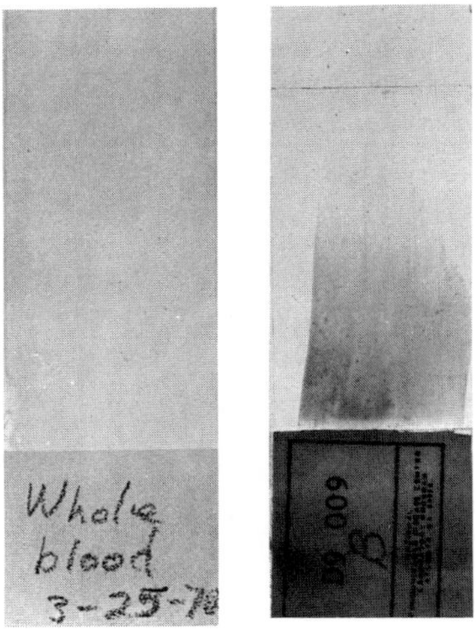

FIGURE 1.15. The preparation of human blood smears for machine reading was revolutionized by the spinning technique invented by Preston and Norgren [1971]. The pictures show typical smears of human blood on microscope slides, produced using both **(a)** mechanical spinning and **(b)** manual smearing.

however, only a few square millimeters need be observed to perform a 100-cell differential count, a region can usually be found which is satisfactory for visual analysis.

The spinning technique of Preston and Norgren [1971] revolutionized blood-smear preparation by borrowing from the semiconductor industry centrifugal methods used to spread photoresist on silicon wafers prior to masking (see Figure 1.15). As in manual smearing, a few drops of blood are placed on the microscope slide, but this time in the center. The spin axis is perpendicular to the slide and spinning is carried out for approximately 1 second after accelerating the slide rapidly to a velocity of about 5,000 rpm. After the initial invention by Preston and Norgren, further modifications were patented by Staunton [1972]. Detailed studies of the morphological properties imparted to cell by spinning were reported by Ingram and Minter [1969], Rogers [1972, 1973], and Bacus [1974]. Various commercial spinners have been produced by Corning Glass, Perkin-Elmer, Coulter Electronics, and Geometric Data.

Automation of Chromosome Analysis

A major effort in the preparation of samples for machine reading for chromosome analysis was funded by the United States National Institute of Child Health and Human Development in a joint venture between Jet Propulsion Laboratory and the City of Hope by Castleman and Melnyk [1975].

This did not lead to commercial instrumentation. Thus, manual preparation is still used. It is time consuming but not particularly difficult. Blood is cultured for several days using biochemicals that cause the cultured cells to be arrested at the metaphase stage of cell division. Next, the specimen is treated with a hypotonic solution that causes the metaphase cells to expand. Then, when they are spread on the microscope slide, the chromosomes are well separated. The original methodology for observing chromosomes in this manner was discovered by Hungerford et al [1964]. In the early 1970s, biochemical staining for chromosome analysis underwent a major change through the discovery and perfection of new treatments by Caspersson et al [1965, 1971, 1973].

Caspersson's treatment radically alters the appearance of the chromosomes by elucidating "banding patterns." The invention of Caspersson made specific identifications of chromosomes possible as well as studies of the translocation of specific genetic components. Additional discussions of these new developments are provided by Breg et al [1974b].

Alternative staining methods are compared by Sumner and Evans [1973]. Lundsteen and Philip [1974] furnish a discussion of the problems of performing microphotometry on such material. Much of the difficulty in the automation of the analysis of chromosomes lies in the initial task of locating suitable metaphase spreads on the microscope slide. As there are only of the order of 10 to 100 spreads per square centimeter with their individual size ranging from 50 to 100 μm, this is a difficult task to automate. Automatic metaphase location carried out by Wiener spectrum techniques is described in Wald and Preston [1968] and Ledley et al [1972]. Much preliminary work has been required prior to full automation as offered by such systems as CYTOSCAN manufactured by Shandon in Great Britain. CYDAC was used for chromosome analysis by Mendelsohn et al [1964, 1968, 1969] and Mendelsohn and Mayall [1974]. Other work in the United States is reported by Neurath et al [1966, 1969, 1972a], Ledley and Ruddle [1966], Ledley et al [1972], Castleman and Wall [1973], Aggarwal and Fu [1973, 1974], Stone et al [1967], and Klinger and Kochman [1971]. In Great Britain the major effort in automated chromosome analysis was that of Rutovitz [1967, 1968, 1970] using software described by Hilditch [1970] and Hilditch and Rutovitz [1969]. Work in Italy was done by Gallus et al [1968, 1971]. These efforts were closely coordinated through the World Health Organization, and there has been a rigorous interchange of both data and software.

With the introduction in the 1970s of new cytochemical-staining techniques elucidating "banding" patterns, extensive revisions were required in the digital processing methodology. A trace of optical density along its major axis may be expanded in a Fourier series and the coefficients utilized as a feature vector for purposes of classification. Work in this area has been undertaken by Caspersson et al [1971], Lundsteen and Philip [1974], and Thomas at Case-Western Reserve [unpublished].

Other workers such as Wald and Fatora [1974] devised software for analyzing chromosome damage produced by radiation in which chromosomes having multiple centromeres (dicentrics), those with circular (ring) configurations, as well as chromosome fragments are located and tabulated.

At Jet Propulsion Laboratory and the City of Hope, Castleman and Melnyk [1975], under contract to the National Institute of Child Health and Human Development, produced an automated system using the Digital Equipment Corporation PDP11/45 minicomputer capable of locating metaphase spreads and producing a computer-generated karyogram. Typical results are given in Figure 1.16. The entire cycle took about 8 minutes

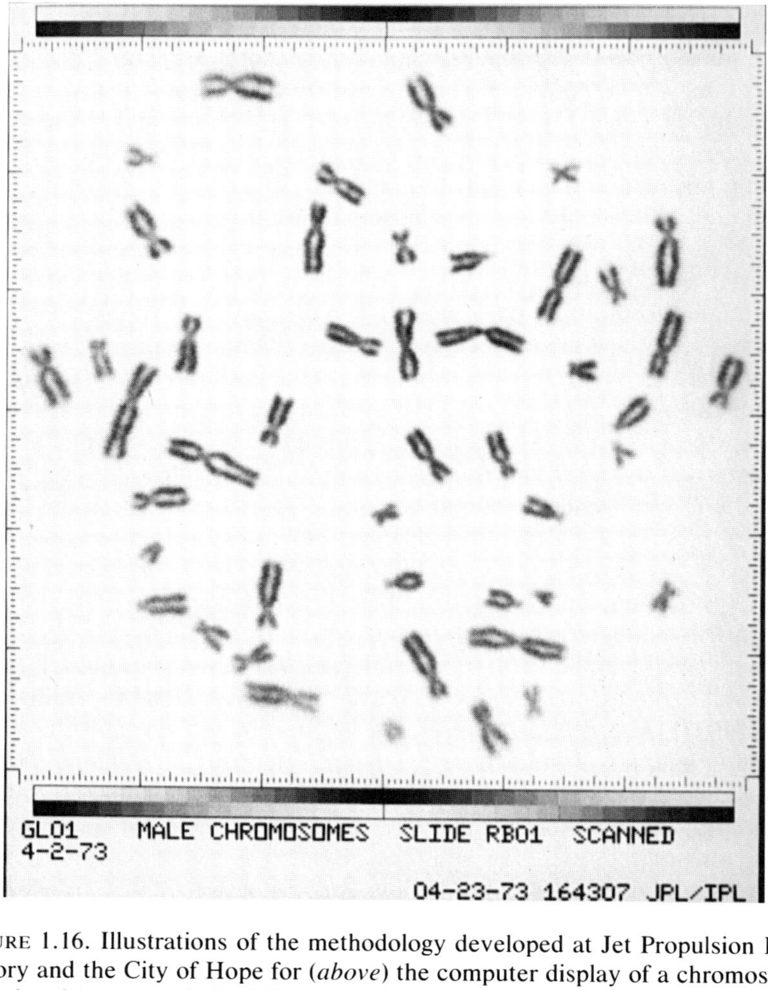

FIGURE 1.16. Illustrations of the methodology developed at Jet Propulsion Laboratory and the City of Hope for (*above*) the computer display of a chromosome spread and (*next page*) the fully-automated preparation of the karyogram, in the project reported by Castleman and Melnyk [1975].

per spread. The hope was that all U.S. newborns could be karyogramed using such systems. The system was then transferred for evaluation to Northwestern University and subsequently abandoned. Unfortunately, because of the unattractive economic nature of commercialization, no further efforts are being undertaken in the United States. Work in this area has continued in Europe. An example, due to ten Kate [1985] of the University of Delft is shown in Figure 1.17.

Automation of Cervical-Smear Analysis

Problems inherent to sample preparation disclosed during the Cytoanalyzer program forced the investigation of new methods. Early work by Diacumakos et al [1962] involved mechanical sieving of the material to remove debris and cellular clumps. These studies used aerosol-spray techniques for applying the sieved material to the microscope slide. Diacumakos et al [1962a–b] also investigated the possibility of discarding the original stain devised by Papanicolaou [1954] in preference to the Weigert's hematoxylin-tetrazine stain. Further efforts conducted as part of the Cytoanalyzer program are discussed in Pruitt [1959]. In the decade that followed, Parry et

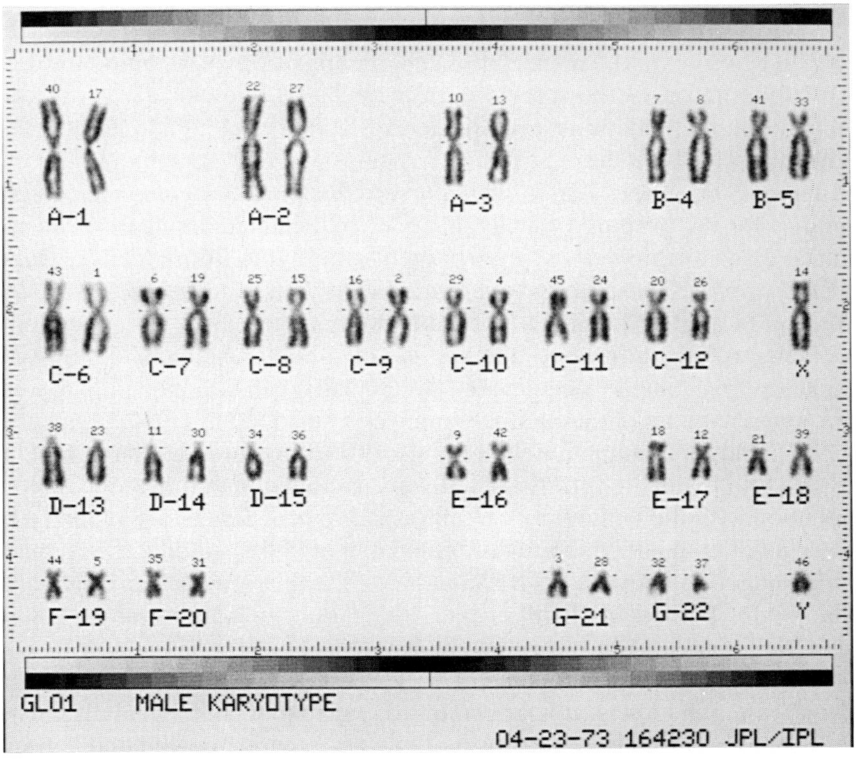

FIGURE 1.16. *Continued.*

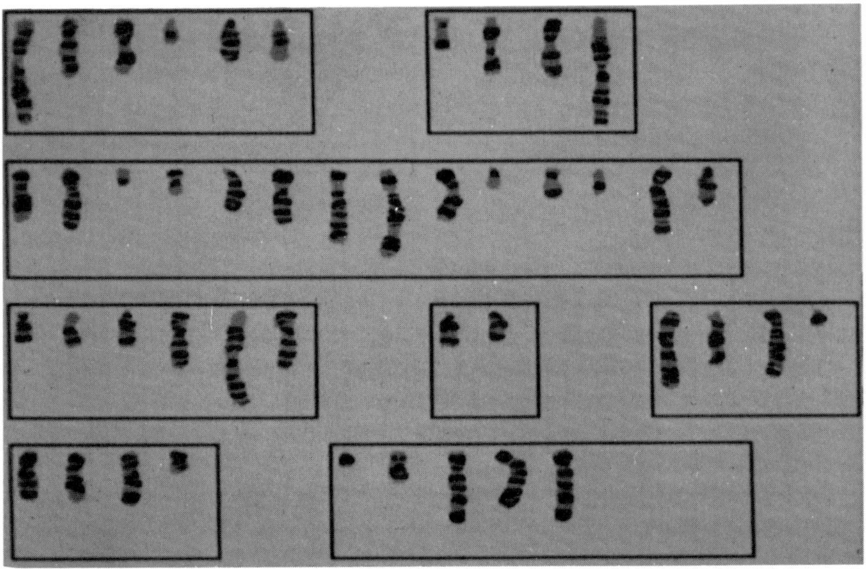

FIGURE 1.17. Chromosomes color-coded according to the size, location, and optical density of each band [TK Ten Kate, 1985]. See Color Plate IV.

al [1971] wrote on the dispersal of cell clumps using ultrasound, and Leif [1970] wrote on methods for separating cells by buoyant density. Other separation methods were investigated by Tolles et al [1978], Tucker and Gresham [1971], Bahr et al [1973], Wheeless and Onderdonk [1973], and Adams [1974]. These last workers pursued the possibility of using acridine orange for the purpose of analyzing epithelial cells in cervical-smear material using a flow-system environment employing fluorescence. Davis [1962] provides a useful survey. More recently, at St. Stephen's Hospital, Husain et al [1978] and Watts et al [1984] pioneered the use of cationic polyelectrolytes to prepare the surface of the microscope slide to assist in generating suitable monolayers for automated cell scanning. Other work on monolayer preparation was done by Schenck et al [1981], Leif et al [1977], and, in conjunction with the BioPEPR project in Holland, by Oud et al [1981]. Further studies by the Jet Propulsion Laboratory in conjunction with the University of California Medical School are reported by Rosenthal et al [1979]. In Japan, Tanaka et al [1981] studied the problem in conjunction with the CYBEST project. Further studies were also conducted by Bahr et al [1978] at the Armed Forces Institute of Pathology and the University of Chicago and by Neugebauer et al [1980] in Germany.

The automated prescreening of cervical smears is still the focus of intense research efforts. It is important to realize that the clinical diagnostic problem is quite different from that encountered in white blood cell differential counting where an exact determination of the proportions of different cell types is the sole objective. In a cervical prescreening device the minimum expected decision is already a diagnostic one. Two questions must be answered: (1) Is this unequivocally a normal specimen, definitely not in need of further human diagnostic assessment? and (2) Is this an adequate clinical sample so that the first decision can be made at all? The cervical prescreening device addresses a search problem—for the presence of tumor cells—and to make matters even more difficult, this is a search problem for what is often a very rare event.

Machines for automated cervical-prescreening devices are presented with several, substantial difficulties:

1. Since the clinical sample may contain very few tumor cells, a large number of cells need to be examined to maintain an adequate probability of detection of the few tumor cells that may potentially be present.
2. The clinical sample must be presented in a "clean" monolayer to prevent cell overlap that might obscure diagnostically important cells.
3. The spatial-sampling density must be high enough to allow an unequivocal rejection of artifacts and to reduce the false-positive rate.

Possibly because the Cytoanalyzer is often cited as an example of a failed venture, one is rarely reminded of the remarkable technical and methodologic achievement that it represented, given the state of technology at that time. It appears today that the nature of the decision problem—ie, the need to recognize normal cells as normal with an extremely high confidence—has not been appreciated and that this required the need to process a very large volume of spatial samples at a very high data rate. The Cytoanalyzer project made it clear, however, that extensive basic research on the information requirements, statistical sampling requirements, sample-preparation technology, and scanning technology remained to be done.

Several studies concern themselves with statistical sampling requirements, estimates of cell and specimen classification error rates, and with the statistics of cellular composition of cervical preparations. Gusberg et al [1969] performed an early feasibility study. Other studies include that of Louis et al [1976], who examined the proportions of free-lying and thus directly measurable abnormal cells in routine cervical smears. Bibbo [1975, 1976a–c] and co-workers conducted comprehensive studies of the cellular composition of routine smears. Neugebauer [1980] compared smear and monolayer preparations in their composition.

A substantial amount of effort was made to present clinical samples in machine-compatible form. This involves several critical steps. Disaggregation techniques were explored, to break up clumps, using ultrasonication and syringing. Methods for the cell deposition on slides in monolayer form were examined and perfected, and finally efforts were made to achieve reducibility of information extraction by better controlled staining. A number of studies were concerned with staining optimization and standardization for automated prescreening, eg, the study of the gallocyanin-chrom-alum stain by Eason and Tucker [1979], a study of different preparation techniques to allow a variety of staining procedures by Van Driel-Kulker et al [1980], an investigation into the suitability of a Thionine-Feulgen, Congo-red stain by Oud et al [1981], and a comprehensive review article by Wittekind [1985].

There is extensive literature on cell dispersal and deposition of monolayers of cervical cells. One of the first publications goes back to Tucker and Gresham [1971], followed by a study by Leif et al [1975] and Otto et al [1976]. Brakey and Rooney report on a vortex technique to break up cell clusters [1976], and Mayall [1976b] provides a general discussion on monodisperse cell samples. Leif et al [1977] report on their efforts for cell dissociation, and Escobar et al [1977] evaluate different chemical agents for the same purpose. Garcia [1977] reports on ultrasonic dispersion. Leif et al [1977] present techniques to optimize cell adhesion to the slide. Mead et al [1978] describe their success with syringing. Otto [1979b] reports on sedimentation velocity separation. The methods used in the CYBEST project are described by Tanaka et al [1981].

Devices and methods for the production of a monolayer of cells are described in a paper by Shibata et al [1972] on the Autosmear device. Wheeless provides an overview of the problem in 1974. Wolley et al [1976] describe the use of a slide spinner. Husain describes the techniques used in his project in 1978. Bahr et al [1978] describe a monolayer device for production of uniformly dense monolayers of epithelial cells. Tolles et al [1978] provide a review of the technology. A simple method for monolayering epithelial cells is reported by Rosenthal et al [1979]. Zahnizer et al [1979] discuss the problem in the context of the BioPEPR project. Hoeffgen et al [1979] report on the quality of morphologic appearance of automatically spread cells and so does Otto et al [1979a, 1980]. Oud et al [1981] report specifically the development of a cervical-smear procedure for the BioPEPR system. Schenck et al [1981] reported their monolayering technique. Bales [1981] writes on the preparation of monolayers of urinary sediment. Watts [1984] discusses the use of cationic polyelectrolytes in the preparation of monolayers.

Bartels et al [1979] provided guidelines for the statistical sampling requirements in the performance testing of prescreening devices and estimates of the frequency of clinical samples for which a very low number of tumor cells could be selected. Moore [1974] published an optimization approach for prescreening frequency. Diegenbach and Baak [1977] studied the differences between nuclei of hyperplastic and malignant-appearing uterine cells. Castleman and White [1980] worked out the trade-offs between cell-recognition error rates and specimen error rates. Bacus et al [1982, 1984] conducted an extensive study of the detectability of premalignant and malignant disease, based on assumptions for cell and specimen error rates.

During the 1970s work on high-resolution, image-analytic systems proceeded with emphasis on improved cell diagnostic capabilities, rather than on the development of clinically feasible prescreening devices. Such systems were really designed more to provide proof of principle, rather than practical clinical prototypes. The great majority of these systems in fact never proceeded beyond that stage. The one notable exception is the Japanese CYBEST system, which evolved through a number of models and is at this time undergoing extensive clinical field trials. These efforts are described in a series of papers by Ishiyama and Hayakawa [1969], Matsumoto et al [1971], Watanabe [1974], Watanabe et al [1976], Tanaka et al [1979], Watanabe et al [1980], Suzuki and Yamamoto [1980], and Mukawa et al [1979, 1983]. An example taken from this project is shown in Figure 1.18. A comprehensive review of the early efforts in this field is found in Wied et al [1981b] and in Pressman and Wied [1979]. Research efforts to construct and design such breadboard models of cervical-cancer-prescreening devices reached an active phase in the late 1970s that has continued into the 1980s.

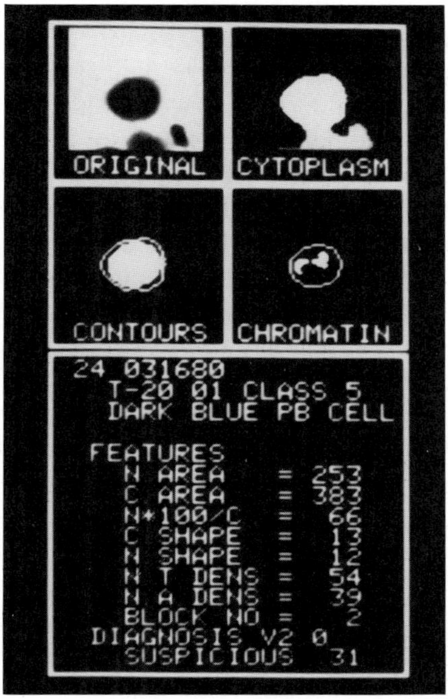

FIGURE 1.18. The CYBEST display shows the original and cytoplasmic image of each cell located in the cervical smear under examination, as well as the contours of the cytoplastmic boundary, the nuclear boundary, and the dense chromatin. Numerical values are computed for eight features and, the diagnosis of the cell is recorded (courtesy Dr. N. Tanaka, Chiba Cancer Center Research Institute, and the Nihon University School of Medicine).

The early work of Kahn et al [1971] and Poulsen [1977] provides another good example of the exploration of the feasibility of *automated microscopy* in cervical-smear analysis. There were numerous other groups engaged in the 1970s. At the GSF in Germany Kinder et al [1979] explored coherent processing techniques, which were also pursued by Kopp et al [1974, 1976]. Husain [1974] worked with a Quantimet system; Al and Ploem [1979] developed the LEYTAS system based on the Leitz TAS processor.

Kishigami et al [1978, 1979] designed and tested an automatic prescreening system based on Feulgen staining of the nuclei, as did Farrow and Tucker [1976]. In Germany Bloss et al [1979] developed the FAZYTAN system based on a very fast videoprocessor. In the Netherlands the BioPEPR system was developed by Voojis et al [see Zahnizer et al, 1980]. Read et al [1979] developed a dual-resolution system at the University of Chicago with a fast low-resolution cell-detection module to preselect potentially abnormal cells prepared in monolayer form. Bengtsson et al [1979] reported a fast diode array scanner for a prescreening device. Kunze [1980] published an account of the ZYPAB system also for use as a prescreening device.

The uncertainties surrounding the clinical use of an automated cervical-cancer prescreening device appear to be as unresolved today as they were 15 years ago. It is clear that there is a clinical need to detect not only malignant, but premalignant lesions as well. It has also become clear that the classification and grading of such precursor lesions by different diagnosticians is far from firmly established [see Wied et al, 1985a]. The question of whether such a device will be able to detect adenocarcinoma of the endometrium remains unsettled. Finally, the performance requirements and methods of ascertaining whether a given device attains them in principle and as a maintenance procedure continue to cause debate.

During this time many research groups were actively engaged in testing flow cytometric devices. Although flow cytometry is not covered in this chapter, it is worth noting studies by Bunnag and Herman [1979] at the National Institutes of Health, Jensen et al [1979] of the Lawrence Livermore Laboratories, Sprenger [1979] in Kiel, and Stoehr and Goerttler [1979]. Wheeless and Onderdonk [1973] pursued the slit scan flow cytometric principle.

DNA Photometry

The pioneering work of Caspersson [1950] and his co-workers in Stockholm first made quantitative cytochemical determinations possible, specifically exact microphotometric determinations of cellular constituents such as DNA and RNA. Mellors [1952] established that squamous cancer cells exhibit increased total absorbance at 260 nm, indicating an increased amount of nucleic acids. Leuchtenberger et al [1954] and Stich and Steele [1962] studied the DNA contents of human tumor cells. The use of the

Feulgen reaction made this methodology accessible to many more laboratories and the 1950s saw intense research aimed at establishing the methodology of quantative microphotometry based on cytochemistry. Excellent reviews of these basic research efforts are given by Swift [1966], Hale [1966], and Deitch [1966].

Quantitative cytochemistry was seen by many as holding the promise of providing reliable markers for tumor cells, and the measurement of total DNA contents per nucleus was especially pursued. It was found that total Feulgen optical density, as a measure of total DNA content/nucleus is not an unequivocal and sufficiently discriminating marker for the recognition of an individual tumor cell. The field then turned to micromorphometric measurement and texture assessment of the nuclear chromatic as a means of attaining reliable machine recognition of abnormal cells. This goal has now been largely achieved, but two subsequent developments deserve further discussion.

Quantitative cytochemistry, and specifically Feulgen DNA photometry applied to a sample of nuclei from a given lesion, has become one of the few research methodologies now entering clinical practice, with the advent of rapid video microphotometers. These instruments make the recording of adequate sample size feasible, and the distribution of DNA values among the nuclei, called the "ploidy pattern," appears to offer both diagnostic and prognostic information, as discussed in the following section.

Another development is the discovery of diagnostically highly discriminating information that can be extracted from digitized imagery by computation only. These clues are visually perceived only poorly, inconsistently, or not at all.

Ploidy Analysis

Research and development in quantitative cytology and histology have indeed led to highly refined methodologies, to advanced instrumentation, and to entirely novel technologies. Practically all of these evolved during the past two decades, following the growth and advances in the microelectronic industry. Yet it must also be said that only a very few of these methodologies have reached the stage of routine clinical application. The basic research underlying ploidy analysis was completed 25 years ago. Clinical feasibility of the recording of ploidy patterns based on a statistically valid sample size has only now become possible. For all of the promise of the methodology, it must be said that clinical data are only now accumulating. It may well be that a straightforward diagnostic, and/or prognostic, interpretation will soon be available.

Generally, six types of ploidy patterns are currently distinguished. These are (1) the normal euploid pattern, with a peak at 2N, a small proportion of cells in the S-phase, and at the 4N peak; (2) a polyploid pattern, with modes at 2N, 4N, 8N, and even at 16N, with different proportions of

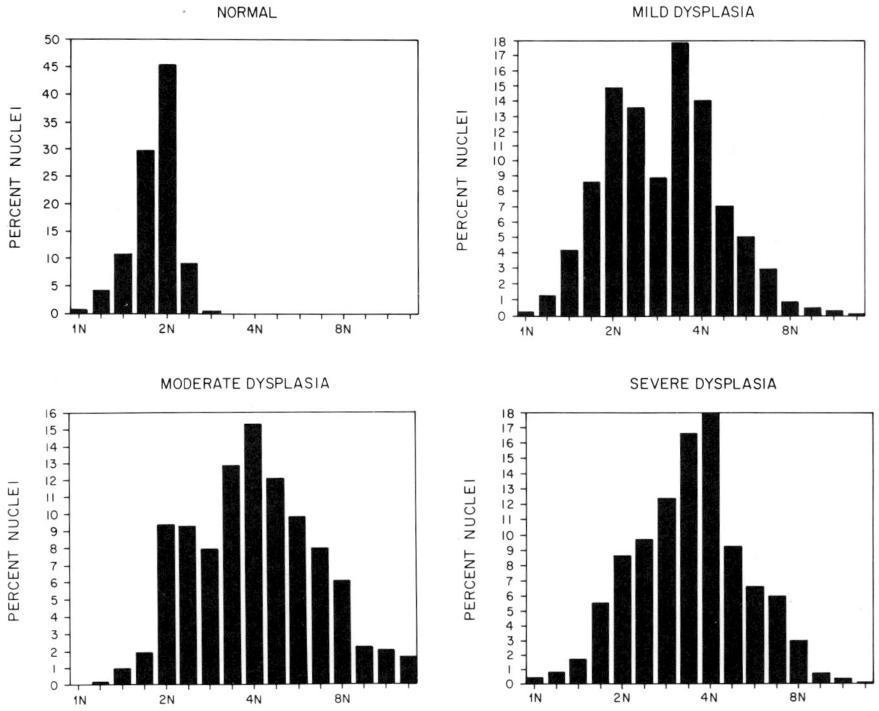

FIGURE 1.19. Ploidy histograms generated over a large population of cells for cases characteristic of normal, mild, moderate, and severe dysplasia.

nuclei at each mode—a pattern that is normal for some tissues and in other cases is indicative of normal tissue repair; (3) aneuploid patterns with nuclei having ploidies spanning a wide range, often with a strong mode in the 3N region; in practice the aneuploid denotation covers all patterns definitely not falling into the euploid, or polyploid tolerance limits; (4) patterns exhibiting a combination of aneuploid and polyploid features; (5) near diploid patterns, with a stemline of nuclei with only slightly increased DNA values, often indicative of a "diploid tumor," or a well-differentiated tumor. Examples are given in Figure 1.19. There has also been a recent, rising interest in hypodiploid patterns, which are believed to have a poor prognosis.

It has been recognized that the interpretation of a ploidy pattern must consider the organ site from which the clinical sample was collected. One also has to consider that cells of various degrees of differentiation, and possible various degrees of chromatin compaction, are affected differently by the various steps in the Feulgen procedure. Specifically, certain tumor cells might be affected in a markedly different fashion from the control cells by the hydrolysis step of the procedure. It is therefore mandatory that the control sample on the basis of which the position of the 2N peak is established be carefully chosen. This differential sensitivity certainly has an affect on which nuclei appear in the aneuploid range.

Multivariate statistical methods for the assessment of ploidy patterns are only now being developed by Weber et al [1985a–b]. Currently, there are no practical ways to set up confidence regions and tolerance regions. For any purpose other than a very coarse visual assessment of the ploidy pattern, statistical considerations demand sample sizes of at least 100 nuclei. This is feasible in a clinical laboratory only by the use of video technology. A number of such systems have been described for ploidy analysis, such as the TAS-based system of Aufferman et al [1984], the Quantimet 720 system of Fisher [1971], and the Saticon color video-microphotometer described by Wied et al [1983a–b].

All of these systems are really scaled-down versions of research instruments and are too expensive for routine use in the clinic. They are gradually being replaced by small, special purpose "ploidy microphotometers." Devices of this kind are essentially an attachment to a laboratory microscope, consisting of a video camera, an image frame buffer, and a microprocessor system. Rapid scanning stages, with computer-controlled positioning, are being added to some of these microphotometers. They offer the great advantage that cells of interest can be identified initially in Papanicolau-stained form by a pathologist. They may then, after destaining and Feulgen processing, be relocated from a computer-stored location map for a correlation between cytodiagnosis, measured image features, and DNA content.

Clinical Studies

The early studies in the 1960s were followed by numerous investigations of material from the female genital tract, constituting a rich literature on ploidy patterns in cervical neoplasia. For the female genital tract it has been established that aneuploid cervical intraepithelial neoplasia lesions have a higher rate of persistence, or recurrence, than those with polyploid patterns. Good references are the classical studies of Jones et al [1970] and Fu et al [1981].

Preinvasive lesions, with modes in the 2N to 3N ploidy range, have been found to be more likely to progress to invasive cancer by Kirkland [1969]. Of the invasive lesions, those of low ploidy have a higher frequency of lymph-node metastasis than carcinomas with high ploidy, as demonstrated in Atkin [1971a] and Fu et al [1980].

Atkin [1971b, 1976a, 1976b] and Auer et al [1984] found that glandular neoplasms of low ploidy from the endocervix and endometrium have a better prognosis than adenocarcinomas of high ploidy. Fu et al [1978, 1979] demonstrated that the ploidy pattern allows one to distinguish active squamous metaplasia from cervical and vaginal intraepithelial neoplasms.

Of particular interest is the differentiation between condylomata and aneuploid lesions of the cervix by Bibbo et al [1983], Weid et al [1984], and Winkler et al [1984]. Atkin [1971, 1979] shows that the radio resistance of cervical squamous carcinomas with low ploidy is greater than that of high ploidy carcinomas. Okagaki et al [1974] demonstrate that postradiation dysplasia with an aneuploid pattern develops recurrent carcinoma more often than dysplasia with euploid or polyploid patterns.

Bibbo et al [1983] used the ploidy pattern to match the cytologic sample—and the histologic biopsy for quality control and diagnostic agreement. Tavares [1966], and Tribukait et al [1979] studied ploidy and proliferation in human bladder tumors. Ploidy measurements were used by Fossa [1977a, 1977b] in a study of transitional cell carcinomas of the human urinary bladder. Ploidy analysis by flow cytometry was found valuable in the differentiation of bladder papillomas from carcinoma in situ of the bladder for the follow-up of conservatively treated low-stage bladder tumors and for detection of carcinoma in situ of the bladder (Devonec et al [1981, 1982]). Nuclear DNA content was established as an important parameter in the prognostic evaluation of prostatic cancer by Ronstrom et al [1981] and Zetterberg et al [1976, 1980]. Bibbo correlated DNA ploidy measurements of fine-needle aspirates of the prostate with clinical diagnosis and staging. (See also Chodak et al [1985].)

The DNA ploidy patterns were studied for bronchial and lung cancer by a number of research groups. Cytomorphologic grading, combined with Feulgen DNA analyses, was used by Nasiell et al [1978] in a study of metaplastic and neoplastic bronchial cells. Auer [1980b] reported DNA analyses of atypical squamous metaplastic cells, carcinoma in situ cells, and cancer cells from bronchogenic cancer. Blondal et al [1980, 1981] studied the DNA contents of nuclei in bronchial carcinoma and examined the prognostic potential of such measurements. Pak [1982] applied quantitative DNA determinations to human pulmonary cytology. Teplitz et al [1983] used DNA values as a marker for bronchogenic cancer.

The DNA cytophotometry of mammary tumors was the subject of studies by Kallenberger et al [1967], Zajicek et al [1970], Atkin et al [1972], Ludwig et al [1973], Sprenger et al [1979], Dziura and Bonfiglio [1979], Cornelisse et al [1979, 1981, 1983], Auer et al [1980a,1980b], and Fallenius et al [1983].

Several studies were aimed at the oral mucosa. Doyle and Manhold [1975] used Feulgen photometry in the study of oral cancer and leukoplakia. Pfitzer and Pape [1973, 1975] measured the DNA contents of leukoplakia cells in the oral mucosa. The DNA contents of the cells in human buccal smears were measured by Rowinski [Rowinski et al 1978]. Feulgen DNA photometry on normal and abnormal oral squamous cells was reported by Cowpe and Longmore [1981].

Ploidy can be predictive of the clinical course of chondrosarcomas, as found by Kreicbergs et al [1980], Kreicbergs and Zetterberg [1980], Cuvelier and Roels [1979], and Auer and Zetterberg [1984].

Differentiated carcinomas of the thyroid have an increased DNA content, especially follicular carcinoma. Benign adenomas show only a moderate increase, as shown by Lukacs et al [1979]. Papillary carcinomas have a diploid ploidy pattern, as reported by Johannessen et al [1981].

In addition, there are studies involving tumors from many other organ sites. Astrogial tumors were studied by De Reucck et al [1979]; lymphoid cell populations, by Braylan et al [1978]; mucinous tumors of the ovary, by Weiss [1969] and Atkin [1971b]; malignant teratoma testicle, by Lederer [1976].

There are numerous studies of DNA contents, differentiation, and cell cycle kinetics. See for example Sandritter et al [1967], Alpen and Johnston [1967], Kiefer et al [1974], Nicolini et al [1977], Kendall and Nicolini [1977], Adelh et al [1978], Doermer and Abmayr [1979], Sans and de la Torre [1979], Sheck et al [1980], Abmayr et al [1982a,b], and Giaretti et al [1983].

The ploidy patterns developing during carcinogenesis of squamous bronchial carcinoma were recorded by Konaka et al [1982]. Nuclear transformations of cells induced by carcinogens and tumor viruses were studied by Romen et al [1979a–b, 1980].

Cell Morphometry

During the past decade, the extraction of quantitative morphometric measurements from digitized cell images has been extended from the work on hematologic and ectocervical materials to cells from nearly every body site. A number of publications serve as a general introduction to this methodology. These include the comprehensive treatment by Bahr et al [1979] and the textbook by Koss and Coleman [1979] on diagnostic cytology. The state of the art of the field in the mid-1970s is reviewed in a paper by Bartels and Wied [1975], which was published in connection with the Los Alamos Life Sciences Symposium, and in the book *Cell Pattern Recognition and Cell Sorting*, edited by Wied et al [1969]. A more recent report is that of Bartels et al [1980], which is a progress report on automated cell diagnosis in clinical cytology, and an updated treatment of the same material by Wied et al [1981]. A still more recent survey of the field can be found in the chapter by Bartels et al [1984] in the proceedings of a NATO Study Institute on the subject. Bartels and Olson [1980] treat the analysis of lymphocytes, and Bartels et al [1980] addresses the major questions concerning information extraction from cell populations.

Basic Methodology

Numerous studies are concerned with basic methodologic questions as, for example, the treatment of chromatin texture. Classic papers in this field are those by Grundmann and Stein [1961] and Sandritter et al [1967], whose group used one of the first stepping stages to derive measures of chromatin texture. This same research group continued research as reported by Kiefer et al [1973, 1974] as well as by DeCampos et al [1973]. Changes in chromatin after infection (in cells from mouse embryos) and during the cell cycle were reported by Rowinski et al [1975] and by Sawicki et al [1974a–b]. Kendall and Nicolini [1976] and Kendall et al [1976] studied nuclear texture during the cell cycle.

Also in the 1970s Stenkvist et al [1978] studied nuclear morphometry in human cancer cell populations. Abmayr et al [1982b] investigated methods of discrimination between nuclei from cells in the G1, S, G2 phases, and Giaretti et al [1983] correlated these findings with autoradiography. Olson and Bartels [1982b] and Bartels and Olson [1984] discovered specificity in chromatin changes for different environmental insults and devised a computer graphic display to present such changes to the microscopist in a visually perceivable manner. Finally, Abmayr et al [1982a] identified cell cycle compartments based on chromatin features.

Several computer software packages were produced for the analysis of cell images. These include the programs of Preston and Onoe et al [1976]; the SCANCANS package by Bengtson et al [1977] and Holmquist et al [1978]; SUPRPIC (Super Picture Computer) by Preston [1983a]; and TICAS by Bartels and Wied [1977].

Morphology of Ectocervical Cells

Extensive collections of images of cells from the ectocervix have been established. Some of the earliest work in this area is represented by the laborious and extensive measurements of nuclear area, cell size, and nuclear/cytoplasmic ratios taken by Reagan et al [1952, 1957].

The TICAS project at the University of Chicago produced the first publication on cervical material wherein Wied et al [1968a] dealt with cells from uterine adenocarcinoma and from normal glandular cells. A number of further papers by Wied et al [1969, 1970b] and Bartels et al [1969, 1970, 1972] reported on ectocervical cells as material. Following this, Bibbo et al [1973] reported classification success rates on cell nuclei from the ectocervix. Two papers by Bahr et al [1973] and Bartels et al [1974a] examined the suitability of the Papanicolaou stain for quantitative image analysis. At the same time Bartels et al [1974b] reported on the evaluation of correlation information in digitized cell images while Taylor et al [1978] gave the design of the first hierarchical classification procedure for cell images from the ectocervix. Automated nucleus finding routines were described by Taylor et al [1975]. Bartels et al [1973] and Sychra et al [1976] provided statistics on the recognition of nuclei from ectocervical cells and on shape features. Bhattacharya et al [1975] and Bibbo et al [1975, 1976a, 1976b, 1977, 1978] reported a series of studies aimed at estimating the numerical composition of cervical smears. Wied et al [1977a] compared classification accuracies obtained for cell images recorded at different spatial resolutions. Wied et al [1977b] also reported on the efficacy of contour and textural features. The question of overlapping cells in an automated recognition system was treated by Sychra et al [1977, 1978]. Bartels et al [1978, 1981] described patient profiles for diagnostic assessment using information from all cells within the clinical sample and also provided an extensive evaluation of features used for the recognition of ectocervical cells, and Wied et al [1981a] discussed the value of spectral information.

At Rush-Presbyterian Hospital Mui et al [1976] described scene segmentation procedures for ectocervical samples. Aggarwal and Bacus [1977] discussed a multispectral approach for this purpose. Galbraith et al [1979] and Marshall et al [1979] examined the spectral characteristics of Papanicolaou-stained cervical cells, and Bacus et al compared the classification success rates achieved by the cytotechnologist versus the automated classifier. A summary of the research done on this project is available in the Final Report to the National Institutes of Health on Contract Number 1-CB-74190 in September of 1982.

In Germany early reports were written by Herrmann et al [1974], Abmayr and Borst [1976], and Abmayr et al [1979]. Abele et al [1977], as well as Tilgner et al [1977], studied edge-detection problems using such material. Kunze et al [1978] and Otto [1978] wrote on methods for automated cytologic prescreening. Borst et al [1979] reported on a new thresholding method for cell-image analysis whose purpose is to achieve reliable image segmentation. Reinhardt et al [1979] reported on feature extraction from ectocervical material and classification success rates. Early work on the TUDAP project was reported in papers by Abmayr et al [1979], Rodenacker et al [1981], and Ott et al [1981].

In Sweden the recognition of ectocervical cells was reported by Holmquist et al [1976], Bengtsson et al [1977], and Holmquist [1978]. In Japan the computer recognition of ectocervical cells and the value of spectral information were discussed by Imsato et al [1975], Noguchi et al [1979], and Noguchi [1982]. In France Meyer [1979] applied iterated image transformations for the purpose of feature extraction, and Laurent et al [1979] gave classification success rates on cervical-cell images. In the Netherlands Smeulders et al [1980] reported on shape parameters for use in cell-classification purposes.

In the United States Stern et al [1982] described automated measurement-making for cervical-cell classification, and Pressman [1976] examined the value of certain textural features for the same purpose. In Canada Poulsen et al [1977] published papers on cell-image segmentation methods and cervical-cell identification.

Bladder Cell Morphology

In the United States Bartels et al [1978], Koss et al [1975, 1977a–b, 1978a–b, 1979, 1980a–e, 1983, 1984b], and Sherman et al [1981] and Sherman and Koss [1983] published extensively on the classification of urothelial cells by computer. After first accomplishing the computer discrimination between normal and malignant urothelial cells, their work led to the successful discrimination of atypical urothelial cells. These studies led to the development of a comprehensive hierarchical classification scheme for cells from the urinary sediment. Stinson and Sporn [1977] and Stinson et al [1977] used automated image analysis to quantitate cellular hyperplasia in urinary bladder epithelium.

Morphology of Cells from Other Tissues

Image analytic studies of bronchial squamous cells have been perfected at Baylor University, as reported by Kimzey et al [1980], Baky et al [1980], Engvall et al [1981], and Winkler et al [1981]. These studies have resulted in the definition of an atypia status index for respiratory cells. Also, Erozan et al [1979] conducted a comparative study of noninvasive and invasive squamous cell carcinoma of the lung. Pasteur et al [1984] reported on the automated cytologic diagnosis of bronchial carcinoma, and Margolick et al [1973] discussed quantitative assessment of pneumocytes.

Regarding normal and malignant gastric epithelial cells, Weinreb et al [1984] disclosed a method for distinguishing between these types using computer-image analysis. Neuronal tissue was studied by Werner and Voss [1979] as well as by Herrmann et al [1979] and Wann et al [1973]. Schoendorf [1981] carried out karyometric studies of the epithelium of the mammary gland as well as for tumor cells of mammary carcinoma. Zajdela et al [1979] correlated prognosis in mammary carcinoma with nuclear diameter, and Cornelisse et al [1981] studied nuclear size variability in mammary cytology specimens. Van Driel-Kulker et al [1982] examined mammary cytology specimens with the LEYTAS system.

Computer recognition of normal and malignant mesothelial cells was studied by Jahoda et al [1973a–b]. Valkov et al [1977] performed a morphometric study of the nucleoli in mesothelial cells, and Kwee et al [1982] discriminated between normal and malignant mesothelial cells in pleural fluids.

A number of studies have been directed toward the quantitation of changes due to various carcinogens or toxic pollutants. Reale et al [1979] and Wied et al [1979] differentiated cell populations from tracheal aspirates of hamsters having experimental squamous cell carcinomas. Nair et al [1980] published a study where subtoxic doses of chlordane were found to cause significant changes in the nuclei of liver cells in animals.

Morphology of Lymphocytes

A substantial amount of work in this field has dealt with the computer analysis of lymphocyte images. Bartels et al [1969] studied lymphocytes in transformation. Zajicek et al [1972] studied lymphocyte populations from adenitis and lymphocytic lymphoma. Bahr et al [1974] measured lymphocytes from patients with dengue and typhoid fever, and Sinski et al [1977] observed changes in lymphocytes in guinea pigs after immunization against coccidiomycosis. McKee [1975] performed a doctoral thesis on changes in lymphocytes induced by immunosuppressive agents, and Jarkowski et al [1971] analyzed lymphocytes from patients with asymptomatic leukemia. Rowinski et al [1972] studied the chromatin in resting and stimulated lymphocytes. Bartels et al [1974] studied learning algorithms for the computer analysis of lymphocyte images. Extensive work was done by Bartels et al [1978], Durie et al [1978], Anderson et al [1975a–b], and Olson et al [1973a, 1974a–c, 1980, 1981] on the differentiation between thymic and bursal lymphocytes. These studies included those on differentiating between murine and human B and T lymphocytes. Olson et al [1980] confirmed the differentiation of these lymphocytes by tagging them with immunoadsorbant beads. Olson and Bartels [1981] also studied splenocytes and peripheral blood lymphocytes in animals and Olson et al [1979] and Anderson et al [1975c] reported on the effect of radiation on lymphocyte populations. Crocker and Curran [1979] measured nuclear diameters in lymph node imprints. Gunzer et al [1981] studied leukocytes in leukemia.

New Morphometric Findings

The TICAS project at the University of Chicago had, as its original goal, the utilization of "computable" image information, ie, of image information providing no (or only very subtle) visual diagnostic clues. Such information could be seen as a true expansion of the ability to perceive diagnostic clues by the use of a computer. There have been a number of situations where such subvisual clues were originally uncovered by the image analytic process, and where subsequently pathologists were able to make the distinction as well. In other situations visual discrimination has remained not possible.

Examples for the utility of such subvisual diagnostic clues are given by Bartels [1980]. Subtle changes in response to the administration of chlordane in subtoxic doses were measured by Nair et al [1980] in rat hepatocytes. It was possible to distinguish B and T lymphocytes with very low error rates, as shown by Olson et al [1974a] and Durie et al [1978]. In bladder cytology the nature of the atypical urothelial cells could be clarified by Koss et al [1977a, 1977b]. Possibly the most startling example was the finding of "markers" in normal-appearing intermediate epithelial ectocervical cells, offering clues to the presence of premalignant and malignant lesions in these patients. So-called malignancy-associated changes (MAC) had been reported before, but were not substantiated by measurement.

In patients with malignant diseases or dysplasia of the ectocervix, normal-appearing intermediate cells were shown to exhibit statistically significant changes in the values of certain of their morphometric features, in their staining properties, and/or in the texture of their chromatin. These changes are small and cannot be detected reliably by the human observer. However, when high-resolution, full-color digital images of these cells are viewed on a computer-graphics display (Figure 1.20) and ordered by the degree to which these changes are expressed, human observers can discern the trend. This phenomenon was first described by Wied et al [1980, 1984] and by his co-workers Bibbo et al [1981] and Bartels et al [1983]. Research teams in Munich made similar findings, as reported in Burger et al [1981], that have been confirmed by Vooijs et al [1982], Reinhardt et al [1982], and Rosenthal et al [1983].

Along similar lines, Kwikkel et al [1985] discovered that slight changes in the nuclear/cytoplasmic ratio are found in intermediate cells from patients with dysplastic or malignant diseases and that these changes can be masked by hormonal contraceptives. Nieburgs [1976] and Nieburgs et al [1959, 1962] reported MAC in seemingly normal cells for many years. Earlier reports of MAC related to peripheral blood smears were made by Johnson and Brady [1970] and in buccal smears by Klawe and Rowinski [1974]. Early papers perhaps did not attract much attention because MAC had not been verified by quantitative measurement.

Marker features in seemingly normal epithelial cells from the urothelium were reported upon by Sherman and Koss [1983]. Also discovered by Wied et al [1982] were statistically significant differences between cells within all of the established atypical cell types, depending on the patient's diagnosis as moderate dysplasia, severe dysplasia, or malignant disease. Morphometric data computed from individual cells may thus provide direct diagnostic clues as to the cytologic condition of the patient and, possibly, eventually to the biologic behavior of the lesion. This is the basis for current studies by Rosenthal et al [1983] on prognostic clues that may possibly be offered by cells in instances of moderate dysplasia. The discovery of these differences make it mandatory that data-bases on cells of a given type be collected according to the clinical diagnostic assessment of the patient. Patient profiles based on these differences may allow potentially better diagnostic and prognostic assessment of the patient.

Tissue-Section Analysis

The analytical evaluation of histopathologic sections has always been an important goal in automated microscopy. To this day the now classic papers by Kirsch [1971] and the studies by Lipkin et al [1966] and Shapiro et al [1974] provide guidance and an excellent introduction to the problems, objectives, and possible solutions in this difficult field. Histopathologic sections do, indeed, pose a difficult problem in quantitative analysis. Nevertheless, they are carried out widely, at greatly varying levels of complexity, and provide valuable diagnostic information. A general solution to the problem, however, is still an elusive goal, and suitable approaches are only in the early stages of exploration.

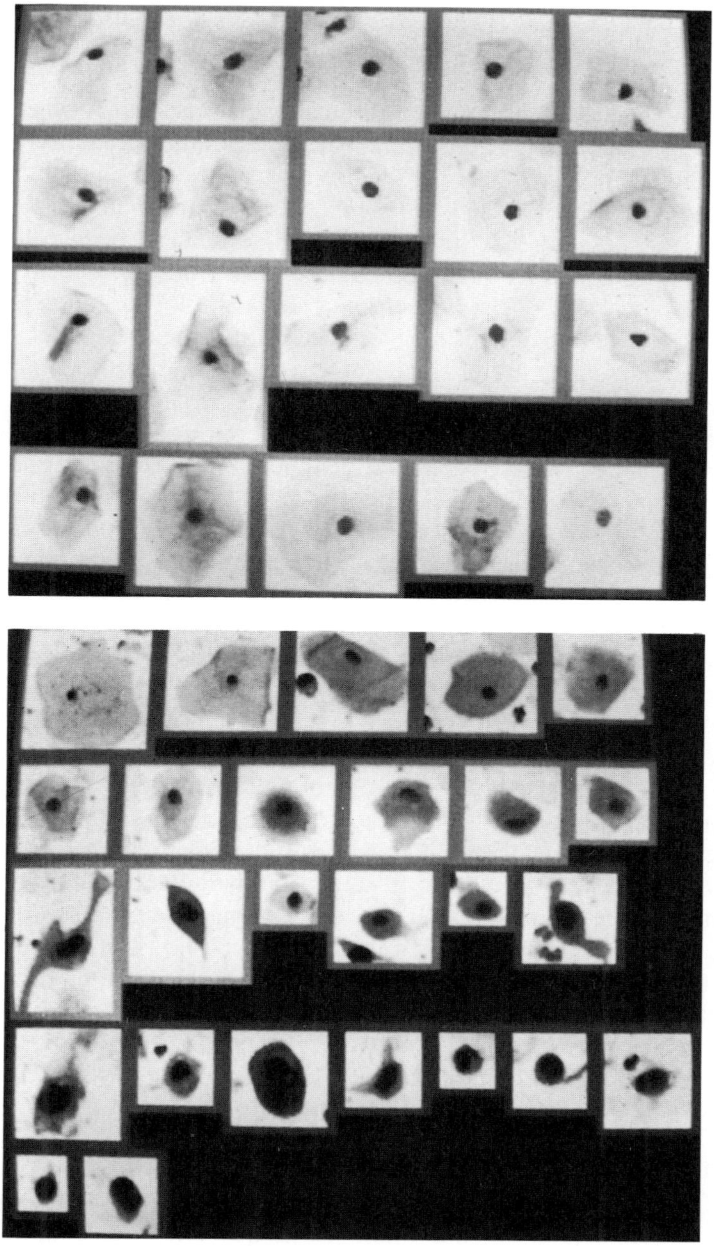

FIGURE 1.20. By ordering cell images on a color-graphics computer display, it is possible to illustrate for the observer the subtle colorimetric, morphologic and textural properties that take place over the range of normal to malignant. *Top* shows normal-appearing intermediate cells; *bottom,* abnormal cells from the ectocervical epithelium. See Color Plate V.

The difficulties arise from several aspects. First, the diagnostic information in a histopathologic section is offered at two levels: (1) the gross histologic structure, or the "architecture" of the tissue, and (2) at the level of the individual cells, nuclei, chromatin, and nucleoli. It is necessary, therefore, to scan a large area at high-spatial resolution. Requirements of up to billions of image points per diagnostic area are not uncommon [Bartels et al, 1984]. Such data volumes present problems in matching partial scan areas necessitated by the restricted field of view of high aperture optics and in the storage requirements. Second, from a technical aspect, computer processing of the imagery in a histopathologic section presents itself as a "closely coupled task." In the processing of a given subarea it may be necessary to refer to and utilize processing results obtained in other subareas of the section. This is in contrast to "loosely coupled processing tasks," or uncoupled tasks, encountered in cytology where one cell image can be processed completely without any need to refer to results obtained from other cell images [Bartels, 1985]. Third, tissue sections are complex pictorial scenes from an image-processing point of view. Scene decomposition is not against a clear background, but from other scene components. Preparation artifacts, defects, and sectioning imperfections all introduce "noise," eg, discontinuities in outlines or boundaries, and a fair amount of image restoration and heuristic decomposition procedures need to be employed.

Finally, there are logical problems in the decomposition and processing of histopathologic sections. The morphology of structures in such sections encompasses a remarkably wide range of patterns, even within the same class of diagnostic tasks. Furthermore, for a given organ site a large number of diagnostic classifications is often required. (For the thyroid there are no less than 32 histopathologic categories.) To map the histomorphometric evaluation into a diagnostic assignment is a formidable task notwithstanding the wealth of diagnostic information offered and, partially, even because of this wealth of data. The problem has not yet found a general solution.

A review of the literature shows several levels of complexity in the extraction of diagnostic information. There are studies where tissue architecture is not considered in the taking of the measurements. It is used only to indicate to the operator where the measurements on individual cells should be taken. The measurement of the total optical density in individual nuclei of a lesion, to establish a ploidy pattern for a tissue sample, is such an example. Then, there are studies where the global texture of the section is recorded and used to derive tissue textural features, which are then used for classification. Examples for such attempts are measurements taken by Olson et al [1973b] on optical density histograms measured in sections of lymphoid tissue. Also see the paper on the extraction of optical density co-occurrence textural features by Pressman [1976]. An extension of this approach are studies where features descriptive of the tissue architecture are extracted and entered into discriminant functions for classification purposes. This approach in many cases provides clinically useful information. An example for such studies is the assessment of the orientation of the major axes of nuclei in bladder epithelium by Prewitt [1978].

This approach can be carried much further, leading to a determination of the proportions of different kinds of cells in a given tissue, of their placement pattern, and, possibly, of abnormality indexes for the nuclei of certain cells. The extensive work done by Preston and Dekker [1980] and his research team [Tourassis et al, 1983, and O'Gorman et al, 1985] provides an excellent example for this approach. Figure 1.21 provides exemplary illustrations of this work.

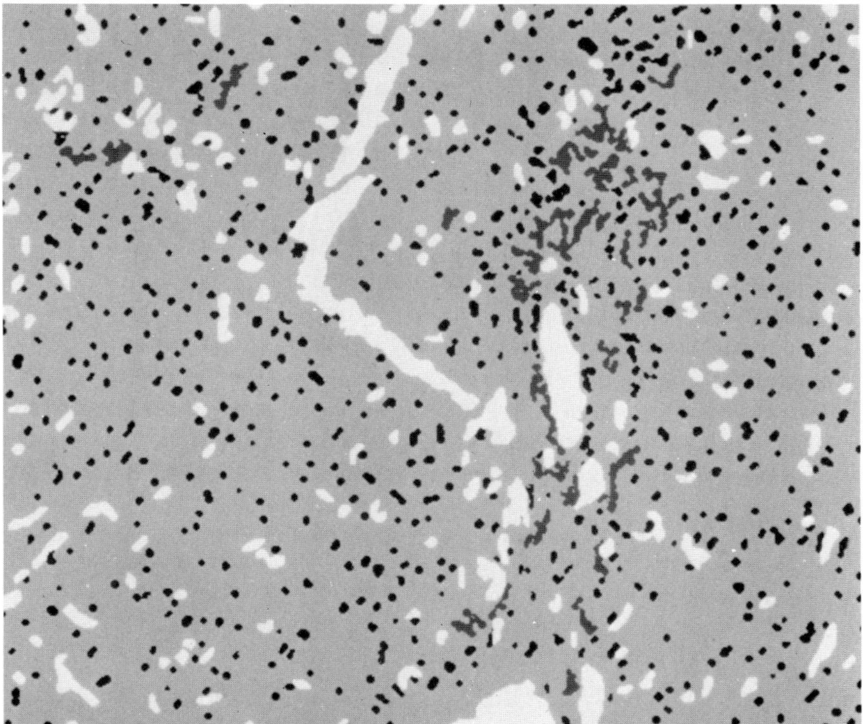

FIGURE 1.21. Computer analysis of the middle right image from Figure 1.8 (*bottom*), which extracts all normal cell nuclei (black), the nuclei of inflammatory cells (red), as well as the cytoplasmic/sinusoid and the cytoplasmic/vessel boundaries. See Color Plate VI.

Finally, there are attempts to endow the processing computer system with an "understanding" of the image structure. This offers two advantages. First, if the analytic system is designed to be image-data driven, an optimized scheduling of the processing tasks can be implemented. Second, image understanding is a requirement for the design of expert systems to aid in the diagnostic assessment of subtle histopathologic changes.

A useful introduction to the methodologies and clinical applications of tissue morphometry (not all of it computer-assisted) is provided in the text by Baak and Oort [1983]. The *Journal for Analytical and Quantitative Cytology* has just expanded its mission to include quantitative histology. Also useful is the *Journal of the International Stereologic Society*. Digital microscopy and morphometry increasingly involve stereologic methods. Texts by Weibel [1979], Aherne and Dunnill [1982], and Oberholzer [1983] summarize these methods.

A somewhat specialized application of digital microscopy is computer-aided three-dimensional reconstruction, which has applications of increasing significance in neuroanatomy and neurobiology. Here, the pioneering work of Sobel et al [1980], Macagno et al [1979], Veen and Peachey [1977], Ware and Lo Presti [1975], Levinthal and Ware [1972] on the CARTOS system is essential reading. Newly emerging applications of this difficult method are studies of invasion patterns in malignant processes [see Suh and Weiss, 1984].

A long-term research effort in the computer-aided assessment of liver sections has been sustained by Simon, Voss, Kunze, Wenzelides, and their co-workers in Berlin. (See Simon et al [1973], Arnold et al [1974], Simon et al [1975], Voss [1974], Barz et al [1977], Voss et al [1977, 1978], Kunze et al [1978], Simon et al [1980, 1981], Kranz et al [1980], Wenzelides et al [1981].) These efforts have now led to the development of an interactive, automated diagnostic system for the assessment of various liver ailments.

Muscle tissue has been studied in various contexts. For example, see Spencer and Eccles [1976] on spinal muscle in scoliosis and Sanfeliu [1980] on the general pattern in muscle-fiber arrangement. The effects of weightlessness on muscle tissues have been studied extensively by NASA (Castleman et al, 1978, 1981; Volker ter Meulen et al, 1972; Chui and Castleman, 1980; Chui et al, 1981). Studies on myocardial fibers are given in Hoyt et al [1984] and Fuster et al [1977]. Piera et al [1979] analyzed the rat myocardium after administration of Isoprenalin. Cornelisse et al [1980] studied type I and type II fibers in human biopsies. In other areas of quantitative histology, Gamel and McLean [1982, 1983, 1984] conducted comprehensive studies of the malignant potential in uveal melanomas. In environmental pathology, toxicology, and carcinogenesis, studies were conducted by Nair et al [1980] on the effects of subtoxic levels of chlordane on rat hepatocytes and by Abmayr et al [1983] on the nuclear morphology of hepatocytes after application of polychlorinated biphenyls. Also of clinical interest are the studies by Boysen and Reith [1980] and by Rigaut et al [1982] on changes in cells of the nasal mucosa. Jordon et al [1982, 1984] conducted studies in detecting early changes due to renal radiation injury, enabling an efficient fractionation scheduling.

Measurements on the parathyroid glands were performed by Grimelius et al [1978]. Quantitative histometric measurements on bone were published by Garrick et al [1972]. Johnston et al [1972] published a computer method for the counting of alveoli in human lung tissue. Grading of astrogliomas based on computerized-image analysis was accomplished by Robertson et al [1978].

The modeling of histologic images by computer by Bartels et al [1976] was a first attempt to find a more general approach to the processing of the complex pictorial scenes. More recent studies by Bartels et al [in press] employ a two-layered analytic procedure. Image decomposition follows a topologic description, which is provided by a knowledge data-base for a given tissue type. The topologic description is discrete in nature and largely invariant against variability in morphology. In its discrete nature it fits exactly into the structure of an expert system, and it is used to direct the image processing. This is followed by a topographic description, which leads to the extraction of a multivariate feature set and the evaluation of dependence schemes and their descriptive statistics in the section. This process is being implemented in an image-data-driven, dynamically reconfigurable multiprocessor computer for the processing of imagery from histopathology, which is under development at the University of Arizona.

Conclusions

Early works in automated biomedical microscopy were applied to the analysis of single cells, with emphasis on blood cells, exfoliated cervical cells, and chromosomes. In the 1970s and 1980s, newer analytical techniques in scene segmentation extended research into the analysis of complex scenes such as images of tissue sections.

For the analysis of single cells, algorithms for measuring shape, size, texture, color, and integrated optical density have been developed. Which of these measurements are useful depends upon the application. In the analysis of blood cells, shape, texture, and color information is employed. In the analysis of cervical cells, DNA photometry, using integrated optical density, is important. Also, spectral information (color) helps in the quantitation of hormonal conditions, keratinization, etc. In the analysis of chromosomes, especially that of banded chromosomes, boundary information plays a major role. Since, in many applications, the response of single cells to environmental stress often expresses itself in the chromatin texture, this feature is important in determining the response to drugs, toxins, low-level radiation, etc. Measures of chromatin texture are also useful for the quantitation of malignancy, as are measures of shape for nucleus, cytoplasm, vacuoles, and nucleoli.

Important discoveries have been made at the University of Arizona and University of New Mexico during studies of the effects of low-level radiation on B and T lymphocytes and of the effects of antineoplastic agents on different subpopulations of lymphocytes from the peripheral blood. It has been found that lymphocytes from different patients have greatly differing radiation sensitivity. In the recognition of B and T cells, error rates of approximately 15% have been achieved in materials stained with the Feulgen, Papanicolau, and Wright-Giemsa stains. By using various fluorescently tagged surface markers, one can further differentiate such cells. Through this mechanism cell-image analysis forms the basis for diagnostic assessment and plays a role in studying the human immune system. Furthermore, the effects of treatment, environmental pollutants, and low-level radiation on a patient may be quantitated in this matter.

Numerous research groups in many countries are actively engaged in research in automated machines for cervical-smear screening. These efforts have been extended to epithelial cells from other body sites. Successful results will undoubtedly be obtained when well-controlled methods of sample preparation and staining are achieved and maintained.

In cytogenetics commercial machines are available that find metaphase spreads automatically and then turn the analysis over to the cytotechnologist. This frees the cytotechnologist from the laborious task of locating the spread and both speeds overall performance and removes the tedious aspects of the work. As sample preparation improves, techniques already developed at several laboratories may be eventually applied to full automation of traditionally stained chromosomes as well as to the analysis of chromosomes stained to exhibit banding patterns. Whether such systems will be cost-effective is as yet unknown. They can, however, provide information about consistent chromosome features that may be difficult to obtain from visual inspection of the complex structure of the banding patterns. Sister chromatid exchanges are also an area of research in cytogenetics.

Because the clinical laboratory is primarily concerned with fixed, nonviable cells and tissue, most image-analytic techniques have been applied to such material. There are, however, exciting prospects for applications of image analysis to living cells in order to characterize their development, metabolic rates, etc. Furthermore, the examination of live tissue is possible. Some work has been done on the examination of the living retina in assessing pigment distributions and in isolating the action of distinct neurotransmitters used in visual-signal analysis. Other research groups have worked with live slices of brain tissue, which are examined by light microscopy in conjunction with recording their neurophysical activity by microelectrodes.

In cancer diagnosis there is growing evidence that DNA photometry and other measures of nuclear characteristics can be correlated with the course of disease. The major goal, of course, is the automated development of prognostic information and the improvement, thereby, of treatment planning. In tissue the groupings of cells that form the tumor can be examined at low power, and the contour of the tumor determined. Some observations indicate that a more rounded border has better prognosis than a tumor with an irregular boundary shape. Dissociation of tumor cells from the main grouping, ie, the tendency of cells to form clumps, is also a prognostic indicator. This is called the "degree of dissociation." By combining descriptions of nuclear abnormalities, the degree of dissociation, and tumor boundaries, these features can then be correlated experimentally with the progress of disease. Three-dimensional reconstruction from consecutive serial sections can overcome the fact that observation of planar slices often leads to an incomplete grasp as to the changes in tissue architecture with disease, age, etc. Such work includes automatically tracking neural linkages using Golgi-stained tissue and tracking pathways. These studies are made possible due to the horseradish-peroxidase reaction.

In summary, image-analysis of cells and tissue is providing new diagnostic clues that will eventually have a significant impact on clinical determinations. By utilizing advanced pattern recognition and image-analytic techniques, further quantitation is being brought into basic studies of cell biology, cell physiology, immunology, and clinical cytology.

Only in hematology has the routine clinical use of computer vision become common and then only at the level of 1,000 machines sold worldwide over a period of 10 years. Automated cervical cytology is being impeded by the difficulty in designing an economic specimen-preparation machine plus the system required for image acquisition and image analysis. This is also true for the analysis of cells from the bladder and lung. In the field of chromosome analysis, the rapid development of new banding techniques has constantly outrun the capability of the engineer to program the new software systems required for image analysis. Standardization of equipment, programming languages and algorithms, and data-bases is lagging. This causes international data exchange to be impeded. However, international groups such as the International Academy of Cytology, and its journal, *Analytical and Quantitative Cytology,* are beginning to provide a common ground for information interchange.

Financial Considerations

The above shows that, to date, successful automation of image analysis for cells and tissues is restricted entirely to the analysis of blood smears. This development represents one of the most exciting technical accomplishments in pattern recognition achieved in this century despite lack of economic viability, at least in the United States. It is the first time that naturally occurring objects (opposed to man-made symbols) have been recognized by computer vision systems with execution rates at thousands of cells per hour and accuracy that equals human performance.

When present commercial equipment was initially marketed, there was considerable question as to its cost-effectiveness. In their initial marketing literature, Geometric Data [1974] pointed out that a technologist can visually inspect 40 smears per hour using a machine but only 10 per hour manually. With expenses at $12.25 per hour, this gives a labor cost saving of $0.92 per smear. For an installation averaging 120 smears per day (365 days per year), this yields an annual saving of $40,296 and a 2.5 year payback on a $90,000 machine (with a $7,200 annual service contract). Opposing this viewpoint, the Chicago Hospital Council [1965] determined that a technologist could read 20 smears per hour at $2.50/hour (mid-1960s wage scale). There were further debates on cost-effectiveness at the 1970 Engineering Foundation Research Conference on Multiphasic Health Testing [Davos] as presented by Preston and Rutovitz [1971]. There it was estimated that the cost of blood-smear examinations were $1.00 per smear. As the cost of technician labor has increased and, more importantly, the difficulty of training technologists in tedious work presents more of a problem, automated computer vision systems for hematology gain further acceptance depending on the degree of competition offered by flow systems. This is also true in Japan and some of the countries in Europe.

This section should not end without mentioning the dispute as to the value of the blood-smear examination itself and as to its statistical validity. The average person has about 10^{10} white blood cells in the peripheral bloodstream at any one moment. Yet the present blood-smear examination catalogs only 10^2 of these cells. This examination used to determine the rate of occurrence of rare cells such as the basophils, which appear at the rate of only one per several hundred cells. For basophils to be properly quantitated, one should examine at least 10^4 cells. To be competitive with flow systems robot microscopes of the future for blood-smear analysis should increase slide transfer velocities and data-processing rates by at least two orders of magnitude. Unfortunately, with the unattractiveness of funding further research and development in this field, this may not occur. In fact, it should be noted that one of the reasons given by Coulter Electronics for abandoning the diff3-series microscopes for blood-smear analysis was that the flow systems are now capable of processing thousands of cells per minute. In the United States at least these systems have displaced the image-analyzing microscope.

Standards and Validation for Current Systems

Automation in any field requires standards for calibration and validation of machine performance. This field often receives little attention. One of the exceptions to this has been the work of Bacus [1973] in investigating the variability of human performance in the visual classification of human white blood cells. The Center for Disease Control also has a continuing program of performance testing of work in clinical laboratories. Hundreds of standard slides are prepared per year and distributed to participating laboratories. The Center for Disease Control is also responsible for standardizing terminology in blood-smear analysis. The current lexicon is given in Table 1.5. Unfortunately, there is no equivalent program at the Center for Disease Control in either chromosome or cervical-smear analysis. However, Bacus [1982] has conducted experiments to compare machine performance with human performance in the classification of individual cervical cells as digitized by the television microscope.

Work is being done on the effect of stains on pattern recognition of cell nuclei as regards size and shape measures. Typical is the work of De-

TABLE 1.5. Red cell descriptive terms (glossary numbers 1–27) and descriptive terms for white blood cells (glossary numbers 35–64).*

1. Cabot rings	35. Alder's granulation
2. Elliptocytes	36. Basophilic bands
3. Hemoglobin c crystals	37. Basophilic metamyelocytes
4. Howell-Jolly bodies	38. Basophilic myelocytes
5. Immature nucleated cells	39. Hyposegmentation (neutrophils)
6. Fragmented red blood cells	40. Auer rods
7. Rouleaux	41. Polyploidy, diploid cells, tetraploid cells, etc.
8. Polymacrocytes	42. Eosinophilic bands
9. Siderocytes	43. Eosinophilic metamyelocytes
10. Malaria parasites	44. Eosinophilic myelocytes
11. Schuffner's dots	45. Immature cells unidentified
12. Malaria crescent-shaped gametocytes	46. Mitosis
13. Maurer's dots	47. Monocytes, atypical
14. Malarial double rings 1-3 trophozoites	48. Monocytes, vacuolated
15. Other malarial forms	49. Neutrophils, vacuolated
16. Polychromatophilia, slight	50. Plasmocytes
17. Polychromatophilia, marked	51. Stem cells
18. Basophilic stippling, slight	52. Tart cells
19. Basophilic stippling, marked	53. Marked leukocytosis
20. Spherocytes, few	54. Toxic granulation, slight
21. Spherocytes, numerous	55. Toxic granulation, marked
22. Target cells, few	56. Smudge cells, few
23. Target cells, many	57. Smudge cells, many
24. Sickle cells, pointed	58. Hypersegmentation of neutrophils, few
25. Sickle cells, blunt end	59. Hypersegmentation of neutrophils, many
26. Burr cells or spinocytes, few	60. Macropolycytes
27. Burr cells or spinocytes, many	61. Atypical lymphocytes occasional (0–5%)
	62. Atypical lymphocytes plasmacytoid type (6–20% of total)
	63. Majority of the lymphocytes atypical
	64. Doehle (RNA) bodies

*Taken from the publication of the Centers for Disease Control, U.S. Department of Health and Human Services.

Campos et al [1973]. The accurate generation of microspheres for spatial calibration has been undertaken by Fulwyler et al [1973] with such workers as Brunsting and Mullaney [1972], Cooke and Kerker [1973], and Phillips et al [1970] working on the light-scattering properties of such spheres. This work has been conducted primarily for the purpose of calibrating flow systems but is also useful in the use of microspheres for calibrating television scanners.

The automation of blood-smear analysis has suffered from the fact that the standard data printout has been established by simply copying earlier manual methods wherein the percentages of the various classes of white blood cells were written on a laboratory "ticket." All feature data—ie, the raw measurement data on such fundamental quantities as nuclear and cytoplasmic size, shape, and color—are discarded. As reported by Mellecker [1972], Preston has made the sole attempt to not only record but visually present patient profiles as derived from automated blood-smear analyzers (see Figure 1.22).

In the future, it should be considered whether the National Bureau of Standards should play a role in the standardization of stains. In the automation of cell and tissue analysis, it may be that the Food and Drug Administration should undertake this activity.

Physical Limitations on Future Systems

In all fields of endeavor, the engineer should predict the limitations that basic physics places on the rate at which systems can operate. If a two order of magnitude improvement is required to produce statistically valid analyses of blood smears, one must examine the basic information carrier that transfers data from the television microscope to the computer. In the case of the television microscope, it is electromagnetic energy in the

CYTOLOGICAL DATA SHEET EXPERIMENT 067

	NEUTROPHIL			LYMPHOCYTE			MONOCYTE		
TRIMMED NUCLEAR AREA	17	22	27	29	38	47	52	80	107
TOTAL CELL AREA	62	93	124	61	89	117	121	177	234
FINE STRUCTURE INDEX	.36	.47	.58	.23	.28	.33	.18	.26	.34
INCLUSION INDEX	.015	.025	.035	–	.008	.018	–	.023	.051
IRREGULARITY INDEX, FINE	–	2.8	13	.11	.25	.60	.66	3.3	5.8
IRREGULARITY INDEX, COARSE	.40	.82	1.2	–	.06	.19	.28	1.4	2.6
ELONGATION INDEX	–	.093	.315	.008	.013	.018	.004	.026	.048
NUCLEAR TRANSMISSION, Y	.29	.38	.47	.29	.35	.41	.44	.45	.46
CYTOPLASMIC TRANSMISSION, Y	.23	.61	1.0	.44	.65	.86	.47	.66	.85
NUCLEAR TRANSMISSION, G	.20	.21	.22	.18	.19	.20	.20	.28	.36
CYTOPLASMIC TRANSMISSION, G	.25	.40	.55	–	.50	–	–	.50	–

FIGURE 1.22. Cytological data sheet for CELLSCAN experiment #067 showing the profile of that specific patient as compared with norms developed from a small population of patients.

optical spectrum. Here the unit of data transfer is the photon having an energy of approximately 0.3 electron volts. To achieve faster scanning rates, the major physical barrier is simply the variability of photon arrival rates (shot noise). The number of photons available is directly dependent upon the color temperature of the illuminating light source. Using non-coherent illumination, a few tens of nanowatts per optical resolution element are available. Lasers provide 100 to 1,000 times more power. This leads to the physical bounds on picture data transfer rates given in Figure 1.23. The number of cell images transferred per minute is dependent upon the fourth power of the picture element size as given in the following equation:

$$R_{\text{scan}} = \frac{(d)^4 \varepsilon(\lambda)\eta(\lambda)(\text{NA})^2}{(S/N)^2 D^2} \int I(\lambda)\, d\lambda \tag{1.9}$$

d = effective detector diameter (related to object plane)
D = object diameter
$I(\lambda)$ = irradiance
NA = numerical aperture in object plane
S/N = acceptable peak signal to root-mean-square (RMS) noise
$\varepsilon(\lambda)$ = optical transfer efficiency
$\eta(\lambda)$ = quantum efficiency
λ = wavelength

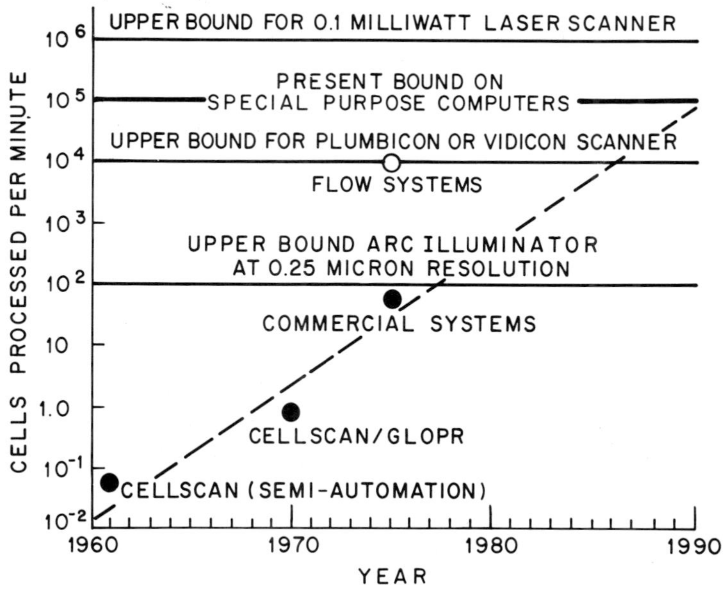

FIGURE 1.23. System performance (as a function of the source of optical energy and of the year the technology was applied) for various types of scanning and digital-picture-processing systems as related to the automated white blood cells differential count.

The fourth-power dependency is due to the fact that, as picture element size increases, the number of photons per picture element increases as the square of the picture element diameter while, at the same time, the number of picture elements per cell decreases as the square of the picture element diameter. Figure 1.23 gives plots for (1) point-scanning devices such as photomultipliers or single semiconductors diodes, (2) light-integrating devices such as television-imaging tubes and diode arrays, and (3) coherent sources (lasers). It is clear that the desired cell transfer rate can be achieved by light-integrating devices but that lasers offer another gain of at least two orders of magnitude. The laser also competes against such phenomena as "image lag" in the light-integrating devices wherein a residual image is retained after readout. The laser exhibits a greater dynamic range. This is the approach being actively pursued by Shack et al [1982], who has developed the ultrafast laser-scanner microscope at the University of Arizona. This scanner is capable of data rates of the order of 100 megawords/s, recording 1.6 billion pixels within 60 seconds, in two wavelength channels. Work is underway to process this data stream in real time, using a multimicroprocessor computer system as reported by Bartels [1986], who projects the need for the processing of pixel volumes of 10^9 to 10^{12} pixels per research project. Examples are studies of malignant cell invasion patterns and three-dimensional tissue reconstruction in neurobiology and neuroanatomy. Applications in the latter area may eventually require the processing of up to 10^{15} pixels, as compared with the 10^8 pixels processed in a typical major research study today. It is work in this direction, where new technology is applied to research in laser-scan computerized microscopes, plus research in the fundamental understanding of how these microscopes may be used to extract significant diagnostic information from carefully prepared samples of cell and tissue that will lead to the reliable, useful, and economic automated microscopes of the future.

Acknowledgments. The authors are indebted to numerous colleagues who furnished material for both text and figures of this manuscript. Where appropriate, these contributions are acknowledged in the figure captions. The authors would also like to thank G. Thomas and his staff at Carnegie-Mellon University for final copy on all illustrations. In particular, the authors would like to acknowledge the many weeks of editorial effort by C. Magdaleno (Executive Suite, Tucson) who proofed the bibliography as well as typed the completed manuscript in its final form.

References

Abele L, Kitahashi T, Wahl F: Ein digitales Verfahren zur Konturfindung und Stoerbeseitigung bei Zellbildern, in *Informatik Fachberichte* (Digitale Bildverarbeitung) GI/NTG, (Nagel HH ed). Springer Verlag, Berlin (1977) 8: 31.

Abmayr W, Bartels PH, Giaretti W, et al: Identification of cell cycle compartments by textural and chromatin features. 1982 Int Conf High Resolution Cell Image Analysis, North Hollywood, Calif (1982a).

Abmayr W, Borst H: Untergrundelemination und merkmalsextration an zellen. Med Physik Forschung Praxis pp 63–73 (1976).

Abmayr W, Borst H, Schwarzkopf P, et al: A feature-selection system for conventional Pap-stained cell-images, in *The Automation of Cancer Cytology and Cell Image Analysis,* (Pressman NJ, Wied GL eds). Tutorials of Cytology, Chicago (1979) pp 258–268.

Abmayr W, Deml E, Oesterle D, et al: Nuclear morphology in preneoplastic lesions of rat liver. Anal Quant Cytol 5: 235–244 (1983).

Abmayr W, Giaretti W, Gais P, et al: Discrimination of G1 S and G2 cells using high resolution TV scanning and multivariate analysis methods. Cytometry 2: 316–326 (1982b).

Adams LR, Kamentsky LA: Machine characterization of human leukocytes by acridine orange fluorescence. Acta Cytol 15:289 (1971).

Adams LR: Acridine orange staining of epithelial cells in strong salt solution. J Histochem Cytochem 22: 492–494 (1974)

Adelh D, Brugal G, Chassery JM, et al: Microscopic image analysis system: Applications to biology: Cell cycle cancerous blood and bone marrow cells. Proc 4th Int Joint Conf on Pattern Recog, Kyoto, Japan (1978) pp 864–871.

Adelh D, Signor G: *Shape Recognition and Artificial Intelligence.* IRIA, Paris (1981).

Aggarwal RK, Bacus JW: A multi-spectral approach for scene analysis of cervical cytology smears. J Histochem Cytochem 25: 668 (1977).

Aggarwal RK, Fu KS: Identification of Abnormal Chromosomes Images. Rep TR-EE 73-12 School Elect Eng, Purdue University (1973).

Aggarwal RK, Fu KS: Automatic recognition of irradiated chromosomes. J Histochem Cytochem 22: 561–568 (1974).

Aherne W, Dunhill MS: *Morphometry,* Edward Arnold, London (1982).

Al I, Ploem JS: Detection of suspicious cells and rejection of artifacts in cervical cytology using the Leyden television analysis system. J Histochem Cytochem 27: 629–634 (1979).

Alexander JC, Thaler AI: The boundary count of digital pictures. J Assoc Comp Mach 18: 105–112 (1971).

Alpen EL, Johnston ME: DNA synthetic rate and DNA content of nucleated erythroid cells. Exp Cell Res 47: 177–192 (1967)

Anderson AC, Fu KS: Design and development of a linear binary tree classifier for leukocytes. Rep TR-EE 79-31 School of Elect Eng, Purdue University (1979).

Anderson RE, Olson GB, Shank C, et al: Computer analysis of defined populations of lymphocytes irradiated in vitro. I Evaluation of murine thoracic duct lymphocytes. Acta Cytol 12: 126–135 (1975a).

Anderson RE, Olson GB, Howarth JL, et al: Computer analysis of defined populations of lymphocytes irradiated in vitro. II Analysis of thymus dependent vs bone marrow derived cells. Am J Pathol 80: 21–32 (1975b).

Anderson RE, Olson GB, Howarth JL, et al: Computer-assisted analysis of irradiated lymphocytes, in *Mammalian Cells: Probes and Problems.* ERDA Symposium Series (1975c) pp 44–54.

Arcelli C, Levialdi S: Neuron counting in three dimensions: A proposal. Acta Cybernetica (Hungary) 5: 65–68 (1973).

Arnold F, Kunze KD, Meyer W, et al: Automatische bildverarbeitung in der pathologie. I. Allgemeine methodische grundlagen. Exp Pathol 9: 16–26 (1974).
Atamer M, Groner W: Investigation of the left shift with peroxidase chemistry of Hemalog D. Adv Auto Anal 3: 33 (1973).
Atkin NB: Cytogenetic factors influencing the prognosis of uterine carcinoma, in *Modern Radiotherapy: Gynaecological Cancer,* (Deeley TJ ed). Butterworths, London (1971a) pp 138–154.
Atkin NB: Modal DNA value and chromosome number in ovarian neoplasia: A clinical and histopathologic assessment. Cancer 27: 1064–1073 (1971b).
Atkin NB: Modal deoxyribonucleic acid value and survival in carcinoma of the breast. Br Med J 1: 271–272 (1972).
Atkin NB: Chromosomal changes in cervical neoplasia, in *The Cervix* (Jordan JA and Singer A eds). WB Saunders, London (1976a) pp 385–393.
Atkin NB: *Cytogenetic Aspects of Malignant Transformation,* Karger, Basel (1976b) pp 73–76
Atkin NB, Richards BM, Ross AJ: The deoxyribonucleic acid content of carcinoma of the uterus: An assessment of its possible significance in relation to histopathology and clinical course. Based on data from 165 cases. Br J Cancer 13: 773–787 (1979).
Auer G, Caspersson T, Wallgren A: DNA content and survival in mammary carcinoma. Anal Quant Cytol 2: 161–165 (1980a).
Auer G, Kato H, Nasiell M, et al: Cytophotometric DNA-analysis of atypical squamous metaplastic cells, carcinoma *in situ* and bronchogenic carcinoma, in *Prevention and Detection of Cancer, Part II: Detection. Second Volume,* (Nieburgs HE ed). Marcel Dekker, New York (1980b) pp 1465–1476.
Auer G, Zetterberg A: The prognostic significance of nuclear DNA content in malignant tumors of breast, prostate, and cartilage, in *Advances in Clinical Cytology, vol. 2,* (Koss LG, Coleman DV eds). Masson, New York (1984) pp 123–134.
Aufferman W, Repges R, Boecking A: Rapid diagnostic DNA cytometry with an automatic microscope and a TV image analysis system. Anal Quant Cytol 6: 179–188 (1984).
Aughey WH, Baum FJ: Angular dependence of lightscattering—A high resolution recording instrument for the angular range 005° to 140°. J Opt Soc Am 44: 833 (1954).
Aus HM, Ruter A, ter Meulen V, et al: Bone marrow cell scene segmentation by computer-aided color cytophotometry. J Histochem Cytochem 25: 662–667 (1977).
Aus HM, ter Meulen V, Kaekell M, et al: Techniques applicable to computer-aided cytophotometry in virology. J Histochem Cytochem 22: 577–582 (1974).
Baak JPA, Kurver PHJ, Boon ME: Computer-aided applications of quantitative microscopy in diagnostic pathology. Pathol Ann 17: 287–306 (Sommers SC, Rosen PP eds) (1982). Appleton-Century-Crofts pp. 287–306.
Baak JPA, Kurver PHJ, Overdiep SH, et al: Quantitative, microscopical, computer-aided diagnosis of endometrial hyperplasia or carcinoma in individual patients. Histopathology 5: 689–695 (1981a).
Baak JPA, Kurver PHJ, Overdiep SH, et al: Quantitative microscopical (computer aided) diagnosis of endometrial hyperplasia or carcinoma in individual patients. Histopathology 5: 689–695 (1981b).

Baak JPA, Oort J: *Morphometry in Diagnostic Pathology*. Springer Verlag, Berlin (1983).

Bacus JW: An Automated Classification of the Peripheral Blood Leukocytes by Means of Digital Image Processing. PhD Dissertation Dept Physiol Univ of Ill, Chicago (1970).

Bacus JW: The observer error in peripheral blood cell classification. Am J Clin Pathol 59: 223 (1973).

Bacus JW: Erythrocyte morphology and centrifugal spinner blood film preparations. J Histochem Cytochem 22: 506 (1974).

Bacus JW: Application of digital image processing techniques to cytology automation. Report prepared for the National Cancer Institute, Contract No. 1-CB-74190 (1982).

Bacus JW, Gose EE: Leukocyte pattern recognition. IEEE Trans Syst Man Cyber SMC-2: 513–526 (1972).

Bacus J, Wiley E, Galbraith W, et al: Cytotechnologist slide screening versus single cell classification accuracy. Anal Quant Cytol 4: 142 (1982).

Bacus JW, Wiley EL, Galbraith MA, et al: Malignant cell detection and cervical cell screening. Anal Quant Cytol 6: 121–130 (1984).

Bahr GF, Bartels PH, Bibbo M, et al: Papanicolaou stain for computer assisted cellular recognition. Acta Cytol 17: 106–112 (1973).

Bahr GF, Bartels PH, Wied GL, et al: Automated cytology, in *Diagnostic Cytology and Its Histopathologic Bases* (Koss LG ed). Lippincott, Philadelphia (1979) pp 1123–1164.

Bahr GF, Bibbo M, Oehme M, et al: An automated device for the production of cell preparations suitable for automatic assessment. Acta Cytol 22: 243–249 (1978).

Bahr GF, Taylor J, Bartels PH, et al: Distinguishing normal human blood lymphocytes from lymphocytes in dengue and typhoid fever. Virchow Arch [Cell Pathol] 16: 205–210 (1974).

Bahr GF, Wied GL: Automation in cytopathologic diagnosis. Multiphasic Health 1: 67–82 (1970).

Baky AA, Winkler DG, Hunter NR, et al: Atypia status index of respiratory cells: A measurement for the early detection and monitoring of neoplastic changes in squamous cell carcino-genesis. Anal Quant Cytol 2: 175–185 (1980).

Bales CE: A semi-automated method for preparation of urine sediment for cytologic evaluation. Acta Cytol 25: 323–326 (1981).

Ball GH, Hall DJ: A clustering technique for summarizing multivariate data. Behav Sci 12: 153–155 (1957).

Barer R: Refractometry and interferometry of living cells. J Opt Soc Am 47: 545 (1957).

Barer R, Wardley J: Ultra-violet television microscopy. Nature 192: 1060 (1961).

Bartels HG, Bartels PH, Bibbo M, et al: Stabilized binary hierarchic classifier in cytologic diagnosis. Anal Quant Cytol 6: 247–261 (1984a).

Bartels PH: Bildgewinnung and Bildverarbeitung. *Symposium on Quantitative Micromorphometry* (Burger G, Goessmer H eds) (1985).

Bartels PH, Abmayr W, Bibbo M, et al: Computer recognition of ectocervical cells—Image features. Anal Quant Cytol 3: 157–164 (1981).

Bartels PH, Bahr GF, Bellamy JC, et al: A self-learning computer program for cell recognition. Acta Cytol 14: 486–494 (1970).

Bartels PH, Bahr GF, Bibbo M, et al: Analysis of variance of the Papanicolaou staining reaction. Acta Cytol 18: 522–531 (1974a).
Bartels PH, Bahr GF, Bibbo M, et al: Objective image analysis. J Histochem Cytochem 20: 239–254 (1972).
Bartels PH, Bahr GF, Griep J, et al: Computer analysis on lymphocytes in transformation. A methodologic study. Acta Cytol 13: 557–568 (1969).
Bartels PH, Bahr GF, Jeter WS, et al: Evaluation of correlation information in digitized cell images. J Cytochem Histochem 22: 69–79 (1974b).
Bartels PH, Bahr GF, Wied GL: Cell recognition from line scan transition probability profiles. Acta Cytol 13: 210–217 (1969).
Bartels PH, Bellamy JC: Self-optimizing, self-learning system in pictorial pattern recognition. Appl Opt 9: 2453–2458 (1970).
Bartels PH, Bibbo M, Bahr GF, et al: Cervical cytology: Descriptive statistics for nuclei of normal and atypical cell types. Acta Cytol 17: 449–453 (1973).
Bartels PH, Bibbo M, Dytch HE, et al: Marker features for malignancy in ectocervical cells-statistical evaluation. Cell Biophys 5: 71–77 (1983).
Bartels PH, Bibbo M, Olson GB, et al: Cell image analysis in quantitative cytology, in *Technical Advances in Biomedical Physics,* (Dendy PP, Ernst DW, Senguen A eds). Martinus Nijhoff, The Hague (1984) pp 59–90.
Bartels PH, Bibbo M, Richards D, et al: Patient classification based on cytologic sample profiles. I. Basic measures for profile construction. Acta Cytol 22: 253–260 (1978).
Bartels PH, Bibbo M, Taylor J Jr, et al: Cell recognition from the statistical dependence of gray values in digitized images. Acta Cytol 18: 165–169 (1974).
Bartels PH, Bibbo M, Wied GL: Modeling of histologic images by computer. Acta Cytol 20: 62–67 (1976).
Bartels PH, Bibbo M, Wied GL: Estimation of the proportion of patients with a very low number of tumor cells in the cervical smear. Anal Quant Cytol 7: 136–142 (1979).
Bartels PH, Buchroeder RA, Hillman DW, et al: Ultrafast laser scanner microscope design and construction. J Anal Quant Cytol 3: 55–66 (1981).
Bartels PH, Chen YP, Durie BG, et al: Discrimination between human T and B lymphocytes by computer analysis of digitized data from scanning microphotometry. II. Discrimination and automated classification. Acta Cytol 22: 530 (1978).
Bartels PH, Graham A, Paplanus S, et al: Computer configurations for the processing of diagnostic imagery in histopathology. *Evaluation of Multicomputers for Image Processing,* (Preston K, Uhr L eds). Academic Press, New York (1986).
Bartels PH, Koss LG, Sychra JJ, et al: Indices of cell atypia in urinary tract cytology. Acta Cytol 22: 387–391 (1978).
Bartels PH, Koss LG, Wied GL: Automated cell diagnosis in clinical cytology, in *Developments in Clinical Cytology,* (Koss LG, Coleman M eds). Butterworths, London (1980) pp 314–342.
Bartels PH, Layton J, Shoemaker RL: Digital Microscopy, in *Survey and Synthesis of Pathology Research,* (Greenberg SD ed). Karger, Basel (1984) pp 28–61.
Bartels PH, Olson GB: Computer analysis of lymphocyte images, in *Methods of Cell Separation,* (Catsimpoolas N ed). Plenum Press, New York (1980) pp 1–99.

Bartels PH, Olson GB: Computer graphic displays for microscopists assisting in evaluating radiation-damaged cells, in *Biological Dosimetry*, (Eisert WG, Mendelsohn ML eds). Springer-Verlag, Berlin (1984) pp 295–297.

Bartels PH, Olson GB, Jeter WS, et al: Evaluation of unsupervised learning algorithms in the computer analysis of lymphocytes. Acta Cytol 18: 376–388 (1974).

Bartels PH, Olson GB, Lockart R, et al: Cytophotometric studies of cell populations. Cell Biophys 2: 339–351 (1980).

Bartels PH, Wied GL: Performance testing for automated prescreening devices in cervical cytology. J Histochem Cytochem 22: 660–662 (1974).

Bartels PH, Wied GL: Extraction and evaluation of information from digitized cell images, in *Mammalian Cells: Probes and Problems*, (Richmond CR, Peterson DF, Mullaney PF, Anderson EC eds). US Energy Res Dev Admin (1975) pp 15–28.

Bartels PH, Wied GL: An image analyzing software system for cytology. Proc of COMPSAC 1977, IEEE Comp Soc 1st Intl Comp Software and Appl Conf, Chicago (1977) pp 282–284.

Barz H, Kunze KD, Voss K, et al: Automatische bildverarbeitung in der pathologie. IV. Altersabhangige Veranderungen morphologischer Parameter der Leberzellkerne im Biopsiematerial. Exp Pathol 14: 55–64 (1977).

Bateman JB: Osmotic responses and light scattering of bacteria. J Colloid Interface Sci 27: 458 (1968).

Beadle C in: The Quantimet image analyzing computer and its applications, in *Advances in Optical and Electron Microscopy*, (Coslett V, Barer R eds). Academic Press, New York (1971) p 361.

Beard JD: Computerized biopsy system, Comput News Physicians C-12 (1984).

Bengtsson E, Eriksson O, Holmquist J, et al: Low resolution segmentation of cervical cells. Report 77-3 Dept Clin Cytol, University Hospital, Uppsala, Sweden (1977).

Bengtsson E, Eriksson O, Holmquist J, et al: Implementation and evaluation of a diode array scanner for digitizing microscopic images, in *The Automation of Cancer Cytology and Cell Image Analysis*, (Presman NJ, Wied GL eds). Tutorials of Cytology, Chicago (1979) pp 269–286.

Bennett W: Spectra of quantized signals. Bell Syst Tech J 27: 446–472 (1948).

Benno RH, Tucker LW, Joh TH, et al: Quantitative immunocytochemistry of tyrosine hydroxylase in rat brain. I. Development of a computer-assisted method using the peroxidase-antiperoxidase technique. Brain Res 246: 225–231 (1982).

Bentley SA, Lewis SM: The use of an image analysing computer for the quantitation of red cell morphological characteristics. Br J Haematol 29: 81–88 (1975).

Berkley C: Electronic counting, sorting, and sizing, Proc 2nd Int Conf Med Electron, Iliffe, London (1959).

Berkman RM, Wyatt PJ: Differential light scattering measurements of heat-treated bacteria. Appl Microbiol 20: 510 (1970).

Bhattacharya PK, Bartels PH, Taylor J Jr, et al: A decision procedure for automated cytology: Test statistic for detecting sample abnormality and inadequacy. Acta Cytol 17: 538–552 (1973).

Bhattacharya PK, Bartels PH, Bahr GF, et al: A test statistic for detecting the presence of abnormal cells in a sample. Acta Cytol 15: 533–552 (1971).

Bhattacharya PK, Bartels PH, Bibbo M, et al: Estimation procedure for the cellular composition of cervical smears. Acta Cytol 19: 366–373 (1975).

Bibbo M, Alenghat E, Bahr GF, et al: A quality-control procedure on cervical lesions for the comparison of cytology and histology. J Reprod Med 12: 811–822 (1983).

Bibbo M, Bartels PH, Bahr GF, et al: Computer recognition of cell nuclei from the uterine cervix. Acta Cytol 17: 340–350 (1973).

Bibbo M, Bartels PH, Chen M, et al: The numerical composition of cellular samples from the female reproductive tract: I. Carcinoma *in situ*. Acta Cytol 19: 438–447 (1975).

Bibbo M, Bartels PH, Chen M, et al: The numerical composition of cellular samples from the female reproductive tract: II. Cases with invasive squamous carcinoma of uterine cervix. Acta Cytol 20: 249–253 (1976a).

Bibbo M, Bartels PH, Chen M, et al: The numerical composition of cellular samples from the female reproductive tract: III. Cases with mild and moderate dysplasia of the uterine cervix. Acta Cytol 20: 565–572 (1976b).

Bibbo M, Bartels PH, Chen M, et al: The numerical composition of cellular samples from the female reproductive tract, in *Automation of Uterine Cancer Cytology*, (Wied GL, Bahr G, Bartels PH eds). Int Acad Cytology, Chicago (1976c) pp 26–53.

Bibbo M, Bartels PH, Chen M, et al: The numerical composition of cellular samples from the female reproductive tract: IV. Carcinoma *in situ* cases exhibiting other than normal vaginal flora. Acta Cytol 21: 705–709 (1977).

Bibbo M, Bartels PH, Chen M, et al: The numerical composition of cellular samples from the female reproductive tract: V. Cell cluster patterns in cases of invasive squamous carcinoma of uterine cervix. Acta Cytol 22: 250–252 (1978).

Bibbo M, Bartels PH, Sychra JJ, et al: Chromatin appearance in intermediate cells from patients with uterine cancer. Acta Cytol 25: 23–28 (1981).

Bins M, Landeweerd GH, Gelsema ES, et al: Texture of white blood cells expressed by the counting densitogram. Cytometry 1: 321–324 (1981).

Blondal T, Bengtsson A: Nuclear DNA measurements in squamous cell carcinoma of the lung: A guide for prognostic evaluation. Anticancer Res 1: 79–86 (1981).

Blondal T, Bengtsson A, Arnorsson T: Nuclear DNA in Bronchial Carcinoma. Presented at 2nd World Congress Int Assoc for the Study of Lung Cancer, Copenhagen, June 9–13 (1980).

Bloss WH, Erhardt R, Reinhardt ER: High resolution image processing system for fast cell analysis, in *The Automation of Cancer Cytology and Cell Image Analysis* (Presman NJ, Wied GL eds). Tutorials of Cytology, Chicago (1979) pp 217–220.

Blum HA: A transformation for extracting new descriptors of shape. Symp Models for Percept Speech Visual Form, Boston (1964).

Boehm N, Sprenger E: Fluorescence cytophotometry: A valuable method for the quantitative determination of nuclear Feulgen-DNA. Histochemie 16: 100 (1968).

Boon ME, Kurver PJH, Baak JPH, et al: The application of morphometry in gastric cytological diagnosis. Virchow Arch [Pathol Anat] 393: 159–164 (1981).

Boon ME, Lowhagen T, Willems JS: Planimetric studies on fine needle aspirates from follicular adenoma and follicular carcinoma of the thyroid. Acta Cytol 24: 145–148 (1980).

Borst H, Abmayr W, Gais P: A thresholding method for automatic cell image segmentation. J Histochem Cytochem 27: 180–187 (1979).

Bostrom RC, Holcome WG: CYDAC-A digital scanning cytophotometer. IEEE Int Conf Rec,Pt 9,110 (1963).

Bostrom RC, Sawyer HS, Tolles WE: The Cytoanalyzer—An automatic prescreening instrument for cancer detection. Proc 2nd Int Conf Med Electron (Paris 1959).

Bourk T, Tretiak OJ: Computer measurement of neutrophil nucleus shape. Eng Foundation Conf Automatic Cytology (Asilomar 1973).

Bowie JE: Differential leukocyte classification using an optical processing system. Proc 23rd ACEMB (1970 p 256.

Boysen M, Reith A: A morphometric model for light microscopic analysis of metaplastic, dysplastic and carcinomatous alterations of the nasal mucosa in nickel workers. Pathol Res Pract 166: 362–371 (1980).

Bradbury S: Microscopial image analysis. Problems and approaches. J Microsc 115: 137–150 (1979).

Brain AE: System for differential particle counting. US Patent No 3,824,383 (1974).

Brakey MW, Rooney IA: A vortex whistle for dispersion of aggregates, in Digest, 11th Int Conf on Med and Bio Eng, Ottawa, Canada (1976) pp 80–81.

Braylan RC, Fowckes BF, Jaffé ES, et al: Cell volumes and DNA distributions of normal and neoplastic human lymphoid cells. Cancer 41: 201–209 (1978).

Breg WR, Miller DA, Allderdice PW, et al: Quinacrine fluorescence patterns and terminal DNA labeling of human C group chromosomes. Nature 236: 76 (1974a).

Breg WR, Miller DA, Allderdice PW, Miller OJ: Identification of translocation chromosomes by quinacrine fluorescence. Am J Dis Child 123: 561 (1974b).

Bremeskow V, Kaden P, Mittermayer C: DNA synthesis during the life cycle of L cells: Morphological histochemical and biochemical investigations with arabinosylcytosine and thiorabinosycytosine. Eur J Cancer 6: 379 (1970).

Brenner JF, Gelsema ES, Necholes TF, et al: Automated classification of normal and abnormal leukocytes. J Histochem Cytochem 22: 797 (1974).

Brenner JF, Nechels TF, Tonacossa IA, et al: Scene segmentation techniques for the analysis of routine bone marrow smears from acute lymphoblastic leukemia patients. J Histochem Cytochem 25: 601–613 (1977).

Briski KP, Baker BL, Cristensen AK: Effect of ovariectomy on the hypothalmic content of immunoreactive gonadotropin-releasing hormone in the female mouse as revealed by quantitative immunocytochemistry and radioimmunoassay. Am J Anat 166: 187–196 (1983).

Broullette JW Jr: Pattern Recognition. Proc 4th Natl Convent Milit Electron (1962) p 1790.

Broullette JW Jr: Form Recognition Method and System. US Patent No 2,986,643 (1961).

Bruce JD: Optimum Quantization. MIT Res Lab Elec Tech Rep No 429 (1961).

Brunsting A, Mullaney PF: Light scattering from coated spheres—A model for biological cells. Appl Opt 11: 675 (1972).

Brunsting A: Light scattering from mammalian cells. PhD Dissert, Univ N Mex (1972).

Brunsting A: Can light-scattering techniques be applied to flow-through cell analysis? J Histochem Cytochem 22: 607–615 (1974).

Bryant FD, Latimer P, Seiber BA: Changes in total light scattering and absorption caused by changes in particle conformation—A test of theory. Arch Biochem Biophys 135: 109 (1969).

Bullock B: Complex information processing—A guided bibliography. Res Rep 444: Hughes Aircraft Corp (1971).

Bunnag B, Herman CJ, Fowlkes BJ, et al: Flow microfluorometric approach to cytology automation, in *The Automation of Cancer Cytology and Cell Image Analysis,* (Presman NJ, Wied GL eds). Tutorials of Cytology, Chicago (1979) pp 1–7.

Burger G: Entwicklung und Anwendung eines hybriden Verfahrens zur automatischen Bildanalyse histologischer und zytologischer Praeparate zur Frueherkennung von Tumoren. Final report, TUDAB project. GSF Report S-522, Munich, West Germany (1977).

Burger G, Abmayr W, Gais P, et al: Bildanalysesystem zur Klassifizierung zytologischer Praeparate. BMFT-FB GSF, Munich, West Germany (1980).

Burger G, Jutting U, Rodenacker K: Changes in benign cell populations in cases of cervical cancer and its precursors. Anal Quant Cytol 3: 261–271 (1981).

Burn GP: The extraction of data from an optical image by scanning. Med Biol Eng 9: 87–95 (1971).

Bushor WE: Medical Electronics Pt VI: Observing life processes. Electronics 34: 65–71 (1961).

Butler JW, Butler MK, Stroud A: Automatic classification of chromosomes, in *Data Acquisition and Processing in Biology and Medicine,* vol 3 (Enslein E ed). Pergamon Press, New York (1966) pp 261–275.

Cahn RL, Poulsen RS, Toussaint G: Segmentation of cervical cell images. J Histochem Cytochem 25: 681–688 (1971).

Carman G, Lemkin P, Lipkin L, et al: A real time picture processor for use in biologic cell identification. II. Hardware implementation. J Histochem Cytochem 22: 732 (1974).

Carvalko J Jr: On determining optimum simple Golay marking transforms for binary image processing. IEEE Trans Comput C-21: 1430–1433 (1972).

Caspersson T: *Cell Growth and Cell Function.* WW Norton, New York (1950).

Caspersson T, Lindsten J, Lomakka G, et al: The use of fluorescence techniques for the recognition of mammalian chromosomes regions. Int Rev Exp Pathol 11: 2 (1973).

Caspersson T, Lomakka G, Lorzech J: The 24 fluorescence patterns of the human metaphase chromosomes-distinguishing characters and variability. Hereditas 67: 89 (1971).

Caspersson T, Lomakka G, Miller A: Computerized chromosome identification by aid of the quinacrine mustard fluorescence technique. Hereditas 67: 103 (1971).

Caspersson T, Lomakka G, Rigler R Jr: Registrierender fluoreszenmikrospectograph zur bestimmung der primar und secondarfluoreszenz verschiedener zellsubstanzen. Acta Histochem (Jena) 6: 123 (1965).

Caspersson T, Lomakka G, Svensson G, et al: A versatile ultramicrospectrograph for multiple line and surface scanning high resolution measurements employing automatized data analysis. Exp Cell Res [Suppl] 3: 40–51 (1955).

Castillo X, Yorkgitis D, Preston K Jr: A study of multidimensional multicolor images. IEEE Trans Biomed Eng BME-29: 111–121 (1982).

Castleman KR, Chui LA, Van der Meulen JP: Spaceflight effects on muscle fibers. NASA Tech Memo 78526: 224–289 (1978).

Castleman KR, Chui LA, Van der Meulen JP: Automatic analysis of muscle fibers from rats subjected to spaceflight. NASA Tech Memo 81289: 267–278 (1981).

Castleman KR, Melnyk J: Automated system for chromosome analysis—Final Report, Internal Document No. 1200-240A Jet Propulsion Laboratory (Sept. 1975).

Castleman KR, Van der Meulen JP, Munsat TL, et al: Quantitative muscle biopsy analysis. Proc SPIE 89: 119–121 (1976).

Castleman KR, Wall RJ: Automatic systems for chromosome identification, in *Chromosome Identification,* (Caspersson T, Zech L eds). Academic Press, New York (1973) pp 77–84.

Castleman KR, White BS: Optimizing cervical specimen classifier. IEEE Trans Pattern Anal Machine Intel PAMI-2: 451–457 (1980).

Catsimpoolas N (ed): *Methods of Cell Separation,* Plenum Press, New York (1980).

Causley D, Young JZ: Flying spot microscope—Use in particle analysis. Research 8: 451–457 (1953).

Chalkey JW: Methods for the quantitative analysis of microstructures. J Nat Cancer Inst 4: 47 (1943).

Cheng GC: Color information in blood cells. J Histochem Cytochem 22: 517–520 (1974).

Chermant JL (ed): *Quantitative Analysis of Microstructures in Materials Science, Biology, and Medicine.* Riederer, Stuttgart (1978).

Chicago Hospital Council: Clinical laboratory study (1965).

Chien YP, Fu KS: A decision function method for boundary detection. Comput Graph Image Proc 3: 125–140 (1974).

Chodak GW, Bibbo M, Dytch HE, et al: Correlation between automated DNA ploidy measurements of fine needle aspiration biopsy of the prostate and cytological diagnosis, staging and gleason sum. Abstract presented at the 1985 American Urological Association meeting in Atlanta.

Christov K, Kiefer G, Kiefer R, et al: Changes in the Nuclear Structure During Thyroid Carcinogenesis in Rats—An Image Analysis Study. Inst Pathol, Univ of Freiburg (1974).

Chui LA, Castleman KR: Morphometric analysis of rat muscle fibers following space flight and hypogravity. Physiologist 23: S-76–S-78 (1980).

Chui LA, Castleman KR, Van der Meulen JP: Morphometric analysis of rat muscle fibers following space flight, in Hideg, Gazenko. Gravitational physiology. Adv Physiol Sci vol 19 (1981).

Coates HN: An electronic machine for statistical particle analysis. Proc IEE 103-B: 479 (1956).

Cole M, Bond CP: Recent advances in automatic image analysis using a television system. J Microsc 96: 89–96 (1972).

Cooke DD, Kerker M: Particle size distribution of colloidal suspension by light scattering based upon single particle counts-polystyrene latex. J Colloid Interface Sci 42: 150 (1973).

Cornelisse CJ, Bots GTAM, Wintzen AR, et al: Real time morphometric analysis of type I and type II fibres in cryostat sections of human muscle biopsies. Pathol Res Pract 166: 218–238 (1980).

Cornelisse C, de Koning HR, Arentz PW, et al: Quantitative analysis of the nuclear area variation in benign and malignant breast cytology specimens. Anal Quant Cytol 3: 128–134 (1981).

Cornelisse CJ, Tanke HJ, de Koning H, et al: DNA ploidy analysis and cytologic examination of sorted cell populations from human breast tumors. Anal Quant Cytol 5: 173–183 (1983).

Cornelisse CJ, Venekamp HW, Smeulders AWM, et al: Cytometric analysis of DNA content, nuclear size, and chromatin texture in benign breast lesions, in *Flow Cytometry IV: Proceedings 4th Int Symp on Flow Cytometry*, (Thorud E, Laerum OD eds). Bergen, Universitets Forlaget, Voss, Norway (1979) pp 373–377.

Collste LG, Darzynkiewicz Z, Traganos F, et al: Flow cytometry in bladder cancer detection and evaluating using acridine orange metachromatic nuclei acid staining of irrigation cytology specimens. J Urol 123: 478–485 (1980).

Cowan WM, Wann DF: A computer system for measurement of cell and nuclear sizes. J Micros 99: 331–349 (1973).

Cowpe JG, Longmore RB: Nuclear area and Feulgen DNA content of normal buccal mucosa smears. J Oral Pathol 10:81–86 (1981).

Crocker J, Curran RC: A study of nuclear diameters in lymph node imprints using the Zeiss Videomat. J Clin Pathol 32: 670–674 (1979).

Cuvelier CA, Roels HJ: Cytophotometric studies of the nuclear DNA content in cartilaginous tumours. Cancer 44: 1363–1374 (1979).

Dahlem Workshop on Biomedical Pattern Recognition and Image Processing (Fu KS, Pavlidis T eds). Verlag Chemie Weinheim (1979).

Davidovits P, Egger MD: Scanning laser microscope for biological investigations. Appl Opt 10: 1615–1619 (1971).

Davidson E: Consecutive cytochemical staining. Adv Auto Anal 1: 447 (1971).

Davis BJ, Ornstein L: High resolution enzyme localization with a new diazo reagent, hexazonium pararosaniline. J Histochem Cytochem 7: 297 (1959).

Davis HJ: The irrigation smear. Acta Cytol 6: 459 (1962).

Dawson IMP, Heanley CP, Heber-Percy AC, et al: CESAR: Cervical smear analyzer and reader. J Clin Path 20: 724–730 (1967).

DeCampos FB, Moore GW, Schulter G: Cell nucleus pattern recognition: Influence of staining. Acta Cytol 17: 510–521 (1973).

Deitch A: Cytophotometry of nucleic acids, in *Introduction to Quantitative Cytochemistry*, (Wied GL ed). Academic Press, New York (1966) pp 327–354.

Deitch A: Microspectrophotometric study of the binding of the anionic dye napthol yellow S by tissue sections and purified proteins. Lab Invest 4: 324 (1955).

Deitch A: A method for the cytophotometric estimation of nucleic acids using methylene blue. J Histochem Cytochem 12: 451–461 (1964).

Delesse MA: Procede mechanique pour determiner la composition de roches. Contes Rendues Acad Sci 25: 544–545 (1846).

Delfiner KP: A generalization of the concept of size. J Microsc 95: 203–216 (1972).

DeReuck J, Roels H, Vander Eecken H: Cytophotometric DNA determination in human astroglial tumors. Histopathology 3: 105–115 (1979).

Deutsch ES: On some pre-processing techniques for character recognition, in *Computer Processing in Communications*. Polytechnic Press (1969) pp 221–234.

Devonec M, Darzynkiewicz Z, Kostryka-Claps ML, et al: Flow cytometry of low stage bladder tumors. Cancer 49: 109–118 (1982).

Devonec M, Darzynkiewicz Z, Whitmore WF, et al: Flow cytometry for follow-up examinations of conservatively treated low stage bladder tumors. J Urol 126: 166–170 (1981).

Dew B, King T, Mighdall D: An automatic microscope system for differential leukocyte counting. J Histochem Cytochem 22: 685 (1974).

Diacymakos EG, Day E, Kopac MJ: Further development in the Cytoanalyzer. Acta Cytol 6: 238 (1962a).

Diacumakos EG, Day E, Kopac MJ: Exfoliated cell studies and the Cytoanalyzer. Ann NY Acad Sci 97: 329 (1962b).

Diacumakos EG, Day E, Kopac MJ: Further development and evaluation of the Cytoanalyzer. Acta Cytol 6: 423 (1962c).

Diegenbach PC, Baak JPA: Quantitative nuclear image analysis: Differentiation between normal, hyperplastic and malignant appearing uterine glands in a paraffin section. II. Computer assisted recognition by discriminant analysis. Eur J Obstet Gynecol Reprod Biol 7: 389–394 (1977).

Diggs LW Sturm C, Bell A: The morphology of blood cells in Wright stained smears of peripheral blood and bone marrow: in *What's New* (Abbott Laboratories 1954).

Dinneen GP: Programming pattern recognition. Proc Western Joint Comput Conf (1955) pp 221–234.

DiStefano HS: A cytochemical study of the Feulgen nuclear reaction. Chromosoma 3: 282 (1948).

Doermer P, Abmayr W: Correlation between nuclear morphology and rate of deoxyribonucleic acid synthesis in a normal cell. J Histochem Cytochem 27: 188–192 (1979).

Doyle JL, Manhold JH: Feulgen microspectrophotometry of oral cancer and leukoplakia. J Dent Res 54: 1196–1199 (1975).

Doyle W: Recognition of Sloppy, Hand Printed Characters. Rep No 54-12 Lincoln Laboratory (1959).

Durie BG, Chen YP, Olson BG, et al: Discrimination between human T and B lymphocytes and monocytes by computer data from scanning microphotometry. Blood 51: 579–589 (1978).

Dziura BR, Bonfiglio TA: Needle cytology of the breast: A quantitative and qualitative study of the cells of benign and malignant ductal neoplasia. Acta Cytol 23: 332–339 (1979).

Eason PJ, Tucker JH: The preparation of cervical scrape material for automated scanning using gallocyanin chrome alum stain. J Histochem Cytochem 27: 25–31 (1979).

Eaves GN: Image processing in the biomedical sciences. *Computers and Biomedical Research,* vol 1. Academic Press, New York (1967) pp 112–234.

Eden M, Green JE, Sun A: Feasibility of computer screening of blood films for the detection of malaria parasites. Bull WHO 48: 211–218 (1973).

Eisert WG, Mendelsohn MC (eds): *Biological Dosimetry*. Springer-Verlag, Berlin (1982).

Engvall J, Greenberg SD, Spjut HJ, et al: Development of a mathematical model to analyze color and density as discriminant features for pulmonary squamous epithelial cells. Pattern Recog 13: 37–47 (1981).

Erozan YS, Pressman NJ, Donavan PA, et al: A comparative cytopathologic study of noninvasive and invasive squamous cell carcinoma of the lung. Anal Quant Cytol 1: 50–56 (1979).

Escobar GI, Todd P, Sattilaro RF: The dispersal of cells from human gynecologie specimens: Chemical agents. J Histochem Cytochem 25: 513–524 (1977).

Evans DMD (ed): *Cytology Automation*. Livingston, Edinburgh and London (1970).

Fallenius A, Svane G, Auer G, et al: Cytochemical classification of nonpalpable breast carcinoma. Anal Quant Cytol 5: 9–12 (1983).

Farrow ASJ, Tucker JH: A real-time TV to computer image input system using run coding. J Histochem Cytochem 24: 111–121 (1976).

Feulgen R, Rossenbeck J: Mikroskopischchemischer nachweiss einer nucleinsaure von typus der thymonucleinsaure und die darauf beruhende elektive farbung va zellkernen in mikroskopisher preparaten. Z Physiol Chem 135: 203 (1924).

Fiel RJ: Small angle light scattering of bioparticles—I Model Systems. Exp Cell Res 59: 413 (1970).

Fiel RJ, Mark EH, Munson BR: Small angle light scattering of bioparticles—III Vaccinia Virus. Arch Biophys 141: 547 (1970).

Fiel RJ, Munson BR: Small angle light scattering of bioparticles. I. Model systems. Exp Cell Res 59: 413 (1970).

Fiel RJ, Scheintaub HM: Small angle light scattering of bioparticles—IV Spleen cells and liver nuclei. Arch Biochem Biophys 158: 164 (1973).

Fischler MA, Elschlager RA: The representation and matching of pictorial patterns. IEEE Trans Comput C-22: 67–92 (1973).

Fisher C: The new Quantimet 720. Microscope 19: 20 (1971).

Fisher C, Bond CP: The Quantimet 720D for densitometry in the life sciences. Microscope 17: 1 (1971).

Fisher RA: The use of multiple measurements in taxonomic problems. Ann Eugenics 7: 179–188 (1936).

Flory LE: Scanning microscopy in medicine and biology. Proc IRE 47: 1889–1894 (1959).

Fossa SD, Kaalhus O, Scott-Knudsen O: The clinical and histopathological significance of Feulgen DNA-values in transitional cell carcinoma of the human urinary bladder. Eur J Cancer 13: 1155–1162 (1977a).

Fossa SD, Kaalhus O: Computer assisted image analysis of Feulgen-stained cell nuclei from transitional cell carcinoma of the human urinary bladder. Acta Pathol Microbiol Scand 86: 590–602 (1977b).

Frasch ACC, Itoiz ME, Cabrini RL: Microspectrophotometric quantitation of the diamimbenzidine reaction for histochemical demonstration of cytochrome oxydase activity. J Histochem Cytochem 26: 157 (1978).

Freeman H: On the encoding of arbitrary geometric configurations. IEEE Trans Electron Comput EC-10: 260–268 (1961).

Freeman H: Techniques for the digital computer analysis of chain-encoded arbitrary plane curves. Proc Nat Electron Conf 17: 421 (1961).

Freere RH, Weibel ER: Stereologic techniques in microscopy. J R Microsc Soc 87: 25–34 (1967).

Fu KS (ed): Special Issue on Pattern recognition and image processing. Proc IEEE 67 (1979).

Fu KS, Pavlidis T (eds): *Biomedical Pattern Recognition and Image Processing,* Verlag Chemie, Weinheim (1979).

Fu YS, Reagan JW, Richart RM: Definition of precursors. Gynecol Oncol 12: 220–231 (1981).

Fu YS, Reagan JW, Richart RM, et al: Nuclear DNA and histologic studies of genital lesions in diethylstilbestrol-exposed progeny. 1. Intraepithelial squamous abnormalities. Am J Clin Pathol 72: 503–514 (1979).

Fu YS, Robboy SJ, Prat J: Nuclear DNA study of vaginal and cervical squamous cell abnormalities in DES exposed progeny. Obstet Gynecol 52: 129–137 (1978).

Fu YS, Temmim L, Olaizola Y, et al: Nuclear DNA characteristics of microinvasive squamous carcinoma of the uterine cervix, in *Progress in Surgical Pathology,* (Fenoglio CM, Wolff MW eds). Masson, New York, (1980) pp 233–244.

Fulwyler MJ, Perrings JD, Cram LS: Production of uniform microspheres. Rev Sci Instr 44: 204 (1973).

Fuster V, Danielson MA, Robb RA, et al: Quantitation of left ventricular myocardial fiber hypertrophy and interstitial tissue in human hearts with chronically increased volume and pressure overload. Circulation 55: 504–508 (1977).

Galbraith W, Marshall PN, Lee ES, et al: Studies on Papanicolaou staining: I. Visible-light spectra of stained cervical cells. Anal Quant Cytol 1: 160–168 (1979).

Gallus G: Contour analysis in pattern recognition for human chromosome classification. Appl Biomed Calcolo Elettronico 2: 95 (1968).

Gallus G, Montanaro N, Maccacaro GA: An application of pattern recognition in biology. Proc 14th Cong Electronics, Rome (1967).

Gallus G, Montanaro N, Maccacaro GA: A problem of pattern recognition in the automatic analysis of chromosomes—Locating the centromere. Comput Biomed Res 2: 187 (1968).

Gallus G, Morabito A, Regoliosi G: On a complete chromosome recognition scheme. Proc Int Fed Auto Contr Symposium (1971) pp 111–117.

Gallus G, Neurath PW: Improved computer chromosome analysis incorporating pre-processing and boundary analysis. Phys Med Biol 15: 435–455 (1970).

Gallus G, Regoliosi G: A decisional model of recognition applied to the chromosome boundaries. J Histochem 22: 546–553 (1974).

Gamel JW, McLean IW: Computerized histopathologic assessment of malignant potential: A method for determining the prognosis of uveal melanomas. Hum Pathol 13: 834–837 (1982).

Gamel JW, McLean IW: Computerized histopathologic assessment of malignant potential: II. A practical method for predicting survival following enucleation for uveal melanoma. Cancer 52: 1032–1038 (1983).

Gamel JW, McLean IW: Computerized histopathologic assessment of malignant potential: III. Refinements of measurement and data analysis. Anal Quant Cytol 6: 37–44 (1984).

Garbay C, Brugal G, Choquet C: Application of colored image analysis to bone marrow cell recognition. Anal Quant Cytol 3: 272–280 (1981).

Garcia GI, Tolles WE: Ultrasonic disaggregation of cell clusters. J Histochem Cytochem 25: 508–512 (1977).

Garrick R, Doman P, Posen S: Quantitative histology of bone: The use of a computer program and results in normal subjects. Clin Sci 43: 789–797 (1972).

Gauvain C, Seigneurin D, Blugal G: A quantitative analysis of the human bone marrow granulocytic cell image using SAMBA 200 cell image processor. Anal Quant Cytol 6: 168–178 (1984).

Gaven JV Jr, Tavitian J, Harakendian A: The informative value of sampled images as a function of the number of gray levels used in encoding the images. Phot Sci Eng 14: 16–20 (1970).

Gelsema ES, Powell BW: Colour measurement and white blood cell recognition. CERN DD 72: 24 (1972).

Gelsema ES, Powell BW, Berrini PL: Procedures for the automation of the white blood cell differential count. Lect Notes Econ Math Syst 83: 237 (1973).

Geometric Data Corp: How to buy a differential counter (1974).

Giaretti WA, Abmayr W, Dormer P: Multiparameter nuclear morphology of the G_v-G_1 transition. Anal Quant Cytol 5: 90–98 (1983).

Giaretti WA, Gans P, Jutting U, et al: Correlation between chromatin morphology as derived by digital image analysis and autoradiographic labeling pattern. Anal Quant Cytol 5: 79–89 (1983).

Gilbert CW: A computer programme for the analysis of human chromosomes. Nature 212: 1437–1440 (1966).

Gilbert H, Ornstein L: A new method of basophil counting using alcia blue. Blood 46: 279–286 (1975).

Gill JE, Jotz MM: Desoxyribonucleic acid cytochemistry for automated cytology. J Histochem Cytochem 22: 470–477 (1974).

Glagoleff AA: On the geometrical methods of quantitative mineralogic analysis of rocks. Trans Inst Econ Min 59: 1 (1933).

Golab PJ: MACAC—An inexpensive and complete biomedical input and output display system. Proc 23rd ACEMB (1970) p 316.

Golay MJE: Apparatus for counting binucleate lymphocytes in blood. US Patent No 3,214,574 (1965).

Golden JF, West SS: Fluorescence spectroscopic and fading behavior of Erhlich's hyperdiploid mouse ascites tumor cells supravitally stained with acridine orange. J Histochem Cytochem 22: 495–505 (1974).

Goodman JW: *Introduction to Fourier Optics*. McGraw-Hill, New York (1968).

Gravatt CC: The applications of light scattering. Appl Spectrosc 25: 509 (1971).

Gray JW: Cell cycle analysis from computer synthesis of desoxyribonucleic acid histograms. J Histochem Cytochem 22: 642–650 (1974).

Greanias EC, Hoppel CJ, Kloomok M, et al: Design of logic for recognition of printed characters by simulation. IBM J Res Dev 1: 8–18 (1957).

Green DK, Neurath PW: The design, operation and evaluation of a high speed automatic metaphase finder. J Histochem Cytochem 22: 531–535 (1974).

Green JE: Analysis method and apparatus utilizing color algebra and image processing techniques. US Patent No 3,851,156 (1974).

Green JE: Method and apparatus for dual resolution analysis of a scene. US Patent No 3,970,841 (1976a).

Green JE: Method and apparatus utilizing color algebra for analyzing scene regions. US Patent No 3,999,047 (1976b).

Green JE: A practical application of computer pattern recognition research. The Abbott ADC-500 differential classifier. J Histochem Cytochem 27: 150–163 (1979a).

Green JE: Video analysis of hematology image data. The ADC-500 preprocessor. J Histochem Cytochem 27:163–171 (1979b).

Green JE: Sample preparation and its effect on automated blood cell differential analysis. Anal Quant Cytol 1: 187–201 (1979c).

Green JE: The Abbott Laboratories ADC-500, in *Real-Time Medical Image Processing*, (Onoe M, Preston K Jr, Rosenfeld A, eds). Plenum Press, New York (1980) pp 119–148.

Grimelius L, Akerstrom G, Johansson H, et al: Estimation of parenchymal cell content of human parathyroid glands using the image analyzing computer technique. Am J Pathol 93: 793–800 (1978).

Gross DS, Rothfeld JM: Quantitative immunocytochemistry of hypothalmic and pituitary hormones. Validation of an automated computerized image analysis system. J Histochem Cytochem 33: 11–20 (1985).

Grundmann E, Stein P: Untersuchungen ueber die kernstrucktur in normalen geweben und im carcinom. Beitr Pathol Anat 125:54–76 (1961).

Gunzer U, Harms H, Haucke M, et al: Computer-aided analysis for the differentiation of mononuclear cells in peripheral blood smears from leukemic patients. Anal Quant Cytol 3: 26–32 (1981).

Gunzer U, Harms H, Haucke M, et al: *Computer Aided Diagnosis of Leukemic Blood Cells*. IEEE Computer Soc Press, Los Angeles 302–306 (1982).

Gusberg SB, Cohen CJ, Yannopoulis K, et al: Automated cytologic screening for uterine cancer—A feasibility study. Obstet Gynecol 34: 284–287 (1969).

Hafford KJ, Preston K Jr: Three dimensional skeletonization of elongated solids. Comput Vision Graph Image Proc (1984).

Halaby SA, Vance ME: Computer-controlled spectral measurements of blood cells. IEEE Trans Biomed Eng BME-26: 34–42 (1979).

Hale AJ: Feulgen microspectrophotometry and its correlation with other cytochemical methods, in *Introduction to Quantitative Cytochemistry*, (Wied GL ed). Academic Press, New York (1966) pp 183–199.

Hall DJ, Wolf D, Bell GH: PROMONADE—An improved interactive graphics man-machine system for pattern recognition. Rep No RADCA-TR-68-572, US Air Force (1969).

Hall EL, Crawford WD Jr, Preston K Jr, et al: Classification of profusion of black lung disease from chest X-rays. Proc 1st Int Joint Conf Pattern Recog (1973) pp 73–87.

Harms H, Ruter A, Haucke M, et al: Digital picture processing of light microscope cell images: Chromatin arrangement and structure analysis. DECUS 5: 405 Digital Press, Maynard (1978).

Harms H, Gunzer U, Aus HM, et al: Computer aided analysis of chromatin network and basophil color for differentiation of mononuclear peripheral blood cells. J Histochem Cytochem 27: 204–209 (1979a).

Harms H, Ruter A, Haucke M, et al: Digital picture processing of light microscope cell images: Examples of cell classification using chromatin arrangement and structure. DECUS 6: 233, Digital Press, Maynard (1979b).

Harper TA: *The Peripheral Blood Film*. Butterworth, London (1968).

Hawksley PGW, Blackett JH, Meyer EW, et al: The design and construction of a photo-electric scanning machine for sizing microscopic particles. Br J Appl Phys 5: 165–173 (1954).

Henry RL, Nalbandian RM, Nichols BM, et al: An automated specific method for the detection of S Hemoglobin, in *Advances in Automated Analysis*, vol 1. Technicon Int Cong Thurman Assoc, Miami (1971) pp 471–476.

Herrmann WR, Arnold F, Kunze KD, et al: An approach to computer analysis of cervical smears. Proc 2nd Int Symp Cancer Det Prevention, Excerpta Medica, Amsterdam (1974).

Herrmann WR, Voss K, Barz H: Image processing in pathology: IX. A new morphometry system and its application on gangliocytes of the nucleus olivaris. Exp Pathol 17: 374–379 (1979).

Highleyman WK: The design and analysis of pattern recognition experiments. Bell Sys Tech J 41: 723 (1962).

Hilditch CJ: The principles of a software system for karyotype analysis, in *Human Population Genetics*, (Jacobs PA ed). University of Edinburgh Press (1970) pp 198–235.

Hilditch CJ, Rutovitz D: Chromosome recognition. Ann NY Acad Sci 157: 339 (1969).

Hlinka J, Sanders FK: Optical shadow casting of living cells. Trans NY Acad Sci 132: 675–687 (1970).

Hodes L: A programming system for on-line analysis of biomedical images. Comm Assoc Comp Mach 13: 279–283 (1970).

Hoeffgen H, Otto K, Soost HJ: Cytomorphologic results of preparation experiments for monolayer deposition of cervical material. J Histochem Cytochem 27:19–24 (1979).

Holmquist J: Computer analysis of cervical cells: Automated feature extraction and classification. J Histochem Cytochem 26: 1000–1005 (1978).

Holmquist J, Bengtsson E, Eriksson O, et al: Computer grading of cell populations from breast cancer, in *The Automation of Cancer Cytology and Cell Image Analysis,* Tutorials of Cytology, Chicago (1979) pp 159–166.

Holmquist J, Imasato Y, Bengtsson E, et al: A microspectrophotometric study of Papanicolaou-stained cervical cells as an aid in computerized image processing. J Histochem Cytochem 24: 1218–1224 (1976).

Horvath WJ, Tolles WE, Bostrom RC: Quantitative measurements of cell properties on Papanicolaou smears as criteria for screening. Trans 1st Int Cancer Cytol Cong (1956) pp 371–397.

Hotelling H: Analysis of a complex of statistical variables into principal components. J Educ Psychol 24: 417–441, 498–520 (1933).

Hou CY, Veress SA, Prothero JW: A quantitative study of biological form. IEEE Trans Biomed Eng BME-17: 106–108 (1970).

Hoyt RH, Ericksen E, Collins SM, et al: Computer-assisted quantitation of myocardial fibrosis in histologic sections. Arch Pathol Lab Med 108: 280–283 (1984).

Hu MK: Pattern recognition by moment invariants. Proc IRE 49:1428 (1961).

Huang TS, Schreiber WF, Tretiak OJ: Image processing. Proc IEEE 59: 1586–1609 (1971).

Hungerford DA: Observations on the morphology and behaviour of normal human chromosomes, in *Mammalian Cytogenetics and Related Problems in Radiobiology,* (Pavan C ed). Pergamon Press, New York (1964).

Husain OAN, Allen RBW, Hawkins EJ, et al: The Quantimet cytoscreen and the interactive approach to cancer screening. J Histochem Cytochem 22: 678–684 (1974).

Husain OAN, Henderson MW: Observations on the use of the Quantimet image analyzing computer in automatic scanning of malignant cells, in *Cytology Automation,* Livingston, Edinburgh and London (1970) pp 214–227.

Husain OAN, Page-Roberts BA, Millet JA: A sample preparation technique for automated cancer screening. Acta Cytol 22: 15–21 (1978).

Imasato Y, Yoneyama T, Watanabe S, et al: Feature evaluation for automated cancer cell identification. Comput Biol Med 5: 245–255 (1975).

Ingram M, Minter FM: Semi-automatic preparation of coverglass smears using a centrifugal device. Am J Clin Pathol 51:214 (1969).

Ingram M, Norgren PE, Preston K Jr: Automatic differentiation of white blood cells, in *Image Processing in Biological Science,* (Ramsey DM ed). University of California Press, Los Angeles (1968) pp 97–117.

Ingram M, Preston K Jr: Automatic analysis of blood cells. Sci Am 223: 72 (1970).

Ishiyama T, Hayakawa F: A study on automatic cytodiagnosis. Jpn Elect Eng 39–40 (1969).

Izzo NF, Coles W: Blood cell scanner identifies rare cells. Electronics 35: 52–57 (1962).

Jagoe R, Paton K: Measurement of pneumoconiosis in chest x-rays by computer. IEEE Trans Comput C-25: 95–97 (1976).

Jahoda E, Bartels PH, Bibbo M, et al: Computer discrimination of cells in serous effusion. II. Peritoneal fluids. Acta Cytol 17: 533–537 (1973a).

Jahoda E, Bartels PH, Bibbo M, et al: Computer discrimination of cells in serous effusion. II. Pleural fluids. Acta Cytol 17: 94–105 (1973b).

Jarkowski TL, Layton JM, Bahr GF, et al: Computer recognition of cells from asymptomatic lymphocytic leukemia. I. Methodologic study. Acta Cytol 15: 147–153 (1971).

Jayaramamurthy SN: Computer Methods for Analysis and Synthesis of Visual Texture. Dept of Computer Science, Univ of Illinois (1973).

Jensen RH, Mayall BH, King EB: Multiparameter flow cytometry applied toward diagnosis of cervical squamous carcinoma, in *The Automation of Cancer Cytology and Cell Image Analysis* (Pressman NF, Wied GC eds), Tutorials of Cytology, Chicago (1979) pp 95–102.

Johannessen JV, Sobrinho-Simoes M, Tangen KO, et al: A flow cytometric deoxyribonucleic acid analysis of papillary thyroid carcinoma. Lab Invest 45: 336–341 (1981).

Johnson B, Brady JM: Malignancy related changes in peripheral blood smears. Acta Cytol 14: 399–403 (1970).

Johnson ET, Goforth LJ: Metaphase spread detection and focus using closed circuit television. J Histochem Cytochem 22: 536–545 (1974).

Johnston DA: An Interactive Classification System. MD Anderson Hosp Tumor Inst, Houston.

Johnston DA: A General Purpose Image Digitizing System. MD Anderson Hosp Tumor Inst, Houston.

Johnston RB, Levine MD, Thorobeck WM: A computer method for counting the number of alveoli per unit area of human lung tissue. IEEE Trans Biomed Eng BME-19: 422–428 (1972).

Jones HW Jr, Woodruff JD, Davis HJ, et al: The evolution of chromosomal aneuploidy in cervical atypia, carcinoma in situ and invasive carcinoma of the uterine cervix. Johns Hopkins Med J 127: 125–135 (1970).

Jordan SW, Olson GB, Bartels PH, et al: Computer-assisted morphometric analysis of renal radiation response. Anal Quant Cytol 4: 174–180 (1982).

Jordan SW, Brayer JM, Bartels PH, et al: Computer-assisted morphometric analysis of renal radiation response, in *Computer-assisted Image Analysis Cytology* (Greenberg SD ed). Karger, Basel (1984) pp 117–147.

Kallenberger A, Hagmann A, Meier-Ruge A, et al: Beziehunger zwischen Sexchromatinvorkummen, Kerngrosse und DNS-werten in Mammatumoren und ihre Bedentung fur die Ueberlebenszeit. Schweiz Med Wochenschr 97: 678–682 (1967).

Kamentsky LA, Liu CN: Computer-automated design of multifont recognition logic. IBM J Res Dev 7: 2–13 (1963).

Karnes J, Robb RA, O'Brien PC, et al: Computerized image recognition for morphometry of nerve attribute of shape in sampled transverse sections of myelinated fibers which best estimates their average diameter. J Neurol Sci 34:43–51 (1977).

Karush W: Lectures on decision theory. US Dept of Commerce Clearinghouse Document No AD 410789 (1963).

Kaufman GI, Wasserman DE, Nester JF: J Assoc Adv Med Instr 6: 230–234 (1972).
Kendall F, Giaretti W, Swenson RA, Nicolini C: Nuclear morphometry of Hela cells during the cycle. Proc Automated Cytol Conf, Eng Found, Pensacola, Fla (1976).
Kendall F, Nicolini C: Nuclear texture analysis of W138 cells stimulated to proliferate. Proc Automated Cytol Conf, Eng Found, Pensacola, Fla (1976).
Kendall FM, Wu CT, Giaretti W, et al: Multiparameter geometric and densitometric analyses of the G_v-G_1 transition of W138 cells. J Histochem Cytochem 25: 724–729 (1977).
Kiefer G, Kiefer R, Moore GW, et al: Nuclear images of cells in different functional states. J Histochem Cytochem 22: 569–576 (1974).
Kiefer R, Kiefer G, Salm R, et al: A method for the quantitative evaluation of eu- and heterochromatin in interphase nuclei using cytophotometry and pattern analysis. Beitr Pathol Anat 150: 163 (1973).
Kiefer R, Kiefer G, Wolf U: Feulgen hydrolysis kinetics in euchromatin and heterochromatic. Inst of Pathology, Univ of Freiburg (1972).
Kiefer R, Sandritter W: Objectivization of the microscopic image of the cell nucleus. Inst of Pathology, Univ of Freiburg (1970).
Kimzey SL, Greenberg SD, Baky AA, et al: Cell atypia profiles for bronchial epithelial cells: Mathematical evaluation of sputum cellular atypia in squamous cell carcinogenesis. Anal Quant Cytol 2: 186–194 (1980).
Kinder J, Reuter B, Hutzler P: Real-time multichannel processing in a microscope, in *The Automation of Cancer Cytology and Cell Image Analysis,* (Pressman NJ, Wied GL eds). Tutorials of Cytology, Chicago (1979) pp 69–73.
Kirkland JA: The study of chromosomes in cervical neoplasia. Obstet Gynecol Surv 24:784–793 (1969).
Kirsch RA: Computer determination of the constituent structures of biological images. Comput Biomed Res 4: 315–328 (1971).
Kirsch RA, Ray LC, Cahn L, et al: Experiments in processing pictorial information with a digital computer. Proc Eastern Joint Comp Conf (1957) p 221.
Kirsch RA: Computer determination of the constituent structure of biological images-pt. 1. NBS Rep No 10173 (1969).
Kishigami Y, Noda S, Hashimoto Y, et al: The collection of vaginal smears suitable for automatic screening by the Auto-Cyto Screener. J Jpn Soc Clin Cytol 17:12–18 (1978) in Japanese.
Kishigami Y, Noda S, Morishita T, et al: Hashimoto Studies on an automatic cytoscreening system: Auto-Cyto Screener), in *The Automation of Cancer Cytology and Image Analysis,* (Pressman NJ, Wied GL eds). Cancer Tutorials of Cytology (1979) pp 112–122.
Klawe H, Rowinski J: Malignancy associated changes (MAC) in cells of buccal smears detected by means of objective image analysis. Acta Cytol 18: 30–33 (1974).
Kline JC, Serra J: The texture analyzer. Microscopy 95:349–356 (1972).
Klinger A, Kochman A, Alexandridis N: Computer analysis of chromosome patterns—feature encoding for flexible decision making. IEEE Trans Comput C-20: 1014 (1971).
Koch AL, Ehrenfeld E: The size and shape of bacteria by light scattering measurement. Biochem Biophys Acta 165: 262 (1968).
Kolers PA, Eden M: *Recognizing Patterns.* MIT Press, Cambridge, Mass (1968).

Konaka C, Auer G, Nasiell M, et al: Pathogenesis of squamous bronchial carcinoma in 20-methylcholanthrene-treated beagle dogs. Anal Quant Cytol 4: 61–71 (1982).
Kopp RE, Lisa J, Mendelsohn J, et al: Coherent optical processing of cervical cytologic samples. J Histochem Cytochem 24: 122–137 (1976).
Kopp RE, Lisa J, Mendelsohn J, et al: The use of coherent optical processing techniques for the automatic screening of cervical cytologic samples. J Histochem Cytochem 22: 598–604 (1974).
Koss LG: Urinary cytology: The subjective diagnostic clues and their evaluation by computer. Anal Quant Cytol 1: 202–208 (1979).
Koss LG, Bartels PH: Urinary cytology: Device capabilities and requirements. Anal Quant Cytol 2: 59–65 (1980a).
Koss LG, Bartels PH, Bibbo M, et al: Computer discrimination between benign and malignant urothelial cells. Acta Cytol 19:378–391 (1975).
Koss LG, Bartels PH, Bibbo M, et al: Computer analysis of atypical urothelial cells: I. Classification of supervised learning algorithms. Acta Cytol 21: 247–260 (1977a).
Koss LG, Bartels PH, Sherman A, et al: Computer identification of degenerated urothelial cells. Anal Quant Cytol 2: 107–111 (1980b).
Koss LG, Bartels PH, Sherman A, et al: Computer identification of multinucleated urothelial cells. Anal Quant Cytol 2: 112–116 (1980c).
Koss LG, Bartels PH, Sychra JJ, et al: Computer analysis of atypical urothelial cells: II. Classification by unsupervised learning algorithms. Acta Cytol 21: 261–265 (1977b).
Koss LG, Bartels PH, Sychra JJ, et al: Computer discriminant analysis of atypical urothelial cells. Acta Cytol 22: 282–286 (1978a).
Koss LG, Bartels PH, Sychra JJ, et al: Diagnostic cytologic sample profiles in patients with bladder cancer using TICAS system. Acta Cytol 22: 392–397 (1978b).
Koss LG, Bartels PH, Wied GL: Computer-based diagnostic analysis of cells in the urinary sediment. J Urol 123: 846–849 (1980d).
Koss LG, Coleman DV (eds): *Diagnostic Cytology and Its Histopathological Bases*. Lippincott, Philadelphia (1979).
Koss LG, Sherman AB, Adams SE: Interobserver and intraobserver differences in the diagnosis of urothelial cells: Comparison with classification by computer. Anal Quant Cytol 6: 112–120 (1984a).
Koss LG, Sherman AB, Adams SE: The use of hierarchical classification in the image analysis of a complex cell population: Experience with sediments of voided urine. Anal Quant Cytol 5: 159–166 (1983).
Koss LG, Sherman AB, Bartels PH, et al: Hierarchic classification of multiple types of urothelial cells by computer. Anal Quant Cytol 2: 166–174 (1980e).
Koss LG, Sherman AB, Eppich E: Image analysis and DNA content of urothelial cells infected by human poliomavirus. Anal Quant Cytol 6:89–94 (1984b).
Kranz D, Voss K, Wenzelides K, et al: Quantitativemorphologische Untersuchungen an Lebern und Lebertumoren nach Einnahme von Kontrazeptiva. Zentralbl Allg Pathol 124: 441–447 (1980).
Kreicbergs A, Soderberg G, Zetterberg A: The prognostic significance of nuclear DNA content in chondrosarcoma. Anal Quant Cytol 2: 272–279 (1980).
Kreicbergs A, Zetterberg A: Cytophotometric DNA measurements of chondro-

sarcoma. Methodologic aspects of measurements in tissue sections from old paraffin embedded specimens. Anal Quant Cytol 2: 84–92 (1980).
Kujoory MA, Mayall BH, Mendelsohn ML: Focus assist device for a flying spot microscope. IEEE Trans Biomed Eng BE-20: 126 (1973).
Kunze KD, Herrmann WR, Meyer W: An image analyzing method for automated cytologic prescreening of cervix carcinoma and its prestages: Demonstrati and preliminary results. Arch Geschwulstforsch 48: 131 (1978).
Kunze KD, Herrmann WR, Meyer W: The ZYPAB image-processing system for cytologic prescreening for cervical cancer. Anal Quant Cytol 2: 252–256 (1980).
Kunze KD, Herrmann WR, Voss K: Image processing in pathology. Exp Pathol 16: 186–193 (1978).
Kwee WS, Veldhuizen RW, Alons CL, et al: Quantitative and qualitative differences between benign and malignant mesothelial cells in pleural fluid. Acta Cytol 26: 401–406 (1982).
Kwikkel HJ, Boon ME, Rijswijk MV, et al: Masking effect of hormonal contraceptives on discriminating quantitative features of visually normal intermediate cells in positive and negative cervical smears. Anal Quant Cytol 8: 227–232 (1986).
Land EH, Blout ER, Grey DS, et al: A color translating ultraviolet microscope. Science 109: 371–374 (1949).
Landeweerd GH: Pattern recognition of white blood cells. PhD thesis, University of Amsterdam (1981).
Landeweerd GH, Gelsema ES: The use of nuclear texture parameters in the automatic analysis of leukocytes. Pattern Recog 10: 57–61 (1978).
Langston C, Thurlbeck WM: The use of simple image analyzers in lung morphometry. J Microsc 114: 89–99 (1978).
Latt SA: Microfluoremetric analysis of desoxyribonucleic acid replication kinetics and sister chromatid exchanges in human chromosomes. J Histochem Cytochem 22: 478–491 (1974).
Laurent JL, Beraud JC, Antoniadis A, et al: Essai de classification automatiques des cellules cervico-vaginales normales et cancereuses a l'aide d'un analyseur d'images et d'un ordinateur. Arch Anat Cytol Pathol 27: 281–285 (1979).
Lederer B: Statistical analysis of cytophotometric DNA measurements demonstrated on malignant testicular teratoma. Acta Cytol 20: 5–6 (1976).
Ledley RS: High speed automatic analysis of biomedical pictures. Science 146: 216–223 (1964).
Ledley RS, Lubs HA, Ruddle FH: Introduction to automatic chromosome analysis. Comput Biol Med 2: 107 (1972).
Ledley RS, Rotolo LS, Golab TJ, et al: FIDAC—Film Input to Digital Automatic Computer and Associated Syntax Directed Pattern Recognition Programming System, in *Optical and Electro-Optical Information Processing*, MIT Press, Cambridge, Mass (1965) pp 591–613.
Ledley RS, Ruddle FH: Chromosome analysis by computer. Sci Am 213: 40–46 (1966).
Leif RC: Buoyant density separation of cells, in *Automated Cell Identification and Sorting*, (Wied GL, Bahr GF eds). Academic Press, New York (1970) pp 21–96, 131–159.
Leif RC, Gail S, Dunlop LA, et al: Centrifugal cytology: IV. The Preparation of fixed stained dispersions of gynecological cells. Acta Cytol 19: 159–168 (1975).

Leif RC, Ingram D, Clay C, et al: Optimization of the binding of dissociated exfoliated cervicovaginal cells onto glass microscope slides. J Histochem Cytochem 25: 538–543 (1977).

Leif RC, Nordquist S, Clay S, et al: A procedure for dissociating Ayre scrape samples. J Histochem Cytochem 25: 525–537 (1977).

Lemkin P: Bone marrow smear image analysis. PhD Dissertation, Comp Sci Dept, U Maryland College Park (1978).

Lemkin P, Carman G, Lipkin L, et al: A real time processor for use in biologic cell identification—I. System Design. J Histochem Cytochem 22: 725 (1974).

Lemons RA, Quate CF: Gigahertz ultrasound. Appl Phys Lett 24: 163–165 (1974).

Lester JM, Williams HA, Weintraub BA, et al: Two graph searching techniques for boundary finding in white blood cell images. Comput Biol Med 8: 293–308 (1978).

Leuchtenberger C, Leuchtenberger R, Davis A: A microspectrophotometric study of the deoxyribonucleic acid (DNA) content in cells of normal and malignant human tissue. Am J Pathol 30: 65–85 (1954).

Levialdi S: On shrinking binary picture patterns. Comm Assoc Comput Mach 15: 7–10 (1972).

Levialdi S, Montanari U: A gray-weighted skeleton. Inf Control 17: 62 (1970).

Levinthal C, Ware R: Three-dimensional reconstructions from serial sections. Nature 236: 207–210 (1972).

Lewis PM: The Characteristic Selection Problem in Recognition Systems. General Electric Rep 61-RL-2796 E (1961).

Liedtke CE, Aeikens B: Segmenmtierungsversuche un Urothelzellen als Vorstufe zur automatischen Urin-Zytologie-Auswertung. Microscopica Acta [Suppl] 4: 230–234 (1980).

Liedtke CE, Kappei F: Wissensgesteurerte Segmentierung von Urothelzellbildern, in VDE Fachberichte, Mustererkennung (Werner R ed). Springer-Verlag, Berlin (1983) pp 390–395.

Lipkin BS, Lipkin LE: Textural parameters related to nuclear maturation in the granulocytic leukocytic series. J Histochem Cytochem 22: 583–593 (1974).

Lipkin LE, Watt WC, Kirsch RA: The analysis, synthesis and description of biological images. Ann NY Acad Sci 129: 984–1012 (1966).

Louis C, Poulsen R, Marshall KG, et al: The occurrence of isolated dysplastic, carcinoma in situ, and invasive type cells in cervical smears from patients with invasive squamous cell carcinoma: Significance to prescreening using image processing techniques. Acta Cytol 20: 158–161 (1976).

Ludwig AS, Okagaki T, Richart RM: Nuclear DNA content of lobular carcinoma *in situ* of the breast. Cancer 31: 1553–1560 (1973).

Lukacs GL, Balazs GY, Zs-Nagy I: Cytofluorimetric measurements on the DNA contents of tumor cells in human thyroid gland. J Cancer Res Clin Oncol 95: 265–271 (1979).

Lundsteen C, Philip J: Microphotometry of banded human chromosomes. J Med Genet 11: 196–200 (1974).

Lurie OB, Bykov RE, Popechitielev EP, et al: Automatic recognition of leukocyte cells in blood preparations. Proc 8th Conf Eng Med Biol (1969).

Macagno ER, Levinthal C, Sobel I: Three-dimensional computer reconstruction of neuronal assemblies. Ann Rev Biophys Bioeng 8: 323–351 (1979).

Mangasarian OL: Linear and nonlinear separation of patterns by linear programming. Oper Res 13:44 (1956).
Mansberg HP, Ohringer P: Design considerations for electronic and electromechanical flying spot scanners. Ann NY Acad Sci 132: 675–687 (1970).
Mansberg HP, Segarra JM: Counting of neurons by flying spot microscope. Ann NY Acad Sci 99: 309–322 (1962).
Mansberg HP: Automatic particle and bacterial colony counting. Science 126: 823 (1957).
Margolick JB, Azen SP, Sherwin RP: An image analyzer quantitation of type II pneumocytes. Am Rev Respir Dis 108: 704–708 (1973).
Marrill T, Hartley AK, Evans TG, et al: CYCLOPS Syst Res Rep No AFCRL-69-0323 US Air Force (1969).
Marshall PN, Galbraith W, Bacus JW: Studies on Papanicolaou staining: II. Quantitation of dye components bound to cervical cells. Anal Quant Cytol 1: 169–178 (1979).
Mason TE, Phifer RH, Spicer SS, et al: An immunoglobulin-enzyme bridge method for localizing tissue antigens. J Histochem Cytochem 17: 563 (1969).
Matsumoto J, Kishigami Y, Noda S, et al: Automated Feulgen reaction in autosytoscreening. J Jpn Soc Clin Cytol 10: 148–154 (1971) in Japanese.
Mawdesley-Thomas LE, Healey P: The use of an automated image analyzer in the quantitive histologic evaluation of toxicologic material with particular reference to the respiratory tract. 8th Ann Mtg Soc of Toxic, Williamsburg, Virginia (1969a).
Mawdesley-Thomas LE, Healey P: The quantitative evaluation of experimental chronic bronchitis. Am Rev Respir Dis 100: 231–233 (1969b).
Mawdesley-Thomas LE, Healey P: Automated analysis of cellular change in histological sections. Science 163: 1200 (1969c).
Mayall BH (ed): J Histochem Cytochem 22: 451–765 (1974).
Mayall BH (ed): J Histochem Cytochem 24: 1–414 (1976).
Mayall BH (ed): J Histochem Cytochem 25: 479–952 (1977).
Mayall BH (ed): J Histochem Cytochem 27: 1–641 (1979).
Mayall BH: A survey of digital image-processing activities at Lawrence Livermore Laboratory II—Biomedical applications. Computer 7: 5–81 (1974b).
Mayall BH: Monodisperse cell samples: The problem and possible solutions, in *The Automation of Uterine Cancer Cytology*, (Wied GL, Bahr GF, Bartels PH eds). Tutorials of Cytology, Chicago (1976b) pp 61–68.
McKee PH: Computer Analysis of Lymphocyte Alterations Induced by Immunosuppressive Chemical Agents. PhD dissertation, Dept Microbiol, Univ of Arizona (1975).
Mead JS, Horan PK, Wheeless LL: Syringing as a method of cell dispersal: I. Effect on intermediate and superficial squamous cells. Acta Cytol 22: 86–90 (1978).
Megla GK: The LARC analyzer and its acceptability. Eng Found Conf Automatic Cytol, Saxtons River (1972).
Megla GK: The LARC automatic white blood cell analyzer. Acta Cytol 17: 3–14 (1973).
Melamed MR, Adams LR, Traganos F, et al: Blood granulocyte staining with acridine orange: Changes with infection. J Histochem Cytochem 22: 526–530 (1974).

Mellecker J: Automated cytology impact survey. Biocharacterist 2: 127 (1972).

Mellors RC, Glassmann A, Papanicolaou GN: A microfluorometric scanning method for the detection of cancer cells in smears of exfoliated cells. Cancer 5: 458–468 (1952).

Mellors RC, Keane JF, Papanicolaou GN: Nucleic acid content of the squamous cancer cell. Science 116: 265–269 (1952).

Mendelsohn ML, Conway TJ, Hungerford DA, et al: Computer-oriented analysis of human chromosomes. I. Photometric estimation of DNA content. Cytogenetics 5: 223 (1964a).

Mendelsohn ML, Hungerford DA, Mayall BH, et al: Human chromosome analysis. Ann NY Acad Sci 157: 376–392 (1969a).

Mendelsohn ML, Kolman WA, Bostrum RC: Initial approaches to the computer analysis of cytophotometric fields. Ann NY Acad Sci 115: 998 (1964b).

Mendelsohn ML, Mayall BH: Chromosome identification by image analysis and quantitative cytochemistry, in *Human Chromosome Methodology,* (Yunis J ed). Academic Press, New York (1974) pp 311–346.

Mendelsohn ML, Mayall BH, Perry BH: Generalized grayness profiles as applied to edge detection and the organization of chromosome images, in *Advances in Medical Physics,* (Laughlin JS, Webster RW eds). (1971) p 327.

Mendelsohn ML, Mayall BH, Prewitt JMS: Approaches to the automation of chromosome analysis, in *Image Processing in Biological Science,* (Ramsey DM ed). Univ of Calif Press, Los Angeles (1969b) pp 119–136.

Mendelsohn ML, Mayall BH, Prewitt JMS, et al: Digital transformation and computer analysis of microscope images, in *Advances in Optical and Electron-Microscopy,* (Cosslett V, Barer R eds). Academic Press, New York (1968) pp 77–150.

Meyer F: Iterative image transformation for an automatic screening of cervical smears. J Histochem Cytochem 27: 128–135 (1979).

Miale JB: *Laboratory Medicine Hematology*. Mosby, St. Louis (1967).

Miller MN: Leukocyte classification by morphological criteria. Eng Found Conf Automatic Cytol, Saxtons River (1972).

Miller MN: Design and clinical results of Hematrak—An Automated Differential Counter. Geometric Data Corp (1973).

Montanari U: A method for obtaining skeletons using a quasi-euclidean distance. J Assoc Comp Mach 15: 600–624 (1968).

Montanari U: On the optimum detection of curves in noise pictures. Comm Assoc Comp Mach 14: 335–345 (1971).

Montgomery POB (ed): *Scanning Techniques in Biology and Medicine*. Ann NY Acad Sci 97: 329–526 (1962).

Montgomery POB, Bonner WA: Ultraviolet flying-spot television microscope. Proc IRE 47: 1889 (1959).

Moore DH: Optimization of cancer detection. J Histochem Cytochem 22: 663–667 (1974).

Moore GA: Recent progress in automatic image analysis. J Microsc 95: 105–118 (1972).

Morton RA: Design of modern image measurement systems, in *Stereology and Quantitative Metallography*. American Society of Testing Material, New York (1972) pp 81–94.

Moses MJ: Quantitative optical techniques in the study of nuclear chemistry. Exp Cell Res 2: 75 (1952).

Mueller W: The LEITZ texture analyzing system. Leitz Sci Tech Inf 1: 101–116 (1974).

Mui JK, Fu KS: Automated Classification of Nucleated Blood Cells Using Digital Image Processing Techniques. TR-EE78-47, Purdue University (1978).

Mui JK, Bacus JW, Fu KS: A scene segmentation technique for microscopic cell images, Proc of the Symp Computer-Aided Diag of Med Images. IEEE Comp Soc, New York (1976) pp 99–106.

Mui JK, Fu KS, Bacus JW: Automated classification of blood cell neutrophils. J Histochem Cytochem 25: 633–640 (1977).

Mukawa A, Kamitsuma Y, Jisaki F, et al: Progress report on experimental use of CYBEST model 2 for practical gynecological mass screening: Alterations of the specimen rejection threshold and specimen preparation. Anal Quant Cytol 5: 31–34 (1983).

Mukawa A, Tanaka N, Ikeda H, et al: A practical evaluation of the CYBEST system in the use of mass-screening on epidermoid carcinoma of the uterine cervix, in *The Automation of Cancer Cytology and Cell Image Analysis* (Pressman NJ, Wied GL eds). Chicago, Tutorials of Cytology (1979) pp 193–199.

Mullaney PF, Dean PN: Cell Sizing: A small-angle light scattering method for sizing particles of low refractive index. Appl Opt 8: 2361 (1969).

Mullaney PF, Steinkamp JA, Crissman HA, et al: Laser flow microphotometer for rapid analysis and sorting of individual mammalian cells, in *Laser Applications in Medicine and Biology* (Wolbarsht ML ed). Plenum Press, New York (1974).

Mullaney PF, Van Dilla MA, Coulter JR, et al: Cell Sizing: A light scattering photometer for rapid volume determination. Rev Sci Instr 40: 1029–1032 (1969).

Nadel EM: Computer analysis of cytometric fields by the CYDAC and its historical evolution from the cytoanalyzer. Acta Cytolog 9: 203 (1965).

Nair KK, Bartels PH, Mahon DC, et al: Image analysis of hepatocyte nuclei from chlordane-treated rats. Anal Quant Cytol 2: 285–289 (1980).

Nasiell M, Kato H, Auer G, et al: Cytomorphological grading and Feulgen DNA-analysis of metaplastic and neoplastic bronchial cells. Cancer 41: 1511–1521 (1978).

Nelson GD, Levy DM: A dynamic programming approach to the selection of pattern features. IEEE Trans Syst Sci SSC-4: 145–151 (1968).

Neugebauer D, Otto K, Soost HJ: Numerical analysis of cell populations in smear and monolayer preparations from the uterine cervix. Anal Quant Cytol 3: 91–95 (1980).

Neurath PW: A review of image processing in biology and medicine. Med Res Eng 9–15 (June–July 1971).

Neurath PW, Brand DH, Schreiner ED: Man-machine interaction for image processing. Ann NY Acad Sci 157: 324–338 (1969).

Neurath PW, Brenner JF: A computer classification of white blood cells into 15 classes using the Tufts PIQUANT System: Proc 25th Ann Conf Eng Med Biol, Bal Hbor, Fla (1972).

Neurath PW, Brenner JF, Selles WD: Computer identification of white blood cells. Int Comput Symp, Davos (1973).

Neurath PW, Enslein K: Human Chromosome Analysis as Computed from Arm Length Measurements. Rep NYO-3748-12, US Atomic Energy Commission (1968).

Neurath PW, Falek A, Bablouzian FL, et al: Human chromosome analysis by computer—An optical pattern recognition problem. Ann NY Acad Sci 128: 1013–1028 (1966).

Neurath PW, Gallus G, Selles WD: Interaction computer-aided chromosome analysis. J Assoc Adv Med Instr 4: 6–14 (1970).

Neurath PW, Gallus G, Vastola E: Computer system for differential white blood cell counting. Phys Med Biol: 17: 716 (1972a).

Neurath PW, Kess B, Low DA: Individual human karyotyping through quantitative analysis. Comput Biol Med 2: 181 (1972b).

Nicolini C, Giaretti W, DeSaive C, et al: The G_{10}-C transition of W138 cells: Geometric and densitometric texture analyses. Exp Cell Res 106: 119–125 (1977).

Nieburgs HE: Malignancy associated changes (MAC), in *Compendium on Diagnostic Cytology,* 4 ed, (Wied GL, Koss LG, Reagan JW eds). Tutorials of Cytology, Chicago (1976) pp 215–231.

Nieburgs HE, Herman BE, Reisman H: Buccal cell changes in patients with malignant tumors. Lab Invest 2: 80–88 (1962).

Nieburgs HE, Zak FG, Allen DC, et al: Systemic cellular changes in material from human and animal tissues in presence of tumors. Proc Int Soc Cytol Council, 7th Annual Meeting (1959) pp 137–144.

Noguchi Y: A segmentation method for cell images based on a two-dimensional histogram. Proc 1st Int Symp Med Imaging Image Interpret. IEEE Comput Soc, New York (1982) pp 134–139.

Noguchi Y, Tenjin Y, Sugishita T: A method of cellular images using color information. Bull Electromech Lab 41: 524–549 (1977).

Noguchi Y, Tenjin Y, Sugishita T: An automated analysis method of overlapped cells based on a set of images observed through three different wavelengths of source light, in *The Automation of Cancer Cytology and Cell Image Analysis* (Pressman NJ, Wied GL eds). Int Acad Cytology (1979) pp 185–192.

Norgren PE, Ingram M, Preston K Jr: Automatic identification of blood cells using morphology and color. Digest 7th Int Conf Med Biol Eng, Stockholm (1967) p 122.

Norgren PE: High speed laser beam detection of chromosome spreads. Ann NY Acad Sci 157: 514–524 (1969).

Oberholzer M: *Morphometrie in der Klinischem Pathologie*. Springer-Verlag, Berlin (1983).

Oberst PW, Crook JW, Swain SF, et al: Toxic effects of high concentrations of bromo-benzylnitrile vapor in various animal species. 8th Ann Mtg Soc of Toxicology, Williamsburg, Virginia (1969).

O'Gorman L, Preston KP Jr, Sanderson AC: A system for automated liver tissue image analysis—Methods and results. IEEE Trans Biomed Eng BME-32: 696–706 (1985).

Okagaki T, Meyer AA, Sciarra JJ: Prognosis of irradiated carcinoma of cervix uteri and nuclear DNA in cytologic postirradiation dysplasia. Cancer 33: 647–652 (1974).

Olson GB, Anderson RE, Bartels PH: Computer analysis of defined populations of lymphocytes irradiated *in vitro*. III. Evaluation of human T and B cells of peripheral blood origin. Hum Pathol 10: 179 (1979).

Olson GB, Anderson RE, Bartels PH: Differentiation of murine thoracic duct lymphocytes into T and B subpopulations by computer cell-scanning techniques. Cell Immunol 13: 327–335 (1974a).

Olson GB, Anderson RE, Bartels PH: Differentiation of T and B lymphocytes by cell image analytical techniques. J Reticuloendothel Soc 15: 23a (1974b).

Olson GB, Bartels PH: Computable image features of thymus and bursal small lymphocytes. Fed Proc 33: 733 (1974c).

Olson GB, Bartels PH: Assessment of environmental-induced insults upon lymphoid cells by computerized morphometric analysis, in *Biological Dosimetry*, (Eisert WG, Mendelsohn MC eds). Springer-Verlag, Berlin (1982a) pp 203–218.

Olson GB, Bartels PH: in *Biological Dosimetry* (Eisert WG, Mendelsohn MC eds). Springer-Verlag, Berlin (1982b) pp 203–318.

Olson GB, Bartels PH: Differentiation of splenocytes and peripheral blood lymphocytes from mice infected with Friend murine leukemia virus. Pattern Recog 13: 57 (1981).

Olson GB, Donovan RM, Bartels PH, et al: Microphotometric differentiation of human T and B cells tagged with monospecific immunoadsorbant beads. Anal Quant Cytol 2: 144–152 (1980).

Olson GB, Wied GL, Bartels PH: Differentiation of chicken thymic and bursal lymphocytes by cell image analysis. Acta Cytol 17: 454–461 (1973a).

Olson GB, Wied GS, Bartels PH: Differentiation of lymphoid tissues by analysis of digitized images. Acta Cytol 17: 89–93 (1973b).

Onoe M, Preston K Jr, Rosenfeld A (eds): *Real-Time Medical Image Processing*. Plenum Press, New York (1980).

Ooms ECM, Essed C, Veldhvizen RW, et al: The prognostic significance of morphometry in T1 bladder tumors. Histopathology 5: 311–318 (1981).

Ornstein L, Ansley HR: Spectral matching of classical cytochemistry to automated cytology. J Histochem Cytochem 22: 453 (1974).

Ornstein L: Pattern recognition morphology and the generation of hypotheses. Am Assoc Adv Sci Conf, Washington DC (1960).

Ornstein L: Computer learning and the scientific method—A proposed solution to the information theoretical problem of meaning. J Mt Sinai Hosp 32: 837 (1965).

Ott R, Schuermann J, Reinhart ER, et al: Automated classification of cytological specimens based on features extracted from nuclei images. Pattern Recog 13: 83–87 (1981).

Otto K: Automatische analyse von zellen der cervix uteri unter veerwendung konventioneller und neuer praparationsverfahren. PhD Dissertation, Tech Univ, Munich (1978).

Otto K: Zelldeposition fur automatisiertes prascreening gynakologischer praparate. Acta Histochem (suppl) 21: 137–145 (1980).

Otto K, Bayer E, Dvorak R, et al: Specimen collection and cell dispersal on slides for automated cell detection, in *Technologies for Automation in Cervical Cancer Screening*, (Goerttler K, Herman CJ eds). Deutsches Krebsforschungszentrum, Heidelberg (1976) pp 120–122.

Otto K, Hoffken H, Soost HJ: Components and results of a new preparation technique for automated analysis of cervical samples. Anal Quant Cytol 1: 127–135 (1979a).

Otto K, Hoffken H, Soost HJ: Sedimentation velocity separation: A preparation method for cervical samples. J Histochem Cytochem 27: 14–18 (1979b).

Oud PS, Zahniser DJ, Harders-Hendriks R, et al: The development of a cervical smear procedure for the BioPEPR image-analysis system. Anal Quant Cytol 3: 73–80 (1981a).

Oud PS, Zahniser DJ, Raaijmakers MLT, et al: Thionine-Feulgen-Congo red staining of cervical smears for the BioPEPR image-analysis system. Anal Quant Cytol 3: 289–294 (1981b).

Pak HY, Ashdjian V, Yokota SB, et al: Quantitative DNA determination by image analysis: I. Application to human pulmonary cytology. Anal Quant Cytol 4:95–98 (1982).

Papanicolaou GN: *Atlas of Exfoliative Cytology*. Harvard University Press, Cambridge, MA (1954).

Parpart AK: Television microscopy in biological research. Science 113: 483–484 (1951).

Parry JS, Cleary BK, Williams AR, et al: Ultrasonic dispersal of cervical cell aggregates. Acta Cytol 15: 1963 (1971).

Pasteur X, Alepee BA, Cava E, et al: Automated cytologic diagnosis of bronchial carcinoma. Anal Quant Cytol 6: 30–36 (1984).

Paton KA: Properties of automatic metaphase selection. Ann Hum Genet (London) 35: 67–77 (1971).

Patrick EA, Fischer FP: Cluster mapping with experimental computer graphics. IEEE Trans, Comput C-18: 987–991 (1969).

Pearse AA, Rost FWD: A microspectrofluorimeter with Epi-illumination and photon counting. J Microsc 89: 321 (1969).

Pfitzer P, Pape HD: Characterization of tumor cell populations by DNA measurements. Acta Cytol 17: 19–25 (1973).

Pfitzer P, Pape HD: Investigation of DNA content of leukoplakia cells of oral mucosa. J Maxillofac Sum 3: 119–124 (1975).

Phillips DT, Wyatt PJ, Berkman RM: Measurement of the Lorenz-Mie scattering of a simple particle: Polystyrene latex spheres. J Colloid Interface Sci 34: 159 (1970).

Piera O, Meyer R, Voss K, et al: Vergleichende quantitative-morphologische untersuchungen (automatic mikroskophildanalyse—mikromorphometrische analyse) des myokard der ratte nach wiederholten gaben von isoprenalin. Zentralbl Allg Pathol 123: 590–593 (1979).

Ploem JS, deSterke JA, Bonnet J, et al: A microspectrofluorometer with Epi-illumination operated under computer control. J Histochem Cytochem 22: 668–677 (1974).

Pollister AW, Moses MJ: A simplified apparatus for photometric analysis and photomicrography. J Gen Physiol 32: 567 (1949).

Pollister AW, Ris H: Nucleoprotein determination in cytological preparations. Cold Spring Harbor Symp Quant Biol 12: 147 (1947).

Poulsen RS, Oliver LH, Cahn RL, et al: High resolution analysis of cervical cells: A progress report. J Histochem Cytochem 25: 689–695 (1977).

Pressman NJ: Markovian analysis of cervical cell images. J Histochem Cytochem 24: 138 (1976).

Pressman NJ: Optical texture analysis for automatic cytology and histology—A Markovian approach. PhD Thesis, Lawrence Livermore Laboratories, Univ of Calif, Los Angeles (1979).

Pressman NJ, Wied GL (eds): *Tutorials of Cytology*, Chicago (1979) pp 95–102

Preston K Jr: The CELLSCAN system: A leukocyte pattern analyzer. Proc Western Joint Comput Conf (1961) p 173.

Preston K Jr: *Coherent Optical Computers*. McGraw-Hill, New York (1972a).

Preston K Jr: Use of the CELLSCAN/GLOPR system in the automatic identification of white blood cells. Biomed Eng 7: 226 (1972b).

Preston K Jr: Use of the Golay logic processor in pattern recognition studies using hexagonal neighborhood logic, in *Computers and Automata*. Polytechnic Press, Brooklyn, NY (1972c) pp 609–623.

Preston K Jr: Applications of cellular automata to biomedical image processing, in *Computer Techniques in Biomedicine and Medicine*. Auerbach 295–331 (1973).

Preston K Jr: Digital picture analysis in cytology, in *Digital Picture Analysis* (Rosenfeld A ed) Springer Verlag, Heidelberg (1976) pp 209–294.

Preston K Jr: Tissue section analysis: Feature selection and image processing. Pattern Recogn 13: 17–36 (1981).

Preston K Jr: Progress in image processing languages, in *Computing Structures for Image Processing*, (Duff MJB ed). Academic Press, London (1983a) pp 195–211.

Preston K Jr: Xi-filters. Proc IEEE Trans Acoustics Speech Signal Proc, ASSP-31: 861–876 (1983b).

Preston K Jr: Introductory comments on high resolution automated microscopy. Anal Quant Cytol Histol 7: 4–5 (1985).

Preston K Jr: High-resolution leukocyte analyzers: Retrospective and prospective. Appl Opti 426: 3258–3265 (1987).

Preston K Jr, Dekker A: Differentiation of cells in abnormal human liver by computer image processing. Anal Quant Cytol 2: 203–220 (1980).

Preston K Jr, Duff MJB: *Modern Cellular Automata*. Plenum Press, New York (1984).

Preston K Jr, Norgren PE: Method and apparatus for detecting mitotic blood cells on a blood cell sample slide. US Patent No 3,503,684 (1970).

Preston K Jr, Norgren PE: Method of preparing blood smears. US Patent No 3,577,267 (1971).

Preston K Jr, Onoe M: *Digital Processing of Biomedical Images*. Plenum Press, New York (1976).

Preston K Jr, Rutovitz D: Automatic Cytology, in *Automated Multiphasic Health Testing*, (Berkley C ed). Eng Found, New York (1971).

Prewitt JMS, Mendelsohn ML: The analysis of cell images. Ann NY Acad Sci 128: 1035–1053 (1966).

Prewitt JMS: Parametric and nonparametric recognition by computer: An application to leukocyte image processing. Adv Comput 12:285–414 (1972).

Prewitt JMS, Wu S: An application of pattern recognition to epithelial tissues. Proc 2nd Ann Symp Comput Appl Med Care IEEE Computer Society, New York (1978) pp 15–25.

Pruitt JC, Ingraham SC II, Kaiser RD, et al: Preparation of vaginal-cervical material for automatic scanning-preliminary results with the cytoanalyzer. Trans 6th Ann Mtg Int Soc Cytol Council (1959).

Ramberg EG: A theoretical analysis of operation of flying spot and camera tube microscopes in the ultra-violet. IRE Trans Med Elec ME-12: 58 (1958).

Ramsey DM (ed): *Image Processing in Biological Science*. Univ Calif Press, (1968).

Rao CR: *Advanced statistical methods in biometric research*. Wiley, New York (1952).

Rauch DJ: Digital matched filter detection and discrimination of a class of lymphocytes. 5th Hawaii Conf on Syst Sci, Univ of Hawaii (1972).

Read JS, Borvec RT, Bartels PH, et al: A fast image processor for locating cell nuclei in uterine specimens, in *The Automation of Cancer Cytology and Cell Image Analysis,* (Pressman NJ, Wied GL eds). Chicago, Tutorials of Cytology (1979) pp 143–155.

Reagan JW, Hamonic MJ, Wentz WB: Analytical study of the cells in cervical squamous-cell cancer. Lab Invest 6: 241–250 (1957).

Reagan JW, Moore RD: Morphology of the malignant squamous cell: A study of six thousand cells derived from squamous cell carcinomas of the uterine cervix. Ann J Pathol 28: 105–127 (1952).

Reale FR, Bartels PH, Bibbo M, et al: Differentiation by TICAS analysis of cell populations of tracheal aspirates from hamsters with squamous carcinoma. Am J Clin Pathol 72: 52–58 (1979).

Regoliosi G: Metodi di elaborazione del contorno nella localizzazione del centromero di cromosomi umani. Mong Centro G Zanbon 5: 136 (1970).

Reinhardt ER, Erhardt R, Schwarzman P, et al: Structure analysis and classification of cervical cells using a processing system based on TV. Anal Quant Cytol 1: 143–150 (1979).

Reinhardt ER, Lenz R, Greiner W, et al: Diagnostic relevance of visually nonsuspect cells for cancer diagnosis: Part I. Results of feature evaluation and their consequences for automated specimen analysis. Anal Quant Cytol 4: 157 (1982).

Reinhardt ER, Schwarzmann P, Kersten M, et al: A new approach for fast feature extraction of cytological samples, in *The Automation of Cancer Cytology and Cell Image Analysis,* (Pressman NJ, Wied GL eds). Chicago, Tutorials of Cytology 213–216 (1976).

Rigaut JP, Margules S, Boysen M, et al: Karyometry of pseudostratified, metaplastic and dysplastic nasal epithelium by morphometry and stereology. Pathol Res Pract 174: 342–356 (1982).

Robb RA, Johnson SA, Greenleaf JF, et al: An operator interactive computer-controlled system for high fidelity digitization and analysis of biomedical images, in *Quantitative Imaging in the Biomedical Sciences*. Soc Photo-Optical Instr Eng (1973) p 11.

Robertson AJ, Anderson JM, Brown RA, et al: Grading of astrocytomas using a Quantimet 720 image-analysing computer. J Clin Pathol 34: 469–474 (1978).

Rodenacker K, Gais P, Jutting U: Segmentation and measurement of texture. Stereol Jugosl (Suppl 1) 3: 165–174 (1981).

Rogers CW: Sample preparation—A quantitative evaluation of different techniques. Eng Found Conf Automatic Cytol, Saxtons River (1972).

Rogers CW: Blood sample preparation for automated differential systems. Am J Med Tech 39: 435–442 (1973).

Romen W, Ruter A, Aus HM: Computer aided cytophotometric analysis of cell transformations induced by carcinogens and tumor viruses, in *The Automation of Cancer Cytology and Cell Image Analysis,* (Pressman NJ, Wied GL eds). Chicago, Tutorials of Cytology (1979) pp 47–51.

Romen W, Ruter A, Harms H, et al: Preneoplastic nuclear alterations in rat's liver: A computer aided cytophotometric analysis. Verh Dtsch Ges Pathol 63: 472–475 (1979).

Romen W, Ruter A, Saito K, et al: Relationship of ploidy and chromatin condensation in rat liver, moreover a comparison of the nuclear texture in sections and touch preparations. Histochemistry 67: 249–256 (1980).
Ronstrom L, Tribukait B, Esposti P-L: DNA pattern and cytological findings in fine-needle aspirates of untreated prostatic tumors. A flow-cytofluorometric study. Prostate 2: 79–88 (1981).
Rosenberg SA, Ledeen KS, Kline T: Automatic identification and measurement of cells by computer. Science 163: 1065–1067 (1965).
Rosenfeld A: Connectivity in digital pictures. J Assoc Comput Mach 17: 146–160 (1970).
Rosenfeld A, Thurston M: Edge and curve detection for visual scene analysis. IEEE Trans Comput C-20: 562–569 (1971).
Rosenquist TH: Use of quantitation in histochemistry: A review of the literature. Histochem J 8: 205 (1976).
Rosenthal DL, Missirlian N, McLatchie C, et al: Predicting cervical epithelial disease states using subcategories of intermediate cells. Anal Quant Cytol 5: 217 (1983).
Rosenthal DL, Stern E, McLatche C, et al: A simple method of producing a monolayer of cervical cells for digital image processing. Anal Quant Cytol 1: 84–88 (1979).
Rost FWD, Pearse AGE: An improved microspectrofluorimeter with automatic digital data logging: Construction and operation. J Microsc 94: 93–105 (1971).
Rothfeld JM, Gross DS: Changes in GnRH within the organum vasculosum lamina terminalis in the ovariectomized estrogen-progesterone treated rat. A quantitative immunocytochemical study using image analysis. Endocrinology 110: 206 (1982).
Rowinski J, Pienkowski M, Abramszuk J: Area representation of optical density of chromatin in resting and stimulated lymphocytes as measured by means of Quantimet. Histochemie 32: 75–80 (1972).
Rowinski J, Sawicki W, Swenson R, et al: Changes in chromatin after infection of mouse embryo fibroblasts with polyoma virus detected by image analysis. Acta Cytol 19: 136–141 (1975).
Rowinski J, Souchier C, Czyba JC: DNA content of cells in human buccal smears: A preliminary study. Acta Histochem 62: 276–281 (1978).
Rueter A, Wittekind D, Harms H, et al: Genormte Farbmessung und automatisierte zytophotometrie in azur B eosin getarbten Preparaten, in *Informatik-Fachberichte*. Springer-Verlag, Berlin (1980) pp 133–145.
Rutovitz D: Machines to classify chromosomes, in *Human Radiation Cytogenetics*. North-Holland, Amsterdam (1967) p 589.
Rutovitz D: Data structures for operations on digital images, in *Pictorial Pattern Recognition*, (Cheng GC, Ledley RS, Pollock DC, Rosenfeld A eds). Thompson (1968a) pp 106–133.
Rutovitz D: Automatic chromosome analysis. Br Med Bull 24: 260–267 (1968b).
Rutovitz D: Centromere finding. Some shape descriptors for small chromosome outlines. Mach Intell 5: 435 (1970).
Salomon LL, Musgrave GH: Automatic counting of bacterial colonies. Ann NY Acad Sci 157: 168–182 (1969).
Sammon JW Jr: A non-linear mapping for data structure analysis. IEEE Trans Comput C-18: 401–409 (1969).
Sammon JW Jr: Interactive pattern analysis and classification. IEEE Trans Comput C-19: 594–616 (1970).

Sandritter W, Bohm N: Ploidy of normal and malignant cells, in *The Automation of Uterine Cancer Cytology*, (Wied GL, Bahr GF, Bartels PH eds). Chicago, Tutorials of Cytology (1976) pp 289–304.

Sandritter W, Carl M, Ritter W: Cytophotometric measurements of the DNA content of human malignant tumors by means of the Feulgen reaction. Acta Cytol 10: 26–30 (1966).

Sandritter W, Kiefer G, Schluter G, et al: DNA in heterochromatin cytophotometric pattern recognition image analysis among cell nuclei in duct epithelium and in carcinoma of the human breast. Beitr Pathol 151: 87–96 (1974).

Sandritter W, Kiefer G, Schluter G, et al: Eine cytophotometrische methode zur objektivierung der morphologie von zellkernen: ein Beitrag zum Problem von Hu- und Heterochromatin. Histochemie 10: 341–352 (1967).

Sanfeliu A: An Application of a Distance Measure Between Grafts of the Analysis of Muscle Tissue Patterns. Rep TR-EE 81-15, Dept Elec Eng, Purdue University (1980).

Sans J, de la Torre C: G_v and G :2 cells as identified by their chromatin pattern in dormant and proliferating meristems. J Cell Biol 19: 294–298 (1979).

Sapp ER: A Technique for Character Recognition by Moments of Arc Length. US Dept of Commerce Clearinghouse Document AD 289676 (1962).

Saraga P, Wavish PR: Edge-coding operators for pattern recognition. Electron Lett 7: 736–738 (1971).

Sawicki W, Rowinski J, Abramczuk J: Image analysis of chromatin in cells of preimplantation mouse embryos. Cell Biology 63: 227–233 (1974a).

Sawicki W, Rowinski J, Swenson R: Change of chromatin morphology during the cell cycle detected by means of automated image analysis. J Cell Phys 84: 423–428 (1974b).

Schaphorst RA: Automatic Video Data Analysis. US Dept of Commerce Clearinghouse Document No AD 322669 (1961).

Schenck U, Neugebauer D, Otto K, et al: Monolayer preparation of cervical samples: Cytomorphologic characteristics and differential cytology. Anal Quant Cytol 3: 285–288 (1981).

Schoendorf H: Karyometric studies of epithelium in mammary glands and tumorous cells of mammary carcinomas. Acta Cytol 25: 63 (1981).

Scott LB, Preston K Jr: Apparatus for Counting Irregularly Shaped Objects. US Patent No 3,408,485 (1968).

Sebestyen GS: *Decision Making Processes in Pattern Recognition*. Macmillan, New York (1962).

Selfridge OG: Pattern recognition and modern computers. Proc Western Joint Comput Conf (1955) pp 91–93.

Selzer RH: Computer processing of the Roentgen image. Radiol Clin North Am 7: 461–471 (1969).

Sencer D: Statement to Senate Subcommittee on Anti-trust and Monopoly. Committee on the Judiciary, Washington DC (1967).

Sengbusch G von: Slit-scan analysis of tape-mounted cells, in *Technologies for Automation in Cervical Cancer Screening*, (Goerttler K, Herman CJ eds). Springer-Verlag, Heidelberg (1976) pp 46–50.

Serra J: Stereology and structuring elements. J Microsc 95: 93–103 (1972).

Shack R, Bell B, Hillman D, et al: Ultrafast laser scanner microscope—first performance tests. Proc Int Workshop Physics and Engineering in Medical Imaging (Nalcioglu O ed). Asilomar, Pacific Grove (1982) pp 49–57.

Shackel B: Medical applications of television—A survey. Med Electron Biol Eng 1: 35–50 (1963).
Shapiro B, Lemkin P, Lipkin L: The application of artificial intelligence techniques to biologic cell identification. J Histochem Cytochem 22: 741 (1974).
Sheck LE, Muirhead KA, Horan PK: Evaluation of the S phase distribution or flow cytometric DNA histograms by autoradiography and computer algorithms. Cytometry 1: 109–117 (1980).
Sherman AB, Koss LG: Morphometry of benign urothelial cells in the presence of cancer. Anal Quant Cytol 5: 221 (1983).
Sherman AB, Koss LG, Adams S, et al: Bladder cancer diagnosis by image analysis of cells in voided urine using a small computer. Anal Quant Cytol 3:239–249 (1981).
Shevechuk MM, Richart RM: DNA content of condyloma acuminatum. Cancer 49: 489–492 (1982).
Shibata H, Kobayashi Y, et al: On the method of specimens for diagnostic cytology by Autosmear. Jpn Soc Cytol 11: 7–14 (1972) in Japanese.
Simon H, Arnold F, Kemmer C, et al: Morphometrische untersuchungen an der mauseleber mittrels automatischer bildverarbeitung nach applikation von oestrastilben. Exp Pathol 8: 48–57 (1973).
Simon H, Kranz D, Voss K, et al: Automated data sampling in slides of selected liver diseases. Pathol Res Pract 170: 388–401 (1980).
Simon H, Kunze KD, Voss K, et al: *Automatische Bildverarbeitung in Medizin und Biologie*. Steinkopff, Dresden (1975).
Simon H, Kranz D, Voss K, et al: Zur methode einer automatischen mikroskopbildanalyse an histologischen schnitten. Zentralbl Allg Pathol 125: 399–405 (1981).
Sinski J, Flynt P, Bartels PH, et al: Cell image analysis of lymphocytes from guinea pigs immunized against coccidiomycosis, in *Coccidiomycosis, Current Clinical and Diagnostic Status*, (Ajello J ed). Symposia Specialists Publ, Miami (1977).
Skeggs TL Jr: An automatic method for colorimetric analysis. Am J Clin Pathol 28: 311 (1957).
Sklansky J, Ballard D, Wechsler H: Computer aided analysis of radiographs. US Public Health Service Grant GM-17632 Univ Calif, Irvine (1973).
Slavin G, Sowter C, Robertson K, et al: Measurement on jejunal biopsies by computer-aided microscopy. J Clin Pathol 33: 254–261 (1980).
Smeulders AWM, Vossepoel AM, Vrolijk J, et al: Some shape parameters for cell recognition, in *Pattern Recognition in Practice*, (Gelsema ES, Kanal LN eds). North-Holland, Amsterdam (1980) pp 131–142.
Smithline ET: Blood Cell Recognizer. US Patent No 3,315,229 (1967).
Sobel O, Levinthal C, Magagno ER: Special techniques for the automatic computer reconstruction of neuronal structures. Ann Rev Bioeng 9: 347–362 (1980).
Soffen GA: Microscopy for the detection of micro-organisms on the surface of Mars. An NY Read Sci 157: 159–167 (1969).
Spencer GSG, Eccles MJ: Spinal muscle in scoliosis. II. The proportion and size of type 1 and type 2 skeletal muscle fibers measured using a computer-controlled microscope. J Neurol Sci 30: 143–154 (1976).
Sprenger E, Sandritter W, Naujoks H, et al: Prescreening in cervical cytology: Progress report, in *The Automation of Cancer Cytology and Cell Image Analysis*, (Pressman NJ, Wied GL eds). Chicago, Tutorials of Cytology (1979) pp 9–12.
Sprenger E, Ulrich H, Schondorf H: The diagnostic value of cell-nuclear DNA

determination in aspiration cytology of benign and malignant lesions of the breast. Anal Quant Cytol 1: 29–35 (1979).

Sprick W, Ganzhorn K: Recognition of Numerals by Contour Following. IEEE Rep 3254 (1959).

Sprick W: Apparatus for Identifying Line Traces. US Patent No 2,738,499 (1956).

Spriggs AI, Diamond RA, Meyer EW: A new principle for automated screening of cervical smears. Lancet 1: 359 (1968).

Staunton JJ: Clinical Spinner. US Patent No 3,705,048 (1972).

Stein PG, Lipkin LE, Shapiro NH: Spectre II: General purpose microscope input for a computer. Science 166: 328–333 (1969).

Steinkamp JA, Crissman HA: Automated analysis of deoxyribonucleic acid, protein, and nuclear to cytoplasmic relationships in tumor cells and gynecologic specimens. J Histochem Cytochem 22: 616–621 (1974).

Steinkamp JA, Romero A, Van Dilla MA: Multiparameter cell sorting: Identification of human leukocytes by acridine orange fluorescence. Acta Cytol 17: 113 (1973).

Stenkvist B, Westman-Naeser S, Holmquist J, et al: Computerized nuclear morphometry as an objective method for characterizing human cell populations. Cancer Res 38: 4688–4697 (1978).

Stern E, Rosenthal DL, McLatchie C, et al: An expanded cervical cell classification system validated by automated measurements. Anal Quant Cytol 4: 110–114 (1982).

Sternberger LA: *Immunochemistry*. Wiley, New York (1979).

Sternberger LA, Harroy PH, Cuculis JJ, et al: The unlabeled antibody enzyme method of immunohistochemistry. Preparation and properties of soluble antigen-antibody complex (peroxidase-antiperoxidase) and its use in the identification of spirochetes. J Histochem Cytochem 18: 315 (1970).

Stich HF, Steele HD: DNA content of tumour cells: III. Mosaic composition of sarcomas and carcinomas in man. J Natl Cancer Inst 28: 1207–1218 (1962).

Stinson SF, Lilga JC, Reese DH, et al: Quantitation with an automated image analyzer of nuclear-cytoplasmic changes induced by hydrocortisone in bladder epithelium. Cancer Res 37: 1428–1431 (1977).

Stinson SF, Sporn MB: Use of an automated image analyzer to quantitate cellular hyperplasia in urinary bladder epithelium. J Microsc 109:329–335 (1977).

Stoehr W, Goerttler K: Dual laser flow analysis and sorting of cells including the uterine cervix, in *The Automation of Cancer Cytology and Cell Image Analysis,* (Pressman NJ, Wied GL eds). Chicago, Tutorials of Cytology (1979) pp 75–79.

Stone SP, Littlepage JL, Clegg BR: Second report on the chromosomes scanning program at the Lawrence Radiation Laboratory. Proc Seminar Soc Photo Optical Eng 157–171 (1967).

Streefkerk JG, van der Ploeg, van Duijn P: Quantitative aspects of cytochemical peroxidase procedures investigated in a model system. J Histochem Cytochem 21: 215 (1973).

Suh O, Weiss L: The development of a technique for the morphometric analysis of invasion of cancer. J Theo Biol 108: 547–562 (1984).

Sumner AT, Evans HJ: Mechanisms involved in the banding of chromosomes with Quinacrine and Giemsa-II. The interaction of dyes with chromosomal components. Exp Cell Res 81: 223 (1973).

Suzuki R, Yamamoto S: Real-time processing in automated cytology, in *Real-*

time Medical Image Processing, (Onoe M, Preston K Jr, Rosenfeld A eds). Plenum Press, New York (1980).
Sweetman F, Ansley HR, Ornstein L: Cell-specific leukocyte esterases. Fed Proc 31: 603 (1972).
Swift H: The quantitative cytochemistry of RNA, in *Introduction to Quantitative Cytochemistry,* (Wied GL ed). Academic Press, New York (1966) pp 355–386.
Sychra JJ, Bartels PH, Bibbo M, et al: Computer recognition of abnormal ectocervical cells: Comparison of the efficacy of contour and textural features. Acta Cytol 21: 765–769 (1977).
Sychra JJ, Bartels PH, Bibbo M, et al: Computer recognition of binucleation with overlapping in epithelial cells. Acta Cytol 22: 22–28 (1978).
Sychra J, Bartels PH, Taylor J, et al: Cytoplasmic and nuclear shape analysis for computerized cell recognition. Acta Cytol 20: 68–78 (1976).
Tanaka N, Ikeda H, Ueno T, et al: Field test and experimental use of CYBEST Model 2 for practical gynecologic mass screening. Anal Quant Cytol 1: 122–126 (1979).
Tanaka N, Ikeda M, Ueno T, et al: CYBEST-CDMS: Automated cell dispersion and monolayer smearing device for CYBEST. Anal Quant Cytol 3: 96–102 (1981).
Tanaka N, Ikeda H, Ueno T, et al: Fundamental study of automatic cytoscreening for uterine cancer: II. Segmentation of cells and computer simulation, in *The Automation of Uterine Cancer Cytology,* (Wied GL, Bahr GF, Bartels PH eds). Chicago, Tutorials of Cytology (1976) pp 228–233.
Tanaka N, Ikeda H, Ueno T, et al: Fundamental study of automatic cytoscreening for uterine cancer: I. Feature evaluation for the pattern recognition system. Acta Cytol 21: 72–78 (1977a).
Tanaka N, Ikeda H, Ueno T, et al: Fundamental study of automatic cytoscreening for uterine cancer: III. New system of automated apparatus CYBEST utilizing the pattern recognition system. Acta Cytol 21:85–89 (1977b).
Tanaka N, Ikeda H, Ueno T, et al: Fundamental study of automatic cytoscreening for uterine cancer: IV. Sample requirements for CYBEST and simulation test of cell dispersion. Acta Cytol 21: 531–535 (1977c).
Tanaka N, Ikeda H, Ueno T, et al: Fundamental study of automatic cytoscreening for uterine cancer: V. Data analysis for improvement of CYBEST. Acta Cytol 21: 536–538 (1977d).
Tanaka N, Ikeda H, Ueno T, et al: Fundamental study on automatic cyto-screening of uterine cancer. Acta Cytol 21: 72–78, 85–89 (1977e).
Tanaka N, Ikeda H, Ueno T, et al: Experimental practical use of automated screening system (CYBEST) for the prescreening of gynecological samples, in *The Automation of Cancer Cytology and Cell Image Analysis,* (Pressman NJ, Wied GL eds). Chicago, Tutorials of Cytology (1979) pp 123–134.
Tanimoto TT: A non-linear model for computer-assisted medical diagnostic procedure. Ann NY Acad Sci 23: 576–578 (1961).
Tanimoto TT, Loomis RG: A Taxonomy Program for the IBM 704. IBM Program Library, New York (1960).
Tavares AS, Costa J, de Carvalho A, et al: Tumor ploidy and prognosis in carcinomas of the bladder and prostate. Br J Cancer 20: 438–441 (1966).
Taylor J, Bahr GF, Bartels PH, et al: Development and evaluation of automatic nucleus finding routines: Thresholding of cervical cytology images. Acta Cytol 19: 288–298 (1975).

Taylor J, Bartels PH, Bibbo M, et al: Automated hierarchic decision structures for multiple category cell classification by TICAS. Acta Cytol 22: 261–267 (1978).

Taylor J, Puls J, Sychra J, et al: A system for scanning biological cells in three colors. Acta Cytol 22: 29–35 (1978).

Taylor WK: Pattern Recognition by Means of Automatic Analogue Apparatus. IEEE Rep 3154 (1959).

Ten Kate TK: Design and implementation of an interactive karyotyping program in C on a Vicom digital image processor. PhD Thesis, Appl Phys Dept, University of Delft (1985).

Teplitz RL, Pak HY, Benfield JR, et al: Quantitative DNA: Comparative studies of a cellular marker for bronchogenic carcinoma. JAMA 249: 1046–1049 (1983).

Thieme GA: A versatile device for microscopic spectrofluoremetry. Acta Physiol Scand 67: 514 (1966).

Tilgner RD, Abele L, Wahl FM: An improved edge detection system applied to cytologic material. Proc Conv Tecn Elaborazione Immagini Interesse Clinico, Pavia (1977) pp 291–302.

Tolles WE: The Cytoanalyzer, an example of physics in medical research. Ann NY Acad Sci 17: 250 (1955).

Tolles WE (ed): *Data Extraction and Processing of Optical Images in the Medical and Biological Sciences*. Ann NY Acad Sci 157: 1–530 (1969).

Tolles WE, Bostrom RC: Automatic screening of cytological smears for cancers: The instrumentation. Ann NY Acad Sci 63: 1211–1218 (1956).

Tolles WE, Garcia GL, Nelson JH: Cytopreparation techniques in support of automated cytology, in *Prevention and Detection of Cancer*. Marcel Dekker, New York (1978) pp 1047–1065.

Tourassis VD, Dekker A, Preston K: Relationship between cell size and weight of the human liver. Anal Quant Cytol 5: 43–54 (1983).

Tretiak OF: Touching and overlapping chromosomes. Ann NY Acad Sci 157: 365–367 (1966).

Tribukait B, Gustafson H, Esposti P: Ploidy and proliferation in human bladder tumors as measured by flow cytofluorometric DNA-analysis and its relations to histopathology and cytology. Cancer 43: 1742–1751 (1979).

Trujillo TT, Van Dilla MA: Adaptation of the fluorescent Feulgen reaction to cells in suspension for flow microfluorometry. Acta Cytol 16: 26 (1972).

Tucker JH: CERVISCAN: An image analysis system for experiments in automatic cervical smear prescreening. Comput Biomed Res 9: 93–107 (1976).

Tucker JH: An image analysis system for cervical cytology automation using nuclear DNA content. J Histochem Cytochem 27: 613–620 (1979).

Tucker JH: Basic performance tests on the CERVIFIP linear array prescreener. Anal Quant Cytol 5: 129–137 (1983).

Tucker JH, Gresham GA: Preparation of cervical scrape materials for automatic scanning. J Obstet Gynaecol Br Commonw 78: 947–953 (1971).

Tycko DH, Anbalagan S, Liu HC: Automated Image Processing Project. Tech Rep No 32 Dept of Comput Sci, SUNY, Stony Brook (1974).

US Army: *Laboratory Procedures in Clinical Hematology*. Tech Manual 8-227-4 (1963).

Uhr L, Vossler C: Pattern recognition program that generates, evaluates, and adjusts its own operators. Proc Western Joint Comput Conf 555 (1961).

Unger SH: A computer oriented toward spatial problems. Proc IRE 46: 1744–1750 (1958).
Valkov I, Topov J, Glukhchev G, et al: Morphometric study on the nucleoli of mesothelial and tumour cells being in pleural cavity effusions. Arch Geschwulstforsch 47: 220–225 (1977).
Van Driel-Kulker AMJ, Ploem JS, Cornelisse CJ: Image analysis of breast cytology specimens using LEYTAS. Anal Quant Cytol 4: 145 (1982).
Van Driel-Kulker AMJ, Ploem JS, Zaaijer JJ, et al: A preparation technique for exfoliated and aspirated cells allowing different staining procedures. Anal Quant Cytol 2: 243–246 (1980).
Vastola EF, Hertzberg RA, Neurath PW, et al: A description by computer of texture in Wright-stained leukocytes. Acta Cytol 18: 231–247 (1974).
Veen A, Peachey LD: Trots: A computer graphics system for three-dimensional reconstruction from serial sections. Comput Graphics Image Proc 2: 135–150 (1977).
Volker ter Meulen, Bartels PH, Bahr, GF: Computer-assisted analysis of a carrier culture infected with Moloney leukemia virus. Acta Cytol 16: 454–463 (1972).
Vooijs PG, Oud PS, Zahniser DJ, et al: Chromatin measurements in Feulgen-stained intermediate cells from normal and abnormal cervical smears. Anal Quant Cytol 4: 154 (1982).
Voss K: Automatische Bildverarbeitung in der Pathologie. II. Objektisolierung und Konturausgleich. Exp Pathol 9: 131–139 (1974).
Voss K, Barz H, Kunze KD: Image processing in pathology. V. Interpretation of feature distributions from liver cell nuclei in histological sections. Exp Pathol 15: 144–152 (1978).
Voss K, Barz H, Wagner H: Automatische Bildverarbeitung in der Pathologie. III. Aufbau und Anwendung eines automatisierten Systems zur Erfassung morphometrischer Parameter von Zellkernen. Exp Pathol 13: 145–161 (1977).
Wald N, Fatora SR: Biomedical significance of human chromosome aberrations: A justification of automated image processing. Proc IEEE Conf Syst Man, Cyber, Dallas (1974).
Wald N, Preston K Jr: Automatic screening of metaphase spreads for chromosome analysis, in *Image Processing in Biological Science,* (D Ramsey ed). University of California Press (1968) pp 9–34.
Walton WH: Automatic counting of microscopic particles. Nature 169: 518 (1952).
Wann DF, Price JL, Cowan WN, et al: An on-line digital computer system for the semi-automatic analysis of Golgi-impregnated neurons. IEEE Trans Biomed Eng BME-20:233–247 (1973).
Ward JE III, Carlson FP, Heywood JD: Coherent optical recognition and counting of reticulated red blood cells. IEEE Trans Biomed Eng BME-21: 12–20 (1974).
Ware RW, Lo Presti V: Three-dimensional reconstruction from serial sections. Int Rev Cytol 40: 325–440 (1975).
Watanabe S: An automated apparatus for cancer prescreening: CYBEST. Comput Graphics Image Proc 3: 350–358 (1974).
Watanabe S, Imasato Y, Genchi H, et al: A Pap smear prescreening system: CYBEST, in *Digital Processing Biomedical Images,* (Preston K Jr, Onoe M eds). Plenum Press, New York (1976).
Watanabe S, Tsunekawa S, Yokomoto Y, et al: The development of a new model cyto-prescreener for cervical cancer, in *Real-Time Medical Image Process-*

ing, (Onoe M, Preston K Jr, Rosenfeld A eds). Plenum Press, New York (1980).

Watts KC, Husain OAN, Tucker JH, et al: The use of cationic polyelectrolytes in the preparation of cell monolayers for automated cell scanning and diagnostic cytopathology. Anal Quant Cytol 6: 272–278 (1984).

Weber JE, Baldessari BA, Bartels PH: Test statistics for detecting aneuploidy and hyperdiploidy. Anal Quant Cytol 7: 131–139 (1985a).

Weber JE, Nielsen T: Multinomial distributions in ploidy analysis. Anal Quant Cytol 7: 140–151 (1985b).

Weibel ER: *Stereological Methods*. Academic Press, London (1979).

Weibel ER, Fisher C, Gahm J, et al: Current capabilities and limitations of available stereological techniques. J Microsc 95: 367–391 (1971).

Weibel ER, Kistler GS, Scherle WF: Practical stereological methods for morphometric cytology. J Cell Biol 30: 23 (1966).

Weinreb M, Zajicek G, Levij IS: Discrimination between normal and malignant gastric epithelial cells by computer image analysis. Anal Quant Cytol 6: 206–210 (1984).

Weiss RR, Richart RM, Okagaki T: DNA content of mucinous tumors of the ovary. Am J Obstet Gynecol 103: 409–424 (1969).

Wenzelides K, Hillmann I, Voss K: Zur Darstellung der Leberzellgrenzen mit Anilinblau und ihre vollautomatische Erfassung mit einem Bildanalysesystem. Acta Histochem 70: 193–199 (1982).

Wenzelides K, Katner M, Roth K: Zur Darstellung der Nukleolen der Leberzellkerne mit einer modifizierten HBFP-Faerbung und ihre vollautomatische Erfassung mit einem Bildanalysesystem. Acta Histochem (in press).

Wenzelides K, Korek G, Voss K: Bestimmung des relativen RNA-Gehaltes von Leberzellen bei Faerbung mit Gallocyanin-Chromalaun. Acta Histochem 69: 307–313 (1981).

Wenzelides K, Kranz D, Simon H, et al: Anwendung der automatischen Mikroskopbildanalyse in der Praxis am Beispiel verschiedener Lebererkrankungen. Zentralbl Allg Pathol 125: 532–538 (1981).

Werner L, Voss K: Klassifizierung von Nervenzellformen der Lamina IV des visuellen Kortex der Albinoratte im Nissl-Praparat mit Hilfe der automatischen Bildverarbeitung. J Hornforsch 20: 467–473 (1979).

Wheeless LL, Onderdonk MA: Preparation of clinical gynecologic specimens for automated analysis: An overview. J Histochem Cytochem 22: 522–525 (1973).

Widrow B: A study of rough amplitude quantization by means of Nyquist sampling theory. IRE Trans Comput CT-3: 266–276 (1956).

Wied GL, Bahr B, Bartels PH: Automated analysis of cellular images by TICAS, in *Automated Cell Identification and Cell Sorting*, (Wied GL, Bahr GF eds). Academic Press, New York (1970a) pp 195–360.

Wied GL, Bahr GF, Bartels PH, et al: Quantitative cytology and histology for improved quality control and diagnostic standardization, in *Cervical Pathology and Colposcopy*, (Kurihara S et al eds). Elsevier Science Publ (1985a) pp 95–99.

Wied GL, Bartels PH, Bahr GF, et al: Computer-assisted identification of cells from uterine adenocarcinoma: I. Clinical feasibility study with TICAS. Measurements at wavelength 530 nm. Acta Cytol 12: 357–370 (1968a).

Wied GL, Bartels PH, Bibbo M, et al: Discriminant analysis on cells from developing squamous cancer of the respiratory tract. Cell Biophys 1: 39 (1979).
Wied GL, Bartels PH, Bibbo M, et al: Rapid DNA cytometry in clinical cytologic diagnosis. Acta Cytol 27: 33–37 (1983a).
Wied GL, Bartels PH, Bibbo M, et al: High resolution color video cytophotometry. Cell Biophys 5: 61–70 (1983b).
Wied GL, Bartels PH, Bibbo M, et al: Computer recognition of ectocervical cells: Classification accuracy and spatial resolution. Acta Cytol 21: 753–764 (1977a).
Wied GL, Bartels PH, Bibbo M, et al: Computer recognition of ectocervical cells: Classification of the efficacy of contour and textural features. Acta Cytol 21: 753–764 (1977b).
Wied GL, Bartels PH, Bibbo M, et al: Cytomorphometric markers for uterine cancer in intermediate cells. Anal Quant Cytol 2: 257–263 (1980).
Wied GL, Bartels PH, Bibbo M, et al: Computer discrimination of ectocervical cells—Assessment of the value of spectral information. Anal Quant Cytol 3: 225–234 (1981a).
Wied GL, Bartels PH, Dytch H, et al: Diagnostic marker features in dysplastic cells from the uterine cervix. Acta Cytol 26: 475–483 (1982).
Wied GL, Bartels PH, Gunter GF, et al: Computer-assisted identification of cells from uterine adenocarcinoma. A clinical feasibility study with TICAS. Acta Cytol 12: 180–204 (1968b).
Wied GL, Bartels PH, Oldfield DG: Taxonomic intracellular analytic system (TICAS) for cell identification. Acta Cytol 12: 177–210 (1968c).
Wied GL, Bibbo M, Bahr GF, et al: Computerized recognition of uterine glandular cells. I. Assessment of a self-optimizing recognition program. Acta Cytol 13: 611–619 (1969).
Wied GL, Bibbo M, Bahr GF, et al: Computerized recognition of uterine glandular cells—III. Assessment of the efficacy of 1um vs. 0.5um spot sizes. Acta Cytol 14: 136–141 (1970b).
Wied GL, Bibbo M, Bartels PH: Computer analysis of microscopic images: Application in cytopathology, in *Pathol Ann,* (Rosen PP ed). Appleton-Century Crofts, New York (1981b) pp 367–409.
Wied GL, Bibbo M, Dytch HE, et al: Computer grading of cervical intraepithelial neoplastic lesions. I. Cytologic indices. Anal Quant Cytol 7: 52–60 (1985b).
Wied GL, Bibbo M, Pishota FT, et al: Intermediate cell markers for malignancy—Consistence of expression. Anal Quant Cytol 6: 243–248 (1984).
Williams GZ: Time-lapse ultraviolet television microscopy—Instrumentation and biological application. IRE Trans Med Electron ME-6: 68–74 (1959).
Winkler DG, Baky AA, Hunter NR, et al: Image analysis of atypical bronchial squamous epithelial cells: A quantitative and qualitative examination of squamous cell carcinogenesis. Anal Quant Cytol 2: 310 (1980).
Winkler B, Crum CP, Fujll T, et al: Koilocytotic lesions of the cervix. The relationship of mitotic abnormalities to the presence of papillomavirus antigens and nuclear DNA content. Cancer 53: 1081–1087 (1984).
Wintrobe MM: *Clinical Hematology*. Lea Febiger, Philadelphia (1956).
Wittekind D: Standardization of dyes and stains for automated cell pattern recognition. Anal Quant Cytol 7: 6–30 (1985).
Wolley RC, Dembitzer HM, Merz HF, et al: The use of a slide spinner in the analysis of cell dispersion. J Histochem Cytochem 24: 11–15 (1976).

Wyatt PJ: Differential light scattering techniques for microbiology, in *Methods in Microbiology*, (Morris JF, Ribbons DW eds). Academic Press, New York (1973) p 183.

Wyatt PJ: Light scattering in the microbial world. J Colloid Interface Sci 39: 479 (1972).

Yasnoff WA, Bacus JW: Scene segmentation algorithm development using error measures. Anal Quant Cytol 6: 45–58 (1984).

Young IT, Paskowitz LI: Localization of cellular structures. Rep NIH Grant GM14940-07 (1974).

Young IT: Automated leukocyte recognition, in *Automated Cell Identification and Cell Sorting*, (Wied GS, Bahr GF eds). Academic Press, New York (1970) pp 187–194.

Young IT: Automated leukocyte recognition. PhD dissertation, MIT, Cambridge (1969).

Young IT: The classification of white blood cells. IEEE Trans Biomed Eng BME-19: 291–298 (1972).

Young JZ, Roberts F: High speed counting with the flying spot microscope. Nature 169: 963 (1952).

Zahniser D, Oud PS, Raaijmakers MCT, et al: Field Test results using the BioPEPR cervical smear prescreening system. Cytometry 1: 200–203 (1980).

Zahniser D, Oud PS, Raaijmakers MCT, et al: BioPEPR: A system for the automatic prescreening of cervical smears. J Histochem Cytochem 27: 635–641 (1979).

Zajdela A, Saravia De La Riva I, Ghossein NA: The relation of prognosis to the nuclear diameter of breast cancer cells obtained by cytologic aspiration. Acta Cytol 23: 75–80 (1979).

Zajicek J, Bartels PH, Bahr GF, et al: Computer analysis of lymphocytes from cases with lymphadenitis and lymphocytic lymphoma. Acta Cytol 16: 284–296 (1972).

Zajicek J, Caspersson T, Jakobsson P, et al: Cytologic diagnosis of mammary tumors from aspiration biopsy smears: Comparison of cytologic and histopathologic findings in 2,111 lesions and diagnostic use of cytophotometry. Acta Cytol 14: 370–376 (1970).

Zajicek G, Meinzer HP, Komitonski D: Automated image processing of colon crypt epithelium. Comput Biomed Res 15: 155–163 (1982).

Zajicek G, Shohat M: On the classification of nucleated red blood cells. Comput Biomed Res 16: 553–562 (1983a).

Zajicek G, Shohat M, Melnik Y, et al: Image analysis of nucleated red blood cells. Comput Biomed Res 16: 347–356 (1983b).

Zeiss C: Universal Micro-Spectrophotometer UMSP I. Pamphlet 40-811/1 e (1964).

Zetterberg A, Esposti PL: Cytophotometric DNA-analysis of aspirated cells from prostatic carcinoma. Acta Cytol 20: 46–57 (1976).

Zetterberg A, Esposti PL: Prognostic significance of nuclear DNA levels in prostatic carcinoma. Scand J Urol Nephrol 55: 53–58 (1980).

Zimmer C: Automatische analyse mikroskopischer bilder. ZEISS-Inform 17: 126–131 (1969).

Zworykin VK, Hatke FL: Ultraviolet color translating microscope. Science 126: 805 (1957).

2
Manipulation of 3D Imagery

GABOR T. HERMAN, SUSHMA S. TRIVEDI, AND
JAYARAM K. UDUPA

Introduction

Our aim is to provide the clinician with computer graphic tools that will enable him to visualize, manipulate, and quantitate the three-dimensional (3D) internal structures of patients.

The necessary information can come from any medical-imaging device that estimates the value of some physical parameter at each of a 3D array of points: our prime example will be the computerized tomography (CT) scanner, which provides us with estimates of the x-ray attenuation of tissue in a sequence of parallel slices. The same graphic procedures can be applied to the output of other devices such as magnetic resonance imagers, positron or single photon emission tomography scanners, or even ultrasound scanners (provided that the data are collected appropriately).

The data from such devices contain detailed information regarding 3D internal structures, but the visualization of the shapes of these structures based on the numbers, or even on the two-dimensional (2D) images of the individual slices, is nontrivial (Figure 2.1). The type of visualization that we have in mind is illustrated in Figure 2.2, which shows images of the skull of a patient before and after craniofacial surgery.

In addition to visualization, the clinician may wish to manipulate the 3D imagery. For example, it may be desirable to plan the surgical procedure on a computer and to produce an image such as the one in Figure 2.2b based on the preoperative CT scan. This way the result of a particular surgical procedure can be observed and evaluated before it is actually performed on a patient. The ability to make quantitative measurements on structures is also of importance to the clinician, as for example, for the preoperative assessment of deformities and the postoperative assessment of results.

In fact, visualization, manipulation, and quantitation all go hand-in-hand. In order to better observe or measure a particular internal feature, we may wish to remove structures that are hiding it in a particular point of

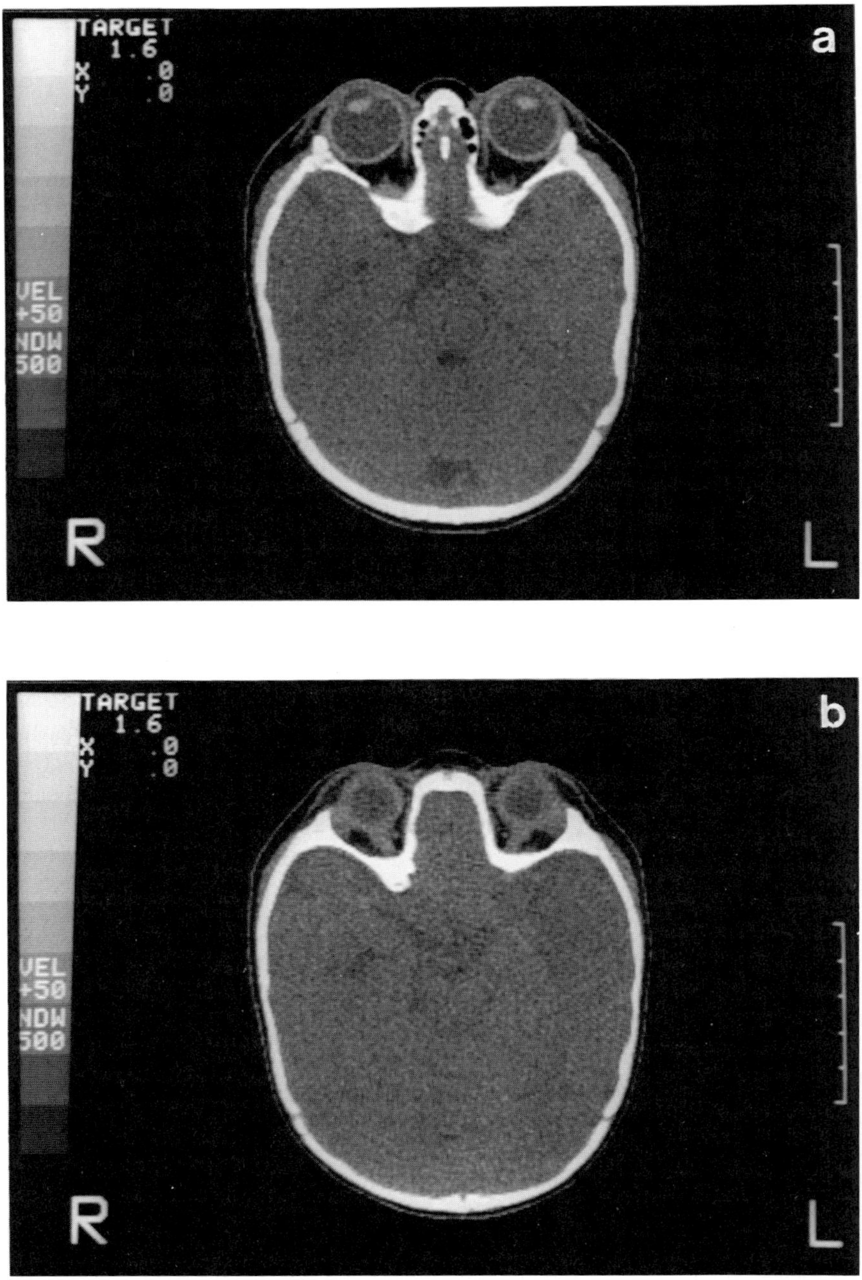

FIGURE 2.1. Four out of the 41 cross-sectional slices of the CT scan of a patient prior to craniofacial surgery. The slices are 6 mm from each other and cover a region from the middle to the roof of the orbits.

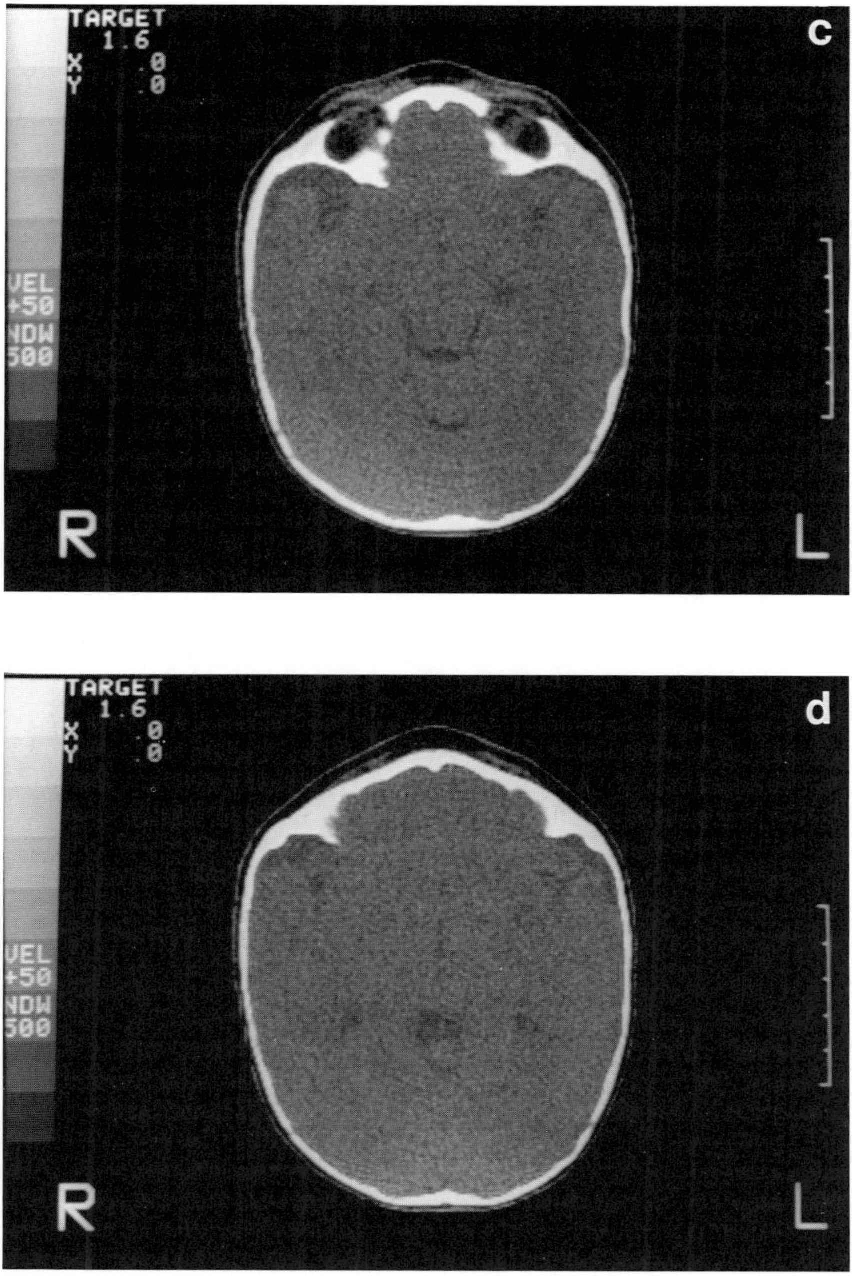

FIGURE 2.1. *Continued.*

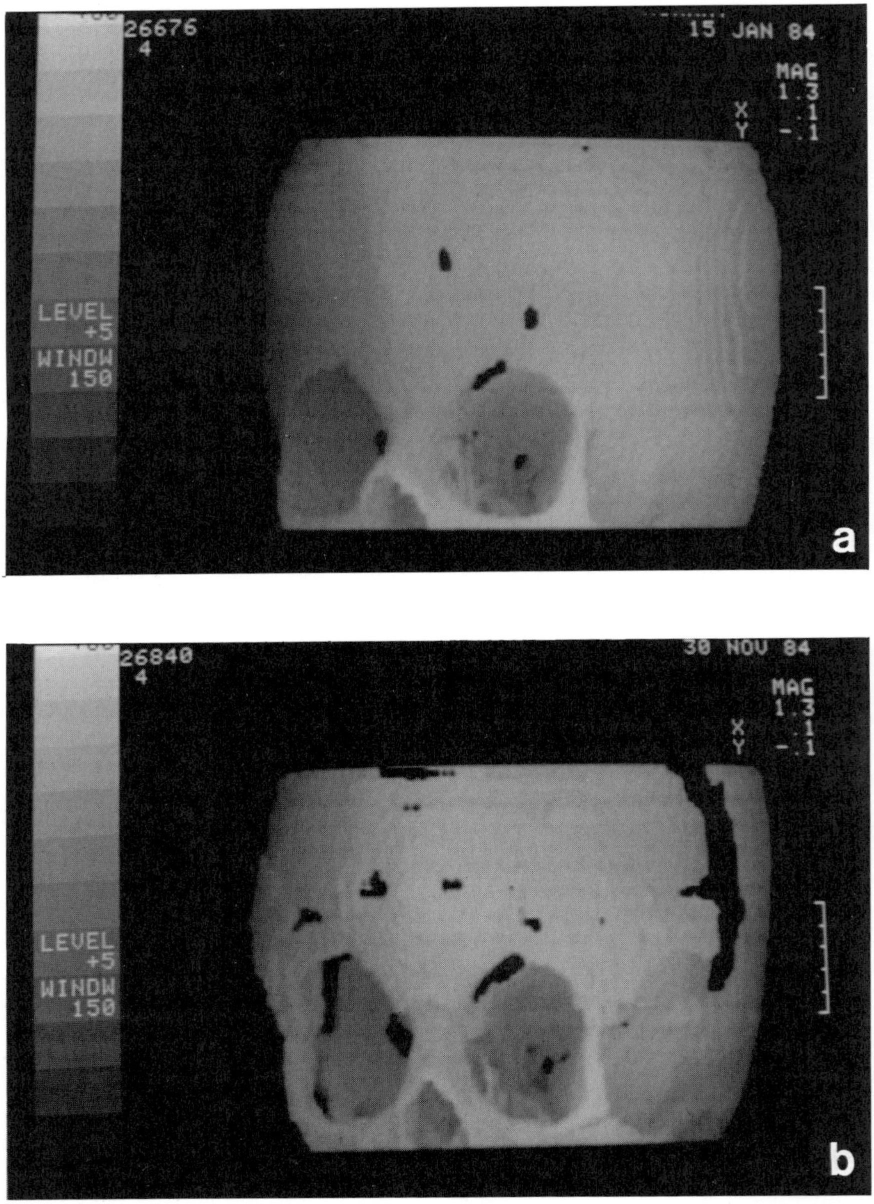

FIGURE 2.2. Displays of the skull of the patient whose CT slices are shown in Figure 2.1. (**a**) 3D display of the skull prior to craniofacial surgery. (**b**) 3D display of the skull after supraorbital reshaping to correct for a developmental anomaly.

view. On the other hand, manipulation of a 3D data set is made much easier by a good display of its relevant content.

In this chapter we discuss computer graphic techniques for display and manipulation of 3D imagery. We pay special attention to the data structures, ie, to the way the information contained in the 3D data is stored in the computer. This is because different types of display and manipulation operations are most efficiently performed using different types of data structures; we discuss this as we go along.

Discrete 3D Scenes and Their Representations

The Binary-Array Representation

In this chapter a *discrete 3D scene* (or *scene* for short) is considered to consist of a rectangular parallelepiped (referred to as the *region* of the scene), which is subdivided by three sets of parallel planes into smaller identical parallelepipeds (referred to as *voxels*), each one of which has a value assigned to it (referred to as the *density* of the voxel). In all applications, the "density" represents some physical property; for example, in CT it is related to the x-ray absorbancy in the voxel. In practice the scene is *digitized* in the sense that the density must be one of the integers between a lower limit L and an upper limit U. If $L = 0$ and $U = 1$, then these are the only possible values, and we say that the scene is a *binary scene*.

We assume that together with the scene, a rectangular coordinate system has been determined that assigns to each voxel a triple (i,j,k) of integers, where $1 < i < I$, $1 < j < J$, and $1 < k < K$. We refer to such a scene as an $I \times J \times K$ scene. The $I \times J \times 1$ scene formed by all voxels (i,j,k), for a fixed k, is called the kth *slice* of the scene.

Scene processing is the 3D analog of picture processing [Rosenfeld and Kak, 1982]. In this article we mention three scene-processing operations: subregioning, interpolation, and segmentation. Each of these operations produces a new scene from another scene.

A scene produced by *subregioning* a set of voxels is a subset of the voxels in the original scene, with densities unchanged. Taking a single slice out of a scene is an example of subregioning. Typically, subregioning is performed as an early step of scene processing: by isolating a region of interest, the size of the data set that needs to be handled by further processing is reduced.

Interpolation produces a scene in which the voxel size is different from that in the original scene. The densities assigned to the new voxels are estimated based on the densities of the original voxels. For example, for reasons to be explained later, we desire to have a *cubic scene*, ie, a scene in which each voxel is a cube. On the other hand, the voxels associated with a CT scanner are typically not cubic; eg, 0.5 mm × 0.5 mm × 3 mm.

In such a case we use linear interpolation to estimate the densities assigned to 0.5 mm × 0.5 mm × 0.5 mm voxels.

Segmentation produces a binary scene. The original and the resulting scenes are of the same size and are subdivided into voxels in the same way. One method of segmentation is *thresholding,* in which a voxel in the binary scene has density 1, if the density in the original scene is above a predetermined (threshold) value. For the illustrations given earlier (Figures 2.1 and 2.2), we were interested in the bony structures of our patient; we therefore chose the threshold so that it was below the densities of the voxels containing mostly bone, but above the densities of the other voxels. Thresholding is the simplest mode of producing from an arbitrary scene a binary scene of the same size. Other techniques are often required to deal with more complicated situations (these often arise when we wish to see soft-tissue surfaces or when the CT scans contain artifacts such as those due to fillings in teeth or breathing motion). Some methods of segmentation in the medical environment are described in Udupa [1982] and Trivedi et al [1985].

The binary scene produced by segmentation is a 3D array of 0's and 1's. This is called the *binary-array* representation of the scene.

For typical medical objects, the binary-array representation seems to be unnecessarily "detailed," since it contains large areas filled with only 0's or only 1's, each explicitly recorded. In the next two sections we discuss more concise ways of representing those binary scenes that are typical in medicine.

Segment Endpoint Representation

The *segment endpoint* (SE, for short) representation is a variant of the run-length representation, which has been used for the efficient transmission of images; see, for example, Foley and van Dam [1983] and Quinn, et al [1984]. The SE is a slice-by-slice, row-by-row representation. Each row of each slice in the binary scene is examined for voxels with density 1. Each contiguous run of 1-voxels in the row is referred to as a segment. The left and right endpoints of all segments are recorded in an increasing order. This representation, owing to its inherent simplicity, can be derived very easily from other representations of the binary scene [see Udupa, 1982; Merrill, 1973].

To be more specific, consider the single binary slice shown in Figure 2.3. Each row of this slice, which contains a 1-voxel, is represented by a sequence of integers as follows:

$$j,\ S_j,\ i_1^L,\ i_1^R,\ i_2^L,\ i_2^R,\ \ldots\ i_{S_j}^L,\ i_{S_j}^R \tag{2.1}$$

where j is the index of the row ($1 < j < J$), S_j is the number of segments in the row ($0 < S_j$), and i_s^L and i_s^R are the indexes of the voxels that are the left-end and right-end voxels of the sth segment, respectively ($1 < i_s^L$

FIGURE 2.3. A slice of the binary scene.

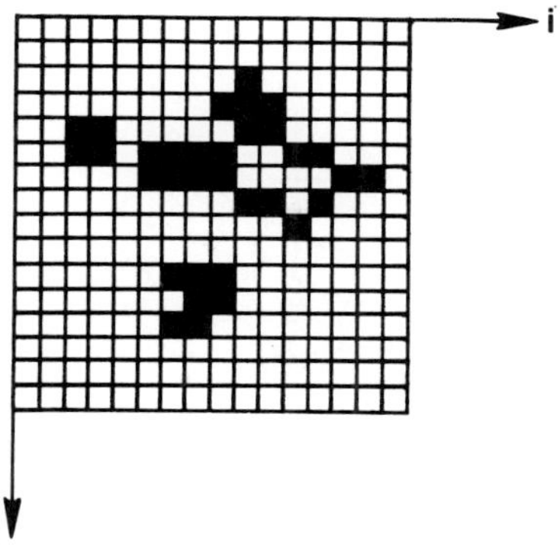

$< I$ and $1 < i_s^R < I$). The whole slice is represented by appending such sequences of integers in their natural row order. Thus, for the slice in Figure 2.3, the SE representation is

3, 1, 10, 10, 4, 1, 9, 11, 5, 2, 3, 4, 10, 11, 6,
3, 3, 4, 6, 9, 12, 13, 7, 2, 6, 9, 14, 15, 8, 2,
10, 11, 13, 13, 9, 1, 12, 12, 11, 1, 8, 10, 12, 1,
9, 10, 13, 1, 8, 9.

Note that a slice with no 1-voxels in it has an empty string of integers as its SE representation.

The SE representation of an $I \times J \times K$ binary scene is obtained by appending, for $k = 1, 2, \ldots, K$, the SE representations of the slices. This is preceded by a sequence of $K + 3$ integers

$$I, J, K, T_1, T_2, \ldots, T_K, \tag{2.2}$$

where I, J, K describe the size of the scene and T_k indicates where the information regarding the kth slice begins in the following sequence. Assigning zero to T_k is used to indicate that the kth slice has no 1-voxels in it.

It is clear that every binary scene can be described by such a sequence of integers and that if the binary array representation is full of long sequences of 0's and 1's, then the SE representation is much more compact. The SE representation has been used for various scene-processing operations such as region search, scene display, and scene manipulation as well as for quantitative analysis of scene regions [Trivedi, 1985b].

The Directed-Contour Representation

In this representation, objects are described by their borders in a slice-by-slice fashion. In each slice, each border of a 1-region is represented by a sequence of voxels, with the convention that as the border is traversed, the 1-region lies to the left (Figure 2.4). We call such a sequence

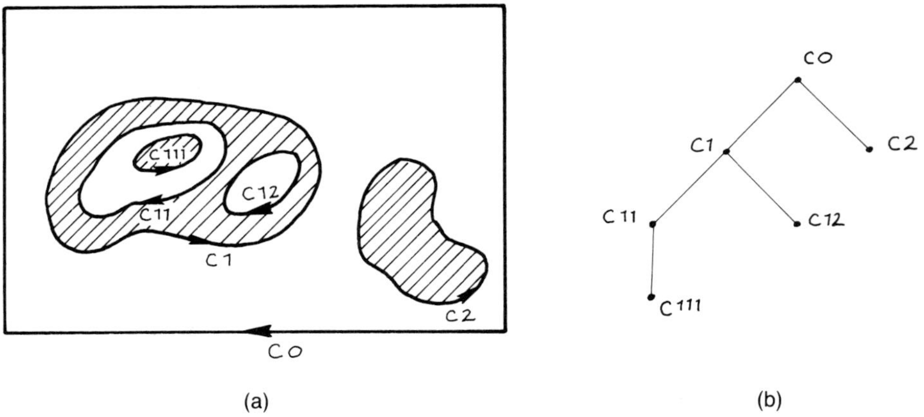

FIGURE 2.4. Illustration of the DC representation. **(a)** In each slice l-regions (hatched) are represented by directed contours such that the l-region lies to the left of the contour. **(b)** A containment tree is associated with the contours to describe the relationship of containment of contours inside contours.

of voxels with a directionality assigned to it a directed contour (DC). The DCs are chain-encoded [Freeman, 1974] so that they can be stored compactly. With each slice we associate a *containment tree* in which each node represents a contour, and its sons represent the contours that are "immediately inside." In the example in Figure 2.4, the border C0 of the slice represents the root node, and C1, C2 are its sons representing contours that are immediately inside C0.

In the previous description, the contours in the DC representation are defined to be the borders of regions. There are situations in which the user has to supply this information by drawing contours; for example, when the result of segmentation is unreliable, or when a part of the object has to be removed for better visualization. To be able to uniquely define the region represented by a set of contours in such situations, the contours should satisfy certain properties. Each contour should be "closed and non-self-intersecting" (see Figure 2.5), and no two contours in the same slice can intersect each other [Udupa, 1982]. Since the user-drawn contours do not always satisfy these properties, they should be appropriately processed to yield valid contours.

In producing the DC representation of a given binary array, one may choose to generate the nodes of the containment tree in either preorder or breadth order. Both approaches require two basic processes: (1) searching for a contour within as well as outside the region enclosed by a given contour, and (2) following a contour in a specified direction once its presence is detected. With reference to the example in Figure 2.4 in the preorder approach, first C1 is detected (assuming search for a contour

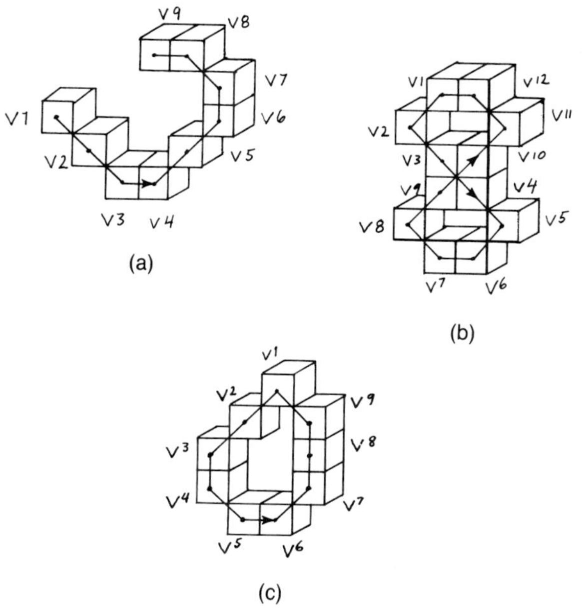

FIGURE 2.5. Illustration of the **(a)** "nonclosed," **(b)** "self-intersecting," and **(c)** "closed and non-self-intersecting" properties of DCs.

is made row-by-row starting from the top), then search is restricted to the region inside C1 to detect C11, and then to the region inside C11 to detect C111, etc. The recursion stops when a leaf node (C111) is reached. Nodes C12 and C2 are then expanded similarly. Roughly speaking, the complexity of the algorithm for the preorder and breadth-order approaches is a function of the height and breadth of the containment tree, respectively. Since, for medical objects, the height of the tree is likely to be much smaller than its breadth, the preorder algorithm is to be preferred [Tuy and Udupa, 1983; Chen et al, 1985a].

Given a DC representation, we can easily produce the equivalent SE representation as follows. For each voxel v in each contour we know, the relative location (ie, the chain code) of the previous voxel v' and the next voxel v''. The voxel v is called a v^+ voxel if its neighbor to the right lies outside the object region locally and a v^- voxel if its neighbor to the left lies outside the object region locally. These definitions are illustrated in Figure 2.6. In Figure 2.6a, v is v^+, since the 1-region that lies to the left of the contour segment $v'vv''$ does not include the voxel to the right of v. The contour may eventually turn around and include this voxel. Locally, however, this voxel does not belong to the object region. In Figure 2.6b, v is v^- and in Figure 2.6c v is both v^+ and v^-. In Figure 2.6d, since the voxels to the right and left of v are both inside the object region locally,

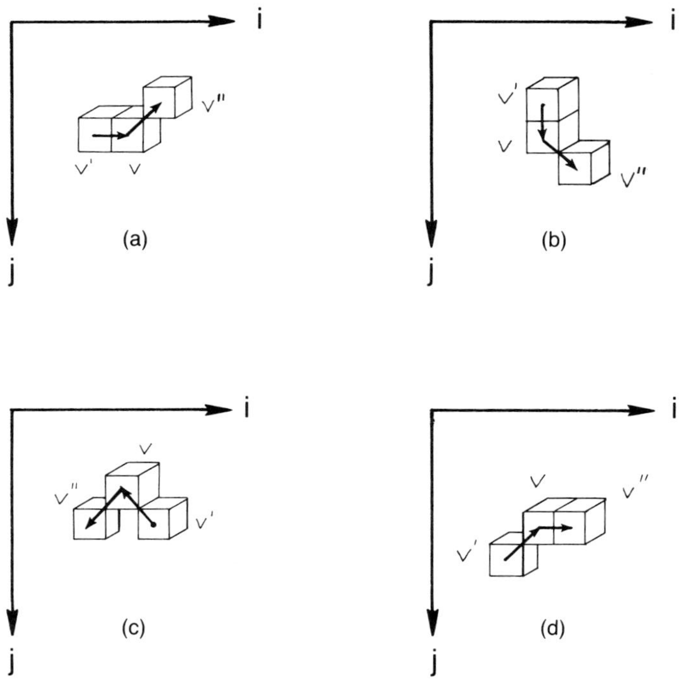

FIGURE 2.6. The voxel v is a v^+ in (**a**), v^- in (**b**), both v^+ and v^- in (**c**), and neither v^+ nor v^- in (**d**).

v is neither v^+ nor v^-. Clearly, since there are only 64 possible configurations of v' and v'' relative to v, whether a voxel in a contour is v^-, v^+, both, or neither can be determined by a table look-up.

To convert a given DC representation of a slice to an SE representation, a list of v^- and v^+ voxels is first prepared as just described. If a voxel is both v^- and v^+, it is entered twice in the list, once identified as v^- and once as v^+. The list is then sorted according to the j (row) coordinate. Within each sublist of voxels with the same j coordinate, sorting is done according to the i coordinate. Within each sorted sublist, the v^- and the v^+ voxels alternate (when a v^- and a v^+ have the same i coordinate within such a sublist, the order of placement is v^- followed by v^+). Within each sublist successive v^- and v^+ pairs represent line segments belonging to the object region. Conversion of such sublists into the form of Equation (2.1) is trivial.

The DC representation is extremely compact. For typical medical objects, it requires about five to eight times less storage than that required by the binary-array representation.

Objects and Their Surfaces

Even though human observers identify objects in 3D scenes without any apparent difficulty, the mathematically precise definitions of an "object" and its "surface" are far from trivial, and equally reasonable definitions can be (and have been) proposed, giving rise to different types of entities being identified as objects. In this section we survey some of the approaches.

First, one has to recognize that in most application areas the discrete 3D scene is considered only to be an approximation to the underlying domain of interest. Usually, we assume that there is a function $f(x,y,z)$ of three real variables, and the density assigned to a voxel is an estimate of the average (or central) value of $f(x,y,z)$ in the voxel. For example, in CT, $f(x,y,z)$ is the x-ray attenuation coefficient of the human body at the point (x,y,z). The object of our interest in this underlying continuous space will occupy some voxels totally, will miss some others, and will partially intersect a third set of voxels. In the process of discretization some information regarding this "true" object is irrevocably lost; nevertheless, recovery of its surface (or a good approximation to it) may be our heuristic aim.

We now discuss an approach to defining the surfaces of an object. Such a definition is unavoidably rather mathematical, since intuitive geometrical notions do not easily translate into precise linguistic descriptions without the aid of mathematics. However, the reader can get a good grasp of what we are aiming at (even without following the details of the next paragraph) by looking at the 2D illustration given in Figure 2.7. Of course, the concept that we are actually defining is three-dimensional (our object is some subset of 3D space); this makes our definition more complex than it would be in two dimensions.

We are dealing with ordinary 3D Euclidean space, R3. We use topological terminology in the standard way [eg, Newman, 1951, or the Appendix]. In particular, we talk about closed, open, and connected sets in R3. The object that we are interested in is a subset of R3. The complement of the object can be divided into a number of components. (*Components* are defined as connected subsets that are not contained in yet larger connected subsets; see Newman, 1951.) The *surfaces* of the object are defined as the frontiers of these components, one surface for each component. (*Frontier* of a set D is the set of points that are closure points both for D and its complement.)

Figure 2.7 gives a 2D example. Consider a subset Q (we assume it to be closed), which is indicated by leaning lines "/". The connected component labeled by E may be an object under a suitable definition. The leaning lines "\" indicate the complement CE of E. This open set CE has two components: the infinite one labeled D and the finite one G en-

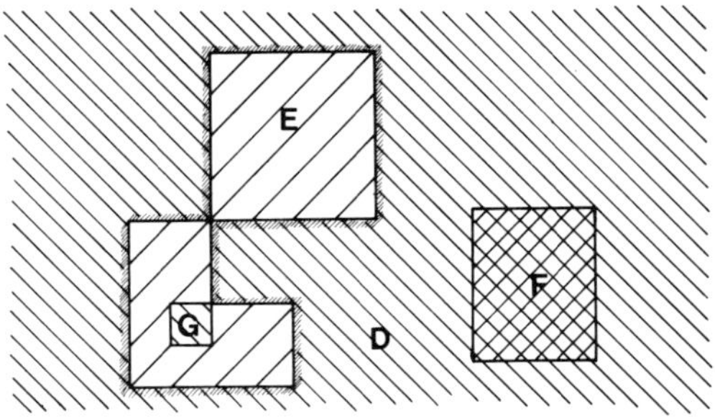

FIGURE 2.7. A 2D illustration of our definition of a surface. The closed set Q is marked by lines "/". It has two components, one is E and the other is F. The complement of E is the open set CE, marked by lines "\". Since F is a subset of both Q and CE, it appears cross-hatched. CE has two components, the open sets D and G. The "exterior surface" of E is the frontier of D, marked by the "hairy" line around E. E also has an "interior surface," namely, the frontier of G. (Based on Fig. 3 in Herman GT: Three dimensional imaging from tomograms, in Hohne KH (ed): *Digital Image Processing in Medicine*, Springer-Verlag, Berlin, 1981, pp 93–118. Reproduced with publisher's permission.)

closed by E. (These are separated, since Q, and hence E, was assumed to be closed.) The surface of E associated with D is the frontier indicated by the "hairy" line around E. The object E has one other surface—not indicated in the figure—associated with G, the other component of CE.

We have now described a definition of a surface that is usable with any definition of an object. This, of course, would lead to some very peculiar surfaces, unless we put some further restrictions on the nature of objects. Note that in the example given above, E is closed, bounded, and connected and that the surface we obtained corresponds to what one intuitively would describe as the "exterior surface" of E.

We now give a precise definition of an object in a discrete scene. We assume that a region is given and is already subdivided into voxels. In what follows we use the word voxel to mean a closed set of points in R3. In other words, a voxel includes all its faces. We put the following restrictions on the notion of an *object in the scene*.

Property 1: An object is the *union of voxels* (ie, for each point p in the object, there is a voxel v entirely contained in the object and containing p).

Property 2: An object is *connected*.

The boundedness of the region of the scene and property 1 imply that an object is a bounded closed subset of R3. Another consequence of property 1 and our definition of a surface is that a surface of an object in the scene is the union of a set of faces of voxels (closed sets including their edges), where each face is contained in exactly one voxel, which is inside the object. Property 2 and standard results of topology [Newman, 1951] imply that a surface of an object in a scene is a connected set. (Recall that as a result of our definition of a surface, one object may have several surfaces.) These ideas are illustrated in Figure 2.8.

These properties, which make no reference to the densities of voxels, can be used to exclude certain subsets of the region as objects in the scene, but tell us nothing about how to find subsets that are objects. Many different approaches can be, and have been, taken.

We restrict our attention to binary scenes. A possible definition is the following. Let Q denote the union of all voxels with density 1. We may then define an object as a connected component of Q. We refer to objects defined this way as *0-objects*. (The set E in Figures 2.7 and 2.8 is a 0-object according to this definition.)

In our own work we have found it more convenient to use the following definition. A *1-object* is a 0-object that cannot be described as the union of two 0-objects U and V whose common points are isolated. (Geometrically this means that we do not allow 1-objects to be connected by vertices only. They must be connected at least by edges, which are one-dimensional entities; hence the name 1-object. Note that the set E in Figures 2.7 and 2.8 is not a 1-object according to this definition, since it is the union of

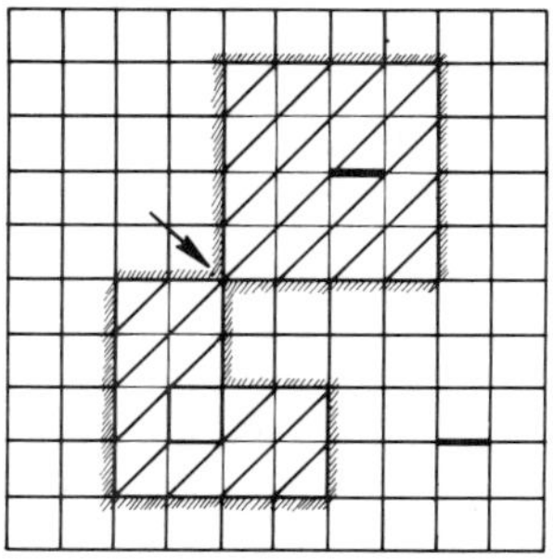

FIGURE 2.8. Two-dimensional illustration of our definition of an object (based on E in Fig. 2.7). The object is indicated by lines "/". Any "voxel" with a line in it belongs to the object entirely (including its edges and corners). In particular, the corner point indicated by the arrow belongs to the object and the whole object is connected (we can get from any point in the object to any other point in the object without leaving the object). Any edge (the 2D equivalent of the notion of a face) marked by the hairy line belongs to the "exterior surface" of the object entirely (cf Fig. 2.7), and the exterior surface is exactly the union of these edges, each one of which is contained in exactly one "voxel" inside the object. The two horizontal edges marked by the heavy lines cannot possibly be in a surface of the object; in one case because it is an edge of two different voxels inside the object and in the other case because it is not an edge of a voxel inside the object at all. Note also that the indicated surface is connected (we can get from any point in the surface to any other point in the surface without leaving the surface).

two 0-objects whose only common point is the one marked by the arrow in Figure 2.8.)

One can similarly define 2-objects (not allowing edge connectivity). For a discussion of such notions see Lobregt et al [1980] and Udupa et al [1982]. Similar distinctions exist in 2D pictures; see Rosenfeld and Kak [1982].

It can be shown [Herman and Webster, 1983] that in a binary scene, for any face F that belongs to exactly one voxel of density 1, there is a unique 1-object O and a unique surface S of O such that S contains F. We now discuss an algorithm that given F, finds S.

There is a complex definition [Herman and Webster, 1983] which, with any face F in S, associates exactly two other faces F1 and F2 in S, to which F is said to be *P-adjacent*. We do not repeat that definition here, but note that given F, it is computationally inexpensive to find F1 and F2.

The following algorithm has been first proposed by Artzy [Artzy et al, 1978]. Its input is a binary scene and a face F0 in a surface S. Its output is a list L of all the faces in S. It makes use of two auxiliary data structures, a queue X, and a list M. We use the verb "mark" to mean "put in the list M" and the verb "queue" to mean "put in the queue X."

```
BEGIN
1. Output and queue F0. Put two copies of F0 in M.
2. While X is not empty,
   a. remove a face F from X;
   b. determine the two faces F1, F2 in S to which F is P-adjacent;
   c. for i = 1, 2,
            if Fi is in M,
                    then remove it from M;
                    else queue, mark, and output Fi;
            end if,
      end for,
   end while.
END
```

It is proved in Herman and Webster (1983) that this algorithm terminates and that at termination the output list L contains precisely one copy of each face in S.

For a detailed discussion of the rather ingenious nature of this algorithm see Artzy et al [1981], Herman and Webster [1983], and Frieder et al [1985a]. The important aspect of it is in step 2c, where membership of Fi in M is checked, and Fi is removed from M if it is found there. This way the size of M is kept small, typically less then 2% of the output list L. The importance of this is easily appreciated when we consider that the size of L in typical medical applications is more than 150,000 and can be greater than 500,000.

This algorithm produces a surface based on the binary-array representation. Alternative algorithms exist for producing surfaces from the SE and DC representations of a binary scene. For a published example, see Udupa [1982].

Display

In the previous section we discussed four different ways of representing the information in binary scenes: the binary array, SE, DC, and as surface(s) of object(s) in the scene. Methods for displaying the information in the scene in a manner similar to that illustrated in Figure 2.2 have been developed for all four representations. In this section we discuss such methods. The relative advantages and disadvantages of the different representations from the display point of view are discussed later. Note also that since techniques exist to translate from one representation into another, under suitable circumstances it may be best to do such a translation prior to display. In what follows we concentrate on direct display based on the different types of representation.

The two basic issues associated with displaying a view of (an opaque diffusely reflecting) 3D object on a 2D screen are (1) ensuring that only that part of the object that is visible from the given view is displayed (visible surface display), and (2) assigning a shade to the visible surface to impart a 3D illusion (shading).

Display from the Binary-Array Representation

Work has been reported on display from binary arrays by Trivedi [1985a], Trivedi [1985b], and Frieder et al [1985b]. In this section we describe the so-called *back-to-front method* for displaying binary arrays.

In this method visible surface display is achieved by painting (and overpainting) on the screen images of individual 1-voxels in an order that is determined by their distance from the screen. To be more precise, we assume that the coordinate system (x,y,z) of the 3D scene is chosen in such a way that the axes are parallel to the planes that subdivide the region of the scene into voxels and that the x, y, and z directions correspond to the i, j, and k indexing, respectively, of the voxels. Thus, for example, the z coordinate is the same for all voxel centers with the same k index. We think of the scene as movable in space and introduce a fixed coordinate system (u,v,w) attached to the 2D screen on which we wish to display (u, horizontal on the screen; v, vertical on the screen; and w, perpendicular to the screen). We call this the *image-space* coordinate system. For any desired view of the 3D scene, there is a 3×4 transformation matrix T (representing translation, rotation, and scaling of the scene) such that any point (x,y,z) in the original scene can be expressed in terms of the image-space coordinates according to

$$\begin{aligned} u &= T_{11}x + T_{12}y + T_{13}z + T_{14}, \\ v &= T_{21}x + T_{22}y + T_{23}z + T_{24}, \\ w &= T_{31}x + T_{32}y + T_{33}z + T_{34}, \end{aligned} \tag{2.3}$$

see Figure 2.9. In our work we display a scene using orthographic projection; this means that the point (x,y,z) projects (if it is visible) onto the point (u, v) on the screen and that its distance from the screen is w.

Eliminating x and y from Equation (2.3), we get

$$w = \alpha_1 + \beta_1 z, \tag{2.4}$$

where α_1 and β_1 are functions of T_{ij} and the position (u,v) of the point on the screen. Thus, for a given point on the screen, the distance w from the screen either decreases or increases with the coordinate z. If the viewing direction is such that w increases with increasing z, then a point in a voxel in slice k cannot hide a point in a voxel in slice k-1, and thus voxels should be painted on the screen in order of decreasing k index to achieve visible surface display. Similar results can be shown for the i and j indexes as

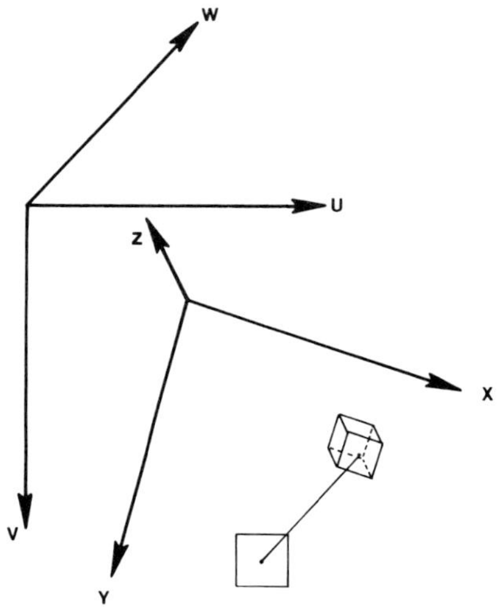

FIGURE 2.9. The 3D scene viewed in image space. A single l-voxel and its "projection" on the screen ($w=0$) are shown.

well. Thus, visible surface display is achieved by three nested loops through the indexes, either in increasing or decreasing order, depending on the desired view. In our implementation the outer loop is through slices (k index), the middle loop is through rows (j index), and in the inner loop we paint onto the screen one-by-one all the 1-voxels in that row (i index).

Painting of a single 1-voxel is done as follows. First note that for a fixed view, the projections of any two voxels onto the screen have the same shape. Using Equation (2.3), we work out the size of a rectangle (with edges parallel to u and v) that just encloses this shape. We paint a 1-voxel by painting on the screen such a rectangle located around the point (u,v) determined by Equation (2.3) and the coordinates (x,y,z) of the center of the voxel (see Figure 2.9). The brightness B assigned to the rectangular area is determined by the value of the distance w obtained from Equation (2.3) according to the following rule (called *distance only shading*):

$$B = \frac{255 * (d_2 - w)}{d_2 - d_1}, \tag{2.5}$$

where d_2 and d_1 are precalculated values such that for all views and all voxels $d_1 < w < d_2$.

This method of display does not require the binary scene be cubic. In Figure 2.10 we show the image of a spine obtained using this method. The CT data provided a 160 × 144 × 30 scene with voxel size 0.4 mm × 0.4 mm × 1.5 mm. The time required to display each 256 × 256 view of this scene is 111 sec (84 sec are needed for a 128 × 128 picture). Image quality is mediocre but acceptable for a rapid preview of the scene. The same procedure can be used to display the object after interpolating the scene so that voxels are cubic. Figure 2.11 shows the same spine displayed after interpolation (scene size now is 160 × 144 × 109), demonstrating a considerable improvement. The time taken for this 256 × 256 picture is nearly 4 min (130 sec are needed for a 128 × 128 picture).

The times quoted here refer to our current implementation on an Eclipse S/200 minicomputer. Since this is a program under development, we expect considerable improvements. For example, it was noticed that a large fraction of the time required to display the object was wasted by the painting of interior voxels (ie, 1-voxels in the scene that are not visible from any direction). Many of the interior voxels can be easily identified by the fact that the six face adjacent neighbors are also 1-voxels. Setting the density of such interior voxels to zero prior to displaying the scene speeds up the process without any loss of image quality. For example, altering the binary scene in this fashion reduced the time required to produce Figure 2.11 from nearly 4 min to less than 1 min, of which only 15 seconds were needed for the display; the rest was needed for the process of identifying interior voxels.

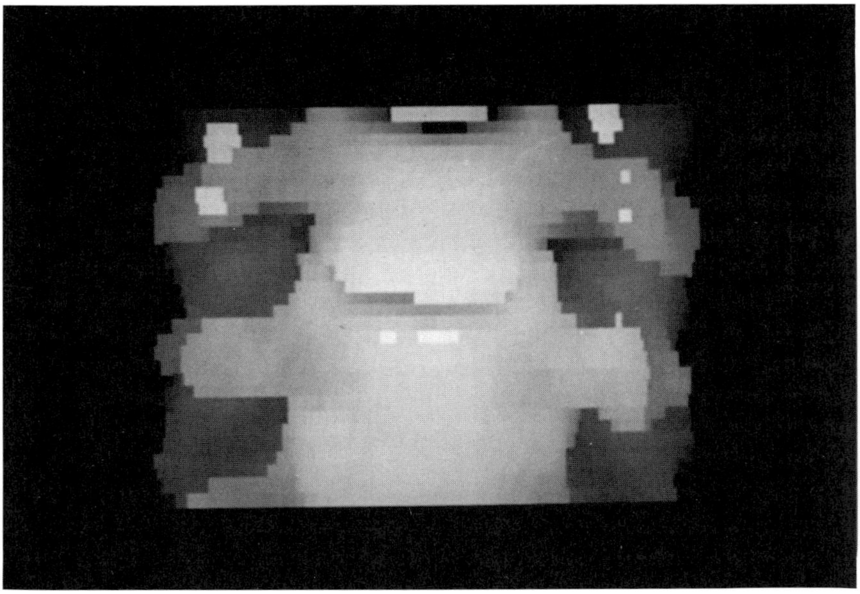

FIGURE 2.10. Display of a human spine from binary-array representation derived from an uninterpolated scene. (Reproduced with permission from Trivedi SS: Interactive manipulation of three-dimensional binary scenes, (in *The Visual Computer,* 2:209–218, 1986.)

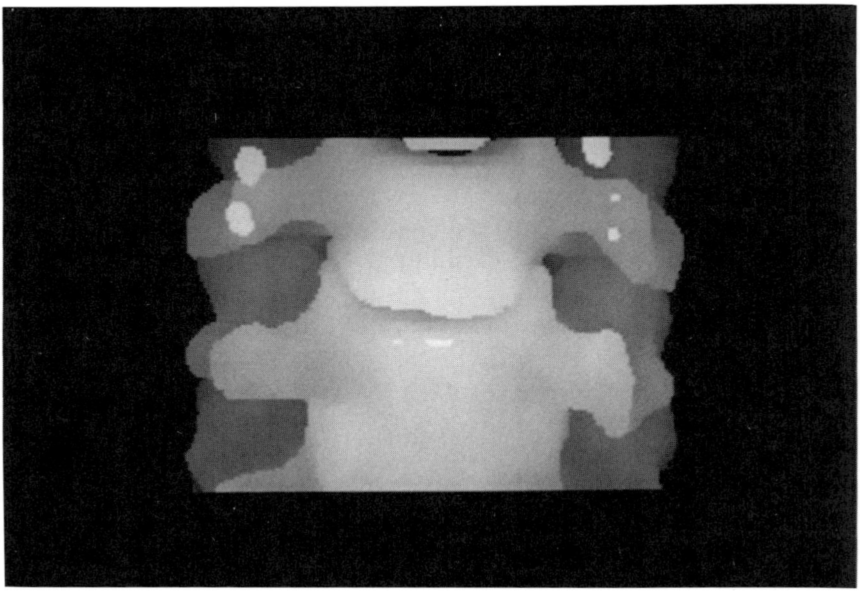

FIGURE 2.11. Display of a human spine from binary array representation derived from an interpolated scene. (Reproduced with permission from Trivedi SS: Interactive manipulation of three-dimensional binary scenes, in *The Visual Computer,* 2:209–218, 1986.)

Display from the Segment End-Point Representation

The back-to-front method can be implemented in the SE representation by simply changing the inner loop on the *i* index. The need for checking the individual voxels is eliminated by simply painting all voxels between the endpoints of the segments. The images obtained using SE representation are identical to those obtained from the binary-array representation and hence are not reproduced here.

An interesting variation of this implementation arises from the fact that by an insignificant restriction on the viewing direction, we can ensure that all voxels in a segment project onto the same scan lines in the image. This means that the geometrical transformation needs to be performed only on the endpoints. The shade assigned to points between the projections of the segment endpoints is obtained by linear interpolation of shades assigned to the two endpoint projections. For details see Reynolds et al [1985].

Display Based on Directed Contours

If high-quality displays or arbitrary views of objects are desired, we translate the DC representation into a surface representation and use the method discussed in the next section. However, that approach is not suitable for an interactive environment because of the lack of immediacy of response. The method to be outlined below achieves speed at the expense of the quality of images and the generality of viewing. Since the display is used here only to guide the user to indicate the nature of manipulation, the quality of display is considered secondary. Our algorithm is based on the following four assumptions and observations.

1. We assume that the viewing direction is obtained by rotation around the axis perpendicular to the slices (Figure 2.12). Under this assumption,

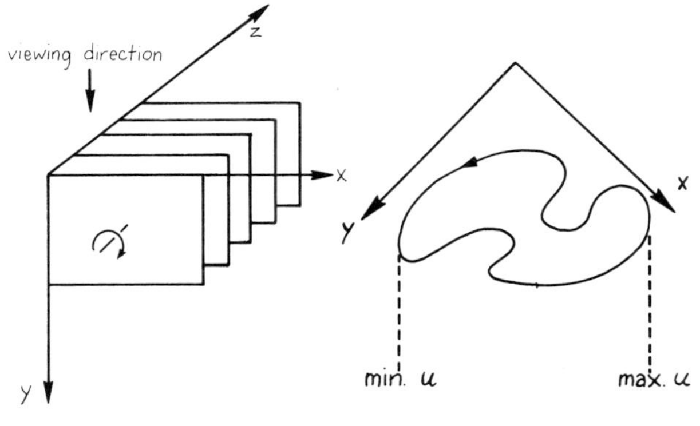

FIGURE 2.12. Illustration of the viewing mode for the DC-based display algorithm. All points with the same value of *z*, project onto a line with the same value of v (cf Fig. 2.9).

only about half of a contour needs to be processed (the part lying between the points where the global maximum and the global minimum u-coordinates occur for the rotated contour). Since the "inside" of the object lies to the left of the directed contour, only those voxels on the contour that are encountered while traversing in the direction of the contour starting from the voxel with the maximum u-coordinate and ending with the voxel with the minimum u-coordinate need to be displayed.

2. Only those contours corresponding to the sons of the root of the containment tree need to be considered for display (cf Figure 2.4).
3. A pair of consecutive voxels of a given contour is considered to represent a segment of the contour. The display algorithm determines visibility in a segment-by-segment fashion. A segment is potentially visible only if the u-component of the direction vector corresponding to the rotated segment is negative. For example, in Figure 2.13, segment v_2v_3 with direction vector 3 is visible, while v_1v_2 with direction vector 1 is not visible. Since for any given orientation of the object there are only eight possible orientations of the segments, their potential visibility can be determined by table look-up.
4. The rendering of the part of the object in two different slices project onto two disjoint sets of scan lines. This implies that the object can be processed slice-by-slice, and at the same time, the image can be generated line-by-line. This nicety completely removes the need for the usually time-consuming memory management required to handle input and output.

The actual rendering algorithm uses a z-buffer mechanism for removing hidden parts [see Herman and Liu, 1979; or Foley and van Dam, 1983]. Shading is assigned by the distance-only formula in Equation (2.5). For an illustration of display using this method see Figure 2.19 and "Manipulation Based on Directed Contours," which appears later in this chapter.

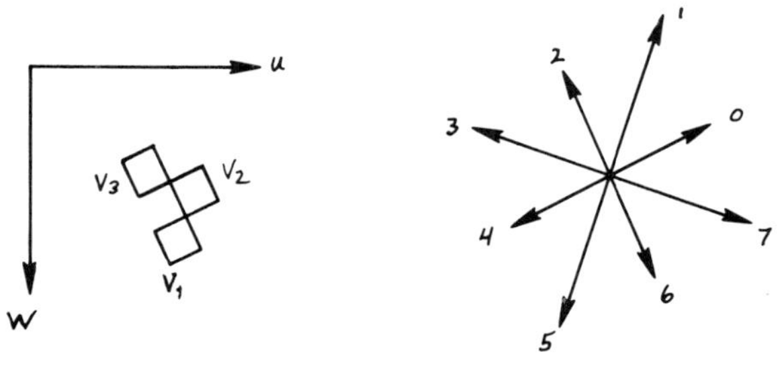

FIGURE 2.13. The potential visibility of contour segments is determined based on the orientation of their direction vectors. For the chosen viewpoint, the segment v_2v_3 with direction vector 3 is visible, while the segment v_1v_2 with direction vector 1 is not visible.

The generation of each image by this approach takes about 2 to 4 sec for small objects, 5 to 14 sec for typical medical objects, and 15 to 35 sec for large objects on an Eclipse S/200 16-bit minicomputer.

Surface Display

Property 1 of objects in a scene results in the surface being a union of voxel faces. Such a surface is unrealistic; "real" objects are not unions of parallelepipeds. One can, in principle, make the voxels arbitrarily small, using interpolation, but that can increase computer costs associated with surface detection and display to an unacceptable level in a medical environment where the patient has to pay for the services rendered.

In our approach to inexpensive but nevertheless realistic surface display in the medical environment, we make use of the fact that the simple nature of the surface elements in a cubic scene allows us to prove powerful mathematical results. Such results lead to efficient computer procedures to handle the large amount of data. An example of such a result, leading to efficient surface tracking, has been given in the section, "Objects and Their Surfaces." Another example (related to surface display) is a theorem [Herman and Liu, 1979] stating that in a cubic scene the distances from an observer of two points A1 and A2 on the surface are ordered in the same way as the distances from the observer of the centers of the faces that contain A1 and A2, respectively. Such a theorem allows rapid determination of the visible part of the surface.

The interpolation that resulted in the cubic scene in our medical examples has been used partly so that we can apply the rapid computer graphic display procedures, which can be applied only to cubic scenes. The other reason is that we believe that it results in a more accurate approximation to the "true" surface than what would have been obtained by using the alternative of thresholding the original scene and then displaying the resulting surface. The surface produced by this alternative method would have a different "resolution" in one direction from that in the other two. The use of original densities for interpolation and the thresholding of the interpolated scene is likely to result in a more accurate approximation to the true organ surface than the alternative. (This has been demonstrated for the binary-array-based display in Figures 2.10 and 2.11.) This is a statement of our personal belief, which is not shared by all workers in this field. Since it is easy to create single objects both for supporting and for refuting this claim, its validity can be resolved only by long, practical experience in the field of application.

At the early stages of our work the quality of the displays based on surfaces in cubic scenes was not quite as good as we desired, especially if they were displayed in a dynamic sequence. A number of techniques have been suggested to deal with this [Herman, 1980]. Recently, we proposed an approach that improves the quality of the displays, but retains those properties of the approach that make it computationally efficient [Herman and Udupa, 1981; Chen et al, 1985b]. In this new method, visibility of a point on the surface is determined in the same fashion as in the previously published approaches [Artzy, 1979; Herman and Liu, 1979]; that is, using a variant of the z-buffer algorithm, which is very rapid for displaying surfaces in a cubic scene. It is the shading of the visible points that is to be done differently. The convention of assigning a single shading value to all points of a face is retained; however, the value depends on the direction cosines of all the faces having an edge in common with the given face. We call this *contextual shading*. Additional computation for contextual shading is minimal, since all necessary information regarding the faces sharing an edge with the given face can be gathered while the boundary surface is tracked and can be stored as attributes of the face in question. Striking improvements in the quality of displays owing to contextual shading have been illustrated by Herman and Udupa [1981] and Chen et al [1985b].

Figures 2.2a and b were produced using contextual shading. A further example is shown in Figure 2.14. The data for this figure are as follows. The 3D binary scene is 141 × 186 × 251. The surface consists of 296,970 elements. Surface tracking took 16 min of elapsed time on our Eclipse S/140 16-bit minicomputer. Displays of the detected surface, such as the one shown in Figure 2.14, took 2 min per image to generate. (Images are

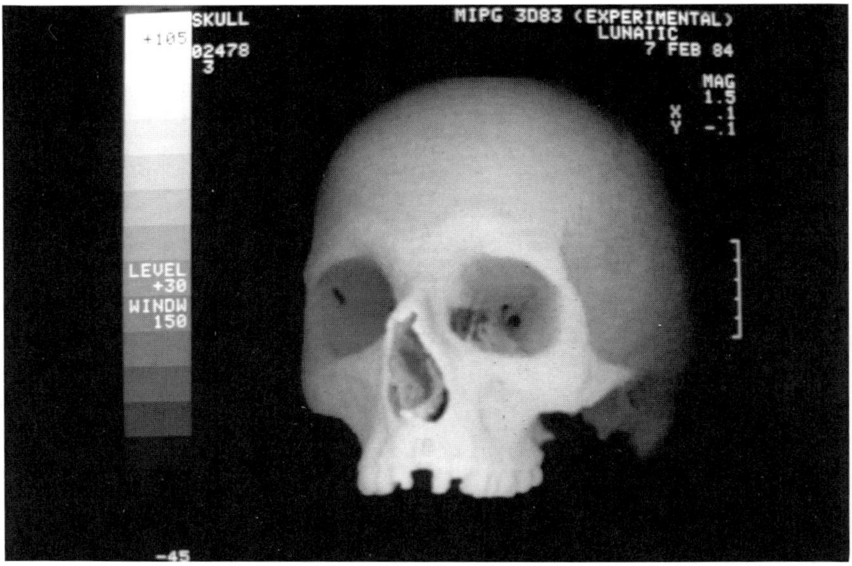

FIGURE 2.14. Surface-based display from x-ray CT data of a human skull specimen using the contextual shading method of Chen et al [1985b].

generated at a 512 × 512 resolution but displayed at 256 × 256 using averaging to reduce sampling affects.)

Manipulation

The result of manipulation is an altered 3D scene. In what follows we describe approaches to manipulation in each of the binary-array, SE, and DC representations. Since in going to a single external surface description of an object (as described earlier) information about the internal structures of the scene is lost, we do not consider the manipulation of objects described by their external surfaces.

Different representations of the scene are best suited for different types of manipulation from the points of view of computational ease and user convenience. In what follows we describe the techniques that we are currently using. Because of the existence of programs to translate one representation into another one, we can move back and forth between the different techniques for manipulation (and display), but this, of course, involves a computational overhead.

As mentioned in the introduction, the intended applications of 3D manipulation are (1) cutting away obscuring structures so as to improve visualization, (2) isolating a specific structure for either independent viewing or quantitative measurements, and (3) simulating specific surgical procedures such as the supraorbital reshaping illustrated in Figure 2.2. As will be seen, our current software fulfills the needs of the first two applications, but not yet of the third. For an alternative approach that has made quite a stride toward the fulfillment of the needs of the third application see Marsh et al [1985].

Manipulation in the Binary-Array Representation

Most of the manipulative operations of interest can be described as one or a combination of the following four operations [Brewster et al, 1984]: addition, subtraction, intersection, and independent motion of object components in the scene. Currently we have working programs to do the first three operations, but not the fourth. We therefore restrict our discussion to addition, subtraction, and intersection.

In each case we use the following terminology. We refer to the 3D scene to be modified as the primary scene P. The modification is done using a secondary scene S, which is a binary scene of the same size as the primary scene. The modified scene M is defined as follows. For addition,

$$M(i,j,k) = \begin{cases} 1, \text{ if } P(i,j,k) = 1 \text{ or } S(i,j,k) = 1, \\ 0, \text{ otherwise,} \end{cases} \tag{2.6}$$

for subtraction,

$$M(i,j,k) = \begin{cases} 1, \text{ if } P(i,j,k) = 1 \text{ and } S(i,j,k) = 0, \\ 0, \text{ otherwise,} \end{cases} \tag{2.7}$$

and for intersection,

$$M(i,j,k) = \begin{cases} 1, \text{ if } P(i,j,k) = 1 \text{ and } S(i,j,k) = 1, \\ 0, \text{ otherwise.} \end{cases} \tag{2.8}$$

Clearly, once the secondary scene S is specified as a binary array, the operation to produce the modified scene is trivial. Also, once the modified scene is produced, it can serve as the primary scene for a further modification, and so on until the final result of the manipulative process is obtained.

We now discuss our way of specifying the secondary scene. We first display the primary scene in an appropriate orientation. (What is "appropriate" depends on the secondary scene we wish to define, as will become clear later.) Let T be the matrix associated with this orientation of the primary scene. Now we draw a simple, closed connected contour on the screen (using a trackball or some other type of input device). Further we specify a minimum and a maximum depth (w_{min} and w_{max}, respectively) in image space (cf Figure 2.9). The secondary scene is defined as follows: a voxel with center at (x,y,z) in object space is assigned the value 1 if after the mapping defined by T, as in Equation (2.3), the value of w is between w_{min} and w_{max} and the point (u,v) falls within the contour we have drawn. In all other cases a voxel in the secondary scene is assigned the value 0.

In our implementation of the binary-array approach, the secondary-scene generation takes as much time as generating one view of the object for display. Once the secondary scene is defined, the production of the modified scene is very inexpensive. This is demonstrated using the primary scene shown in Figure 2.11. Suppose that we wish to remove the overhanging spinous processes for better viewing the neural foramina. (Neural foramina are the openings in the spine through which the central nervous system links to the rest of the body. A bony encroachment into the neural foramina may cause severe discomfort to the patient. In 3D imaging of the spine, the viewing of these openings are obstructed by the overhanging spinous processes.) Using the interactive manipulation techniques described here, we can "scoop out" the overhanging processes, as shown in Figure 2.15. Figure 2.15a shows the overhanging processes that obstruct the viewing of the neural foramina. These processes have been scooped out by subtraction of a secondary object defined by the contour shown in the figure. In the modified scene we have an unobstructed view of the neural foramina as demonstrated in Figure 2.15b.

Secondary object generation in this case took 145 sec and the manipulative operation of subtraction, 25 sec. This is the same as the time that would have been required for an addition or an intersection operation.

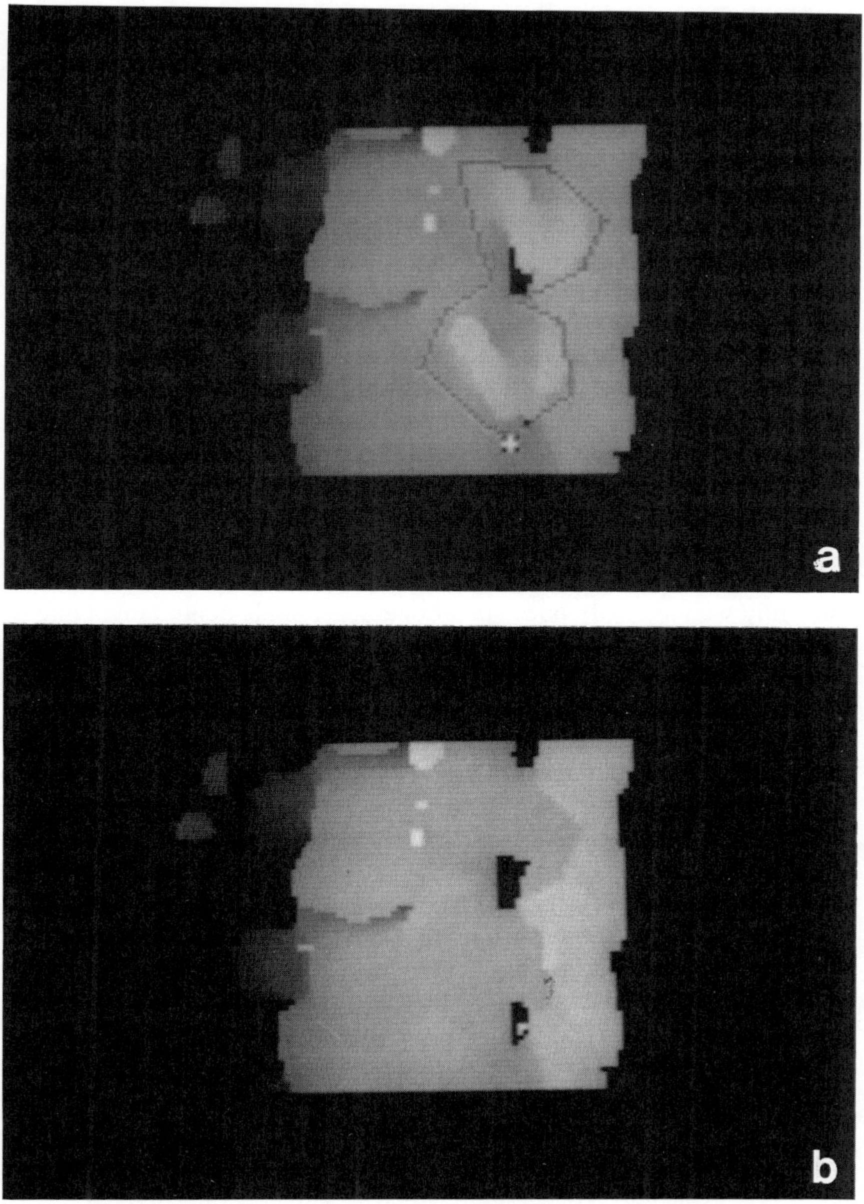

FIGURE 2.15. Demonstration of binary-array-based manipulation: (**a**) The same spine that is shown in Fig. 2.11 is displayed in a different orientation. The region to be "scooped out" is indicated by a contour. (**b**) The frontal bony processes have been "scooped out," leaving us with an unobstructed view of the neural foramina. (Reproduced with permission from Trivedi SS: Interactive manipulation of three-dimensional binary scenes, in *The Visual Computer,* 2:209–218, 1986.)

Manipulation in the Segment Endpoint Representation

We have working programs that given the SE representations of a primary scene *P* and of a secondary scene *S*, produce directly the SE representation of the modified scene *M*. These programs work quite rapidly for objects describable by a few segments in each row, but become time consuming as the number of segments increases. At some point it becomes more efficient to convert the SE representations of *P* and *S* into binary-array representations, to perform the required manipulation to produce the binary-array representation of *M*, and then to convert this into the SE representation of *M*. We have found that the medical objects we have to deal with are complex enough to justify this conversion-based procedure, and so we do not give the details of the direct SE-based operations. These are useful in other areas, such as the design of man-made objects, where the number of segments per row tend to be fewer than that in medicine. For details of direct SE-based manipulation see Trivedi [1985b].

Manipulation Based on Directed Contours

Currently we have only one type of operation implemented involving the DC representation: intersection of the primary scene *P* with a secondary scene *S*. (Subtraction can be effectively achieved by intersecting with the complement of *S*.) We have two different ways of defining *S;* one is using interaction with the slices of *P* (a 2D technique) and one is using 3D displays of P (a 3D technique).

In the *2D technique* [Udupa, 1983] we define the secondary object *S* by actually drawing (using a trackball and a cursor) the DC representation of it onto the display screen, which contains an image of the corresponding slice of the primary object. Each user-drawn contour has to be processed to ensure that it is "closed and nonintersecting." If this processing fails, the user is prompted to redraw the contour. The DC representation of *S* is then translated into a binary-array representation and is intersected with the binary-array representation of *P* according to Equation (2.8).

Two examples of 2D interactive manipulation are given in Figures 2.16 and 2.17. For both examples the data were obtained by x-ray CT. In Figure 2.16 the objective is to examine the internal articular surface of the acetabulum for possible fractures or other abnormality so that a proper surgical corrective procedure can be planned. In Figure 2.17 the objective is to compute the volume of the two orbits. The volume of the abnormal orbit is compared with that of the normal one so as to estimate the surgical correction to be undertaken. Volume computation is repeated postoperatively to judge the effectiveness of the corrective surgical procedure.

The main drawback of 2D interactive manipulation is that drawing contours takes a lot of user time. By carefully designing the software, it is possible to keep the user involvement in the process to a minimum. For

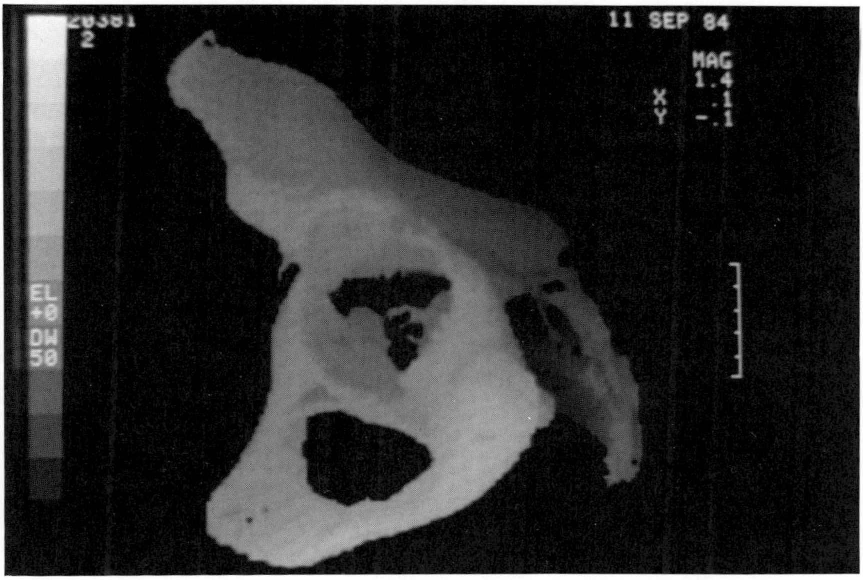

FIGURE 2.16. Visualization of the articular surface of a hip joint separated using 2D interactive manipulation. The acetabulum is rotated after separation so that the internal fracture is clearly visible and the resulting view corresponds closely to what a surgeon gets during the actual operative procedure. (Reproduced with permission from Herman GT, Kemp S: A promising new imaging technique, *Diagnosis* 7:89–91, November 1985.)

example, since the information to be supplied by the user is not likely to change significantly for nearby slices, reproducing automatically the information from the previous slice is often likely to suffice. Ideally, the duration of user involvement should not be more than the time required by a physician to mark the region of interest on a film of a sequence of CT slices. At a loss of generality, it is possible to tune the software to a specific application with significant reduction in the user time requirement. Two-dimensional interaction will probably remain indispensable for applications involving the independent visualization/quantitation of overlapping structures. Applications for which 2D techniques are cumbersome and 3D techniques are more natural and effective are those related to the simulation of surgical procedures.

In the *3D technique* we first display the object at an appropriate orientation on the screen using the direct display method outlined earlier. A "cutting plane" is then indicated by drawing a straight line on the display, the assumption being that the cutting plane is perpendicular to the screen. The secondary scene S is defined by assigning 1 to all the voxels on one side of the cutting plane and 0 to all the voxels on the other side.

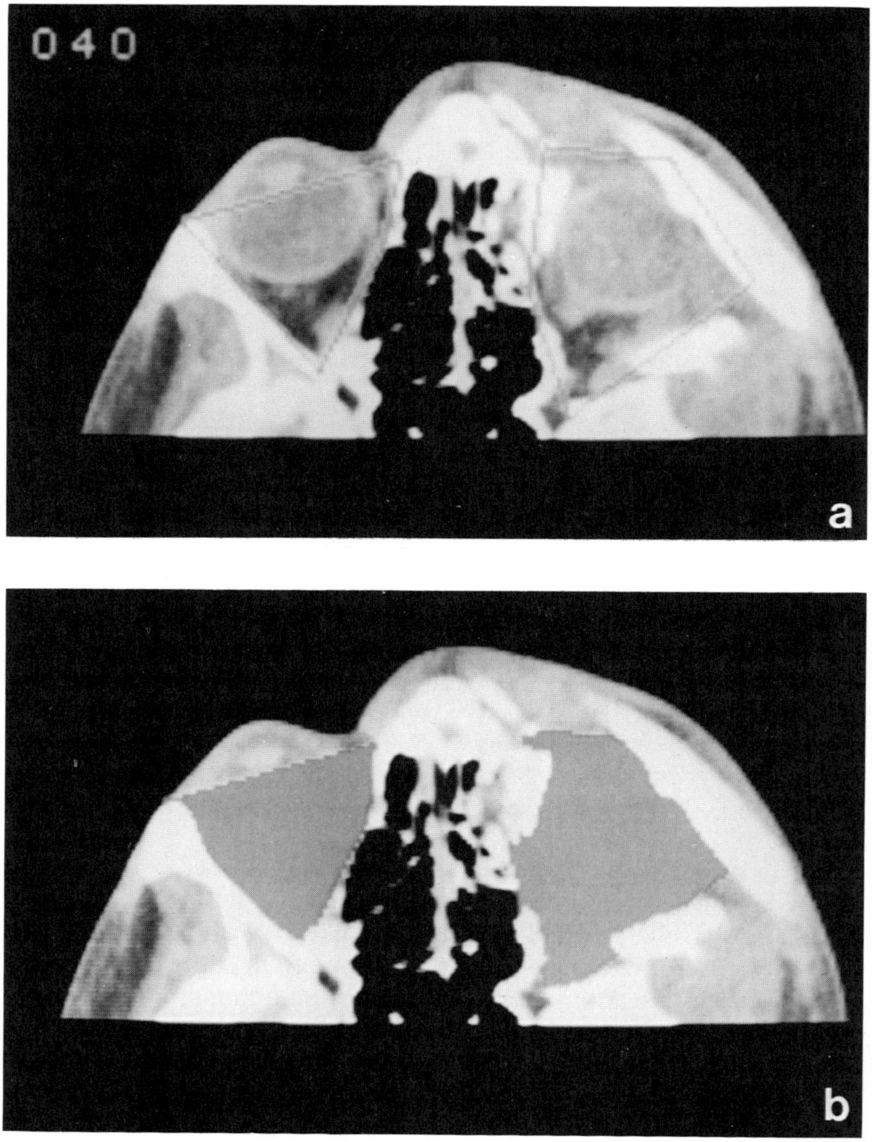

FIGURE 2.17. Computation of the orbital volume using 2D interactive manipulation. **(a)** The user roughly outlines the orbital region on individual slices. **(b)** The interactive software computes the actual orbital region (shown in uniform grey) by restricting thresholding to the user-specified region. The volume of the orbit is output as a by-product when the surface detection algorithm (see text) is applied to the resulting binary array [Udupa, 1981]. See Color Plate VII. (Reproduced with permission of National Computer Graphics Association from Udupa JK: Display and analysis of 3D medical images using directed contours, *NCGA '85 Technical Sessions Proceedings*, Dallas, Texas, vol III, pp 145–155, 1985.)

The DC representation of the modified scene M (defined as the intersection of the primary scenes P and S) is determined from the DC representation of P as follows. Based on the orientation of the scene in the 3D display [the matrix T of Equation (2.3)], we can determine for each slice of P the position of the line L in which the cutting plane meets that slice. Then the modified DCs are determined for that slice based on the points of intersection of L with the contours in the slice. Let P and P' be the points of intersection of L with the border of the slice (see Figure 2.18). Starting from P we traverse along L (towards P') until a point of intersection ($P1$) is reached. Now we traverse along the contour just intersected in the direction of the contour until another point of intersection ($P2$) is reached. Now traverse along L toward $P1$. Continue alternating traversal between L and the contours until $P1$ is reached again. The path traversed from $P1$ back to $P1$ ($P1$ $P2$ $P3$ $P4$ $P1$) represents a modified DC for one side of L. The procedure of traversal should be repeated from other points of intersection (such as $P5$) that have not yet been visited. All contours traversed during such a procedure together with contours not intersected by L ($C2$) and lying on the side of L in question constitute one set of modified DCs for the slice under consideration. To compute the modified contours for the other side of L, the entire procedure should be done commencing the search from P'.

Figure 2.19 illustrates the 3D interactive manipulation procedure on the cranial-vertebral junction of a patient.

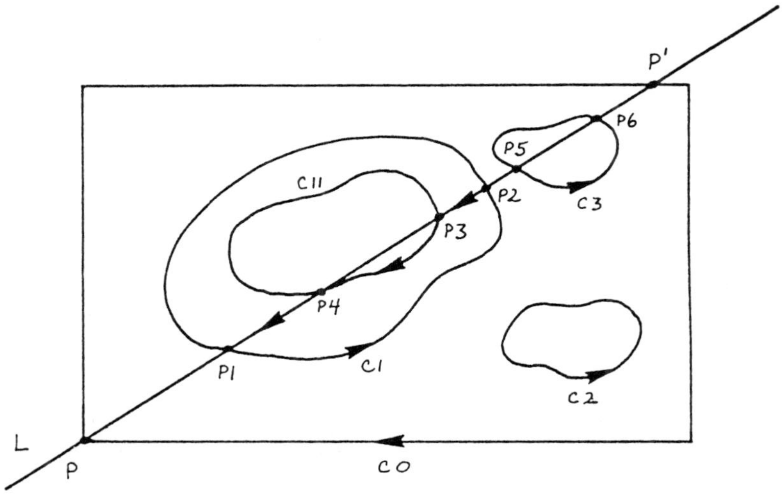

FIGURE 2.18. Illustration of the 3D interactive manipulation algorithm. For each slice the algorithm first determines the line L of intersection of the cutting plane with that slice, and then it computes the modified contours representing the subobjects lying on the specified side of the cutting plane.

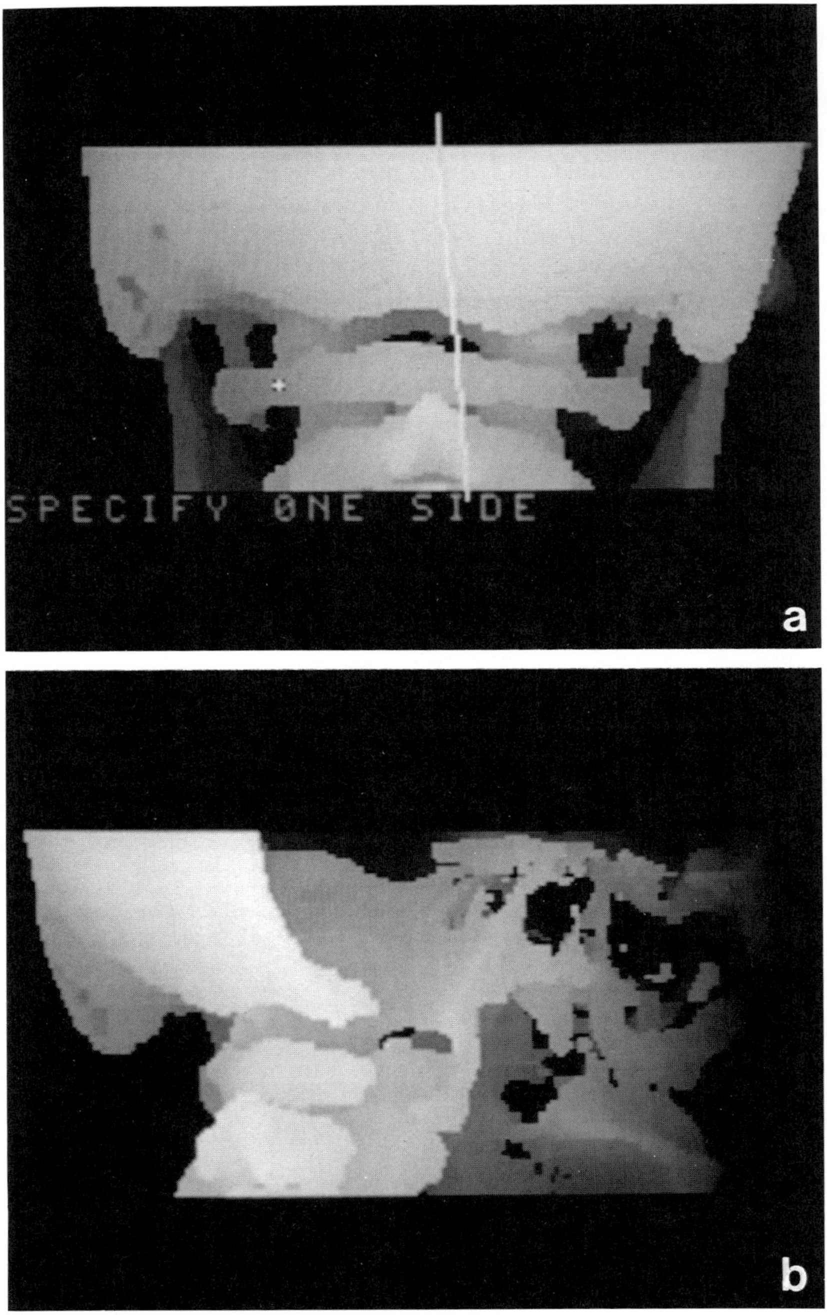

FIGURE 2.19. Illustration of the 3D technique for manipulation in the DC representation. (**a**) A DC-based 3D display of the cranial-vertebral junction of a patient from x-ray CT data, with the cutting plane indicated by the line. (**b**) and (**c**) The resulting sub-objects displayed after a slight rotation in opposite directions so that the internal structures can be examined.

FIGURE 2.19. *Continued.*

Computing time for 3D manipulation is similar to that of the direct display procedure using DCs, ie, 5 to 14 sec for a typical medical object. For example, for the object shown in Figure 2.19a (a 220 × 240 × 100 scene), the DC-based display was created in 11 sec, and the creation of the DC representations of both of the two halves shown in Figure 2.19b took 12 sec.

The major advantages of the 3D technique are its naturalness and speed, which are well suited to surgical simulation studies. An important drawback, however, is that since the original gray-level information is not available in the 3D display, the identification of the boundary between two similar and proximal structures (which appear to be fused in the 3D display because of inadequate resolution, poor segmentation, or partial volume effect) is more difficult than on slices. Hence, the 3D manipulation technique is unlikely to be of much use in the independent visualization of overlapping structures such as the articular surfaces of joints.

Discussion

The four representations of 3D scenes each have their advantages and disadvantages. Table 2.1 summarizes them. The most important comment to make is that the table has to be taken with a very large grain of salt.

In addition to its vagueness, it is based on currently operational software of one particular research group. Our own evaluation of the relative merits of these data structures is constantly changing, and other groups have found innovative techniques that shed new light on their usefulness. We now discuss the merits of each of the data structure in a little more detail and point at samples of relevant work by others.

The binary array, the segment endpoint, and the directed contour data structures are three interchangeable ways of describing the same information. In going from these to a single external surface some of the information is lost, but we make explicit information regarding the connectedness of the surface and the information needed for contextual shading. This is why we stated "not strictly comparable" under storage in the table. For typical medical objects, such as the ones shown in this chapter, our representation of external surface requires as much storage as the binary-array representation.

Even though the binary-array data structure leads to a slow display speed using our software display technique, it is used for real-time display by a specially designed hardware processor [Goldwasser and Reynolds, 1983; Goldwasser, 1985]. Even in the software implementation, one must bear in mind that in most medical applications (eg, CT), the binary array is the natural data structure for the device that collects the information. In such cases, the low speed of the binary-array-based display and manipulation is somewhat compensated for because no overhead (such as surface detection or DC creation) is involved in getting from the binary array to the image. For displaying one view of an object, this approach compares reasonably well with the other techniques. However, it is unlikely that a single view would suffice in the medical situation. The real advantages of the binary-array-based software approach are therefore its power to display arbitrary views and to manipulate the object in more sophisticated ways.

The mediocre quality of the images produced from the binary-array, SE, and DC representations is because these techniques use the distance-only shading rule of Equation (2.5), as opposed to the more sophisticated

TABLE 2.1. Summary of 3D scene representation methods.

	Binary array	Segment endpoint	Directed contour	Object surface
Storage	Inefficient	Efficient	Very efficient	Not strictly comparable
Speed of display	Slow	Medium	Fast	Medium
Image quality	Mediocre	Mediocre	Mediocre	Excellent
Manipulation capability	General but slow	Very inefficient	Limited but fast	None

contextual shading used to display surfaces. Contextual shading of surfaces is made convenient by the fact that the contextual information for a given face in the surface (the relative orientations of the faces in the surface that share an edge with the given one) can be easily obtained during the surface-detection process described earlier. Partial (within slice) contextual information exists in the DC representation and none exists in the other two. One can attempt to improve the shading of displays based on these data structures by making use of the partial contextual information in the DC representation [Heffernan and Robb, 1985] or by postprocessing the distance-only shaded images as in the gradient shading method of Gordon and Reynolds [1985]. However, in our opinion neither of these techniques matches the quality of contextual shading based on the object surface.

The SE representation can be obtained quite efficiently from original medical data (such as CT slices) and, if used appropriately, has a respectable display speed [Reynolds et al, 1985]. The DC representation leads to faster display, but (at least in our current implementation) there is a greater restriction on the viewing directions.

Finally, we mention that there are, of course, other data structures that have been used for manipulation of 3D imagery. One of these, the so-called octree encoding, forms the basis of another very fast hardware device for 3D display and manipulation [Meagher, 1982, 1985].

Acknowledgments. The research of the authors is supported by National Institutes of Health grants HL28438 and RR02546. We are grateful to Ms M.A. Blue for typing, Mr S. Strommer and Mr K. Rude for photography, and Ms M. Kirsch for drawing the illustrations.

Appendix

For the sake of completeness we give here precise definitions of some of the standard topological terminology used in this chapter.

The elements of the *3D Euclidean space* R3 are the *points* (x,y,z), where x, y, and z are real numbers.

A point (x_0, y_0, z_0) is said to be a *closure point* of a set S of points in R3, if for every positive real number ε, there is a point (x,y,z) in S such that

$$(x - x_0)^2 + (y - y_0)^2 + (z - z_0)^2 < \varepsilon.$$

A set of points in R3 is said to be *closed* if it contains all its closure points.

The *complement* of a set S of points in R3 is the set CS of all points in R3 that are not in S.

A set S of points in R3 is said to be *open* if its complement is closed.

A set S of points R3 is said to be *connected* if there do *not* exist two open sets P and Q of points in R3 with all of the following properties:

P and Q have no points in common.
P and S have at least one point in common.
Q and S have at least one point in common.
All points in S are either in P or in Q.

References

Artzy E: Display of three-dimensional information in computed tomography. Comput Graph Image Process 9: 196–198 (1979).

Artzy E, Frieder G, Herman GT: The theory, design, implementation and evaluation of a three-dimensional surface detection algorithm. Comput Graph Image Process 15: 1–24 (1981).

Artzy E, Liu HK, Yau M: A software package for the display of three-dimensional information in medical images. Tech Rep MIPG7, Med Image Process Group, Dept Comput Sci, SUNY at Buffalo, September 1978.

Brewster LJ, Trivedi SS, Tuy HK, et al: Interactive surgical planning. IEEE Comput Graph Appl 4(3): 31–40 (1984).

Chen LS, Herman GT, Reynolds RA, et al: Surface shading in the cuberille environment. IEEE Comput Graph Appl 5(12): 33–43 (1985b).

Chen LS, Hung HM, Udupa JK: Towards real-time interactive display of 3D objects. Proc 18th Hawaii Int Conf Syst Sci, (1985a) pp 160–172.

Foley JD, van Dam A: *Fundamentals of Interactive Computer Graphics*. Addison Wesley Publishing Co, Reading, Massachusetts (1983).

Freeman H: Computer processing of line-drawing images. Comput Sur 6: 57–97 (1974).

Frieder G, Herman GT, Meyer C, et al: Large software problems for small computers: an example from medical imaging. IEEE Software 2(5): 37–47 (1985a).

Frieder G, Gordon D, Reynolds RA: Back-to-front display of voxel based objects. IEEE Comput Graph Appl 5(1): 52–59 (1985b).

Goldwasser SM: The voxel processor architecture for real-time display and manipulation of 3D medical objects. NCGA'85 Tech Sess Proc, Dallas, Texas, (1985) pp 71–80.

Goldwasser SM, Reynolds RA: An architecture for the real-time display and manipulation of three-dimensional objects. Proc Int Conf Parallel Processing, Bellaire, Michigan, (1983) pp 269–274.

Gordon D, Reynolds RA: Image space shading of 3-dimensional objects. Comput Vis Graph Image Process 29: 361–376 (1985).

Heffernan PB, Robb RA: A new method for shaded surface display of biological and medical images. IEEE Trans Med Imaging MI-4(1): 26–38 (1985).

Herman GT: *Image Reconstruction from Projections: The Fundamentals of Computerized Tomography,* Academic Press, New York, (1980).

Herman GT: Three dimensional imaging from tomograms, in Hohne KH (ed): Digital Image Processing in Medicine, Springer-Verlag, Berlin (1981).

Herman GT, Kemp S: A promising new imaging technique. *Diagnosis* 7: 89–91 (November 1985).

Herman, GT, Liu HK: Three-dimensional study of human organs from computed tomograms. Comput Graph Image Process 9:1–21 (1979).

Herman GT, Udupa JK: Display of three-dimensional discrete surfaces. Proc SPIE 283: 90–97 (1981).

Herman GT, Webster D: A topological proof of a surface tracking algorithm. Comput Vis Graph Image Proc 23: 162–177 (1983).

Lobregt S, Verbeek PW, Groen FCA: Three-dimensional skeletonization: principle and algorithm. IEEE Trans Pattern Anal Mach Intell PAMI-2: 75–77 (1980).

Marsh JL, Vannier MW, Stevens WG, et al: Computerized imaging for soft tissue and osseous reconstruction in the head and neck. Clin Plast Surg 12: 279–291 (1985).

Meagher D: Geometric modeling using octree encoding. Comput Graph Image Process 19: 129–147 (1982).

Meagher DJ: Applying solids processing methods to medical planning. NCGA'85 Tech Sess Proc, Dallas, Texas, (1985) pp 101–109.

Merrill RD: Representation of contours and regions for efficient computer search. Commun ACM 16: 69–82 (1973).

Newman MHA: *Elements of the Topology of Plane Sets of Points*, 2nd ed. Cambridge University Press, Cambridge, England (1951).

Quinn JF, Rhodes ML, Rosner B: A technique for locally optimal compression of CT image data. Proc 1984 Joint Alpine Symp Med Comput Graphics, Image Commun Clin Adv Neuro CT/NMR, Innsbruck, Austria, (February 1984) pp 31–36.

Reynolds RA, Gordon D, Chen LS: A dynamic screen technique for shaded graphics display of slice represented objects. Comput Vis Graph Image Proc 38: 275–298 (1987).

Rosenfeld A, Kak AC: *Digital Picture Processing*, 2nd ed. Academic Press, New York (1982).

Trivedi SS: Representation of three-dimensional binary scenes. NCGA'85 Tech Sess Proc, Dallas, Texas, (1985b) pp 132–144.

Trivedi SS: Interactive manipulation of three-dimensional binary scenes. The Visual Computer 2: 209–218 (1986).

Trivedi SS, Herman GT, Udupa JK: Segmentation into three classes using gradients. IEEE Trans Med Imaging MI-5: 116–119 (1986).

Tuy HK, Udupa JK: Representation, display and manipulation of 3D discrete scenes. Proc 16th Hawaii Intl Conf Syst Sci II: 397–406 (1983).

Udupa JK: Determination of 3D shape parameters from boundary information. Comput Graph Image Process 17: 52–59 (1981).

Udupa JK: Interactive segmentation and boundary surface formation for 3D digital images. Comput Graph Image Process 18: 213–235 (1982).

Udupa JK: DISPLAY82—a system of programs for the display of three-dimensional information in CT data. Tech Rep MIPG67, Med Image Process Group, Dept Radiol, Univ Pennsylvania (April 1983).

Udupa JK: Display and analysis of 3D medical images using directed contours, NGCA '85 Tech Sess Proc, Dallas, Texas, (1985) pp 145–155.

Udupa JK, Srihari SN, Herman GT: Boundary detection in multidimensions. IEEE Trans Pattern Anal Mach Intell PAMI-4: 41–50 (1982).

3
Scattered Radiation in Chest Radiography

JAMES A. SORENSON AND LOREN T. NIKLASON

Introduction

Chest radiography is the most frequently performed radiologic study in a typical radiology department. In 1980 more than 60 million chest radiography examinations were performed in the United States, accounting for greater than one third of the total radiologic imaging procedures [Johnson and Abernathy, 1983]. The reasons for this dominant role are twofold: First, the conventional chest radiograph conveys an extraordinary amount of information about a major portion of human anatomy, including lungs, heart, major blood vessels, and substantial portions of the skeleton. These structures, which are involved primarily or secondarily in many diseases, are presented on a single image, with good spatial resolution (several line pairs per millimeter) and with good contrast provided by natural air versus soft tissue versus bone contrast relationships. Second, chest radiography can be performed simply, at relatively low cost, and with low radiation dose to the patient.

Table 3.1 summarizes some of the imaging characteristics and capabilities of chest radiography. Alternative imaging techniques, including computed tomography (CT), magnetic resonance imaging (MRI), ultrasound, and nuclear medicine, though valuable for answering specific diagnostic questions, as yet have been unable to match conventional chest radiography for its current broad spectrum of applications.

In spite of its value for patient care, chest radiography nevertheless suffers from significant well-known limitations. Studies have demonstrated that up to 50% of potentially visible lung nodules on chest radiographs may be missed by experienced observers [Brogdon et al, 1983; Armstrong et al, 1983]. In some cases, nodules representing primary lung tumors are visible in retrospect on previous chest radiographs over periods of months and even years [Muhm et al, 1983]. Given the assumption that earlier detection can improve survivorship, and the central role of chest radiography as a screening test for primary lung cancer, such large error rates pose a significant problem in lung cancer control. In addition, failure to

TABLE 3.1. Imaging capabilities of chest radiography.

Spatial resolution: $\gtrsim$ 5 lp/mm
Contrast resolution: ~ 5%*
Image area: 35 × 43 cm
Entrance radiation dose: ~0.2mSv (20 mrem)
Imaging time: ms

*$\Delta I/I$ for 3-mm-diam soft-tissue nodule (ρ = 1 g/cm^3) surrounded by lung tissue (ρ = 0.3 g/cm^3) at diagnostic x-ray kilovoltages (μ = 0.2 cm^2/g about 70 KeV).
lp = line pairs, mSv = millisievert.

detect metastatic lung nodules arising from primary tumors elsewhere in the body can result in incorrect staging and inappropriate choice of therapeutic measures. Similar detection problems, also involving poorly defined structures of relatively low contrast, occur in the detection and staging of pneumoconiosis (eg, asbestosis) and various infiltrative lung diseases.

Detection and diagnostic errors may arise from three basic causes:

1. The area containing the abnormality is not scanned by the observer.
2. The area is scanned, but the abnormality is not seen.
3. The area is scanned and the abnormality seen, but incorrectly interpreted or identified by the observer.

Eye-movement research suggests that each of these factors contributes significantly, and about equally, to detection errors in chest radiography [Kundel et al, 1978]. Improvement of image quality, in particular of image (nodule) contrast and signal-to-noise ratio (SNR), could decrease each of these types of errors. For example, improved contrast and SNR could improve nodule visibility in the peripheral vision (error #1) as well as in direct vision (error #2). It also could provide more accurate discrimination between subtle low-contrast structure, eg, nodules versus vascular structures (error #3).

For the reasons outlined before, there has been significant recent interest in methods for improving the quality of chest radiographs. Two such approaches are highly efficient x-ray scatter rejection, and image processing and display manipulations. Conventional scatter-rejection techniques used in chest radiography include the use of air gaps and antiscatter grids (Figure 3.1, *top, middle*). Recent research efforts include the development of improved antiscatter grids [Sorenson et al, 1982] and of scanning-slit devices [Sorenson et al, 1980; Tesic et al, 1983] (Figure 3.1, bottom).

Image processing and display manipulations are helpful because in chest radiography there is simultaneously an extremely wide range of x-ray exposure levels transmitted across the chest, eg, lungs versus mediastinum, and inherently low contrast of many normal and abnormal structures, eg, pulmonary vasculature and cancer nodules. As shown in Figure 3.2, the

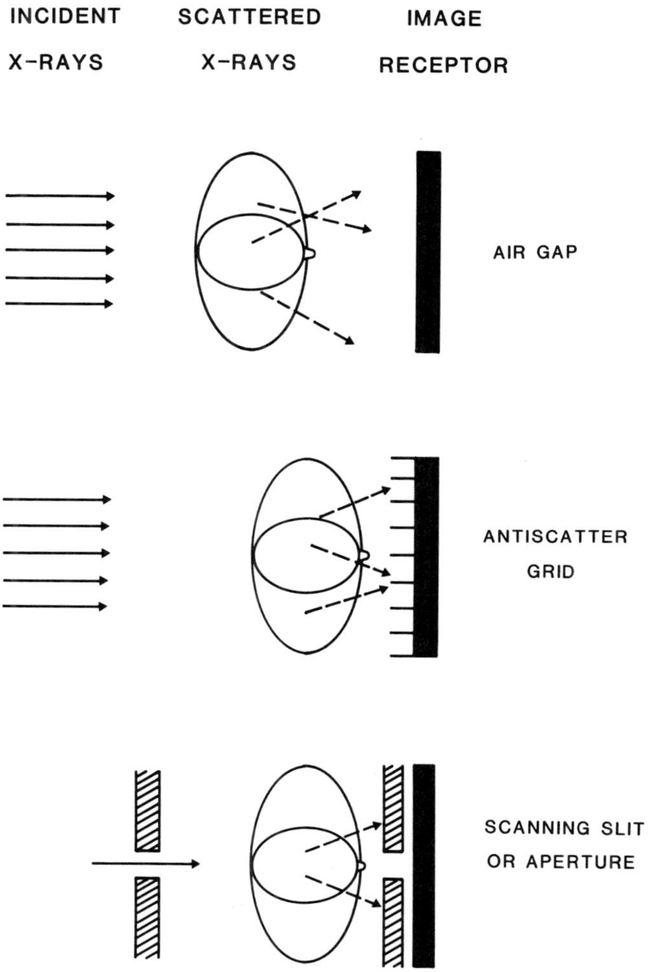

FIGURE 3.1. Scatter-rejection methods used for chest radiography. With the air-gap technique (*top*) the angular deflection of scattered x-rays results in more rapid decrease with distance of scattered x-ray intensity *v* unscattered primary x-ray intensity. Antiscatter grids (*middle*) consist of thin strips of lead or other x-ray absorbent material aligned parallel to the direction of primary x-rays to selectively intercept scattered x-rays. With scanning-slit or aperture devices (*bottom*), x-rays scattered outside of the scanning primary beam are intercepted by layers of lead or other x-ray absorbent materials that form the slit aperture.

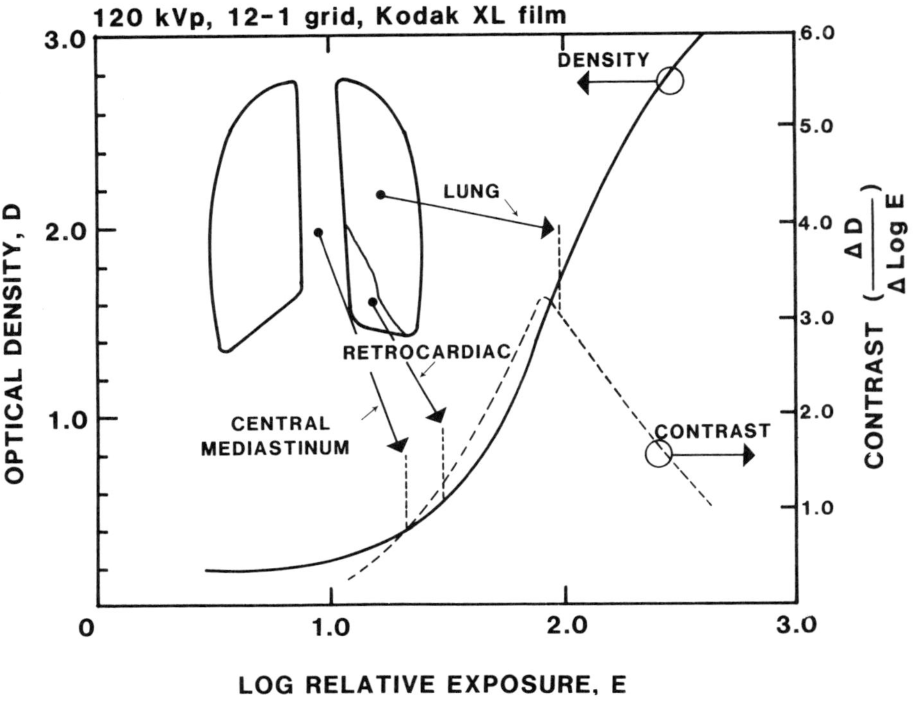

FIGURE 3.2. Sensitometric (H and D) curve for a wide-latitude x-ray film type and conventional chest radiograph. Exposure range across the image exceeds that which can be recorded with good density and contrast. Typically, good contrast is achieved in the lungs only at the expense of poor contrast in the retrocardiac, mediastinum, and other poorly penetrated regions. (Reproduced, with permission, from Sorenson et al, 1981.)

ratio of x-ray intensities transmitted by the lungs, which contain mostly air, versus the mediastinum, which contains mostly bone and soft tissues, can exceed 5:1. With highly efficient scatter rejection, eg, a scanning-slit device, the range is even wider, approaching 10:1 [Tesic et al, 1984]. In conventional chest radiography, it is necessary to use a relatively low-contrast, wide-latitude film type to accommodate the wide range of exposure levels; however, as shown by Figure 3.2, this still is inadequate to encompass the entire range of exposure levels present across the image.

Image-processing techniques can improve the display of chest radiographs. One such technique is unsharp masking, a form of high-pass spatial filtering that suppresses large-area (low-frequency) contrast, eg, lungs versus mediastinum, while preserving or even enhancing detail (high-frequency) contrast, eg pulmonary vasculature or nodules [Ishida et al, 1982; Sonoda et al, 1983; Sorenson, Armstrong et al, 1981; Sorenson, Niklason et al, 1981]. Examples of conventional versus unsharp masked chest radiographs are shown in Figure 3.3. The unsharp masked chest

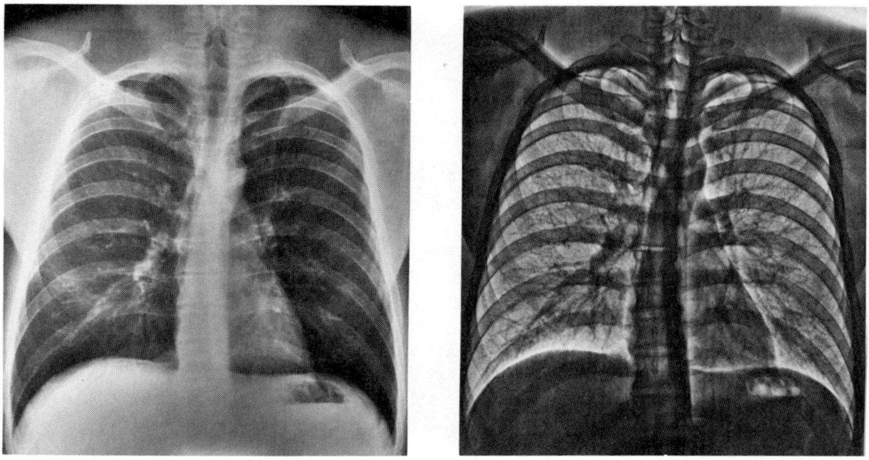

FIGURE 3.3. Examples of conventional chest radiograph (*left*) and unsharp-masked radiograph (*right*) of same subject. (Unsharp-masked image also shown with photographic contrast reversal.) Unsharp masking and other high-pass spatial filtering techniques provide enhanced contrast of fine details, such as pulmonary vasculature, with simultaneous compression of overall exposure range (low spatial frequencies) for improved presentation of structures in mediastinum, below diaphragm, and other areas that normally are poorly presented on conventional chest radiographs.

radiograph in this example was produced by an optical spatial filtration technique [Sorenson & Mitchell, 1985]. The original chest radiograph for this technique was recorded on a very-wide-latitude x-ray film (DuPont WDR) to capture the full range of exposure levels across the chest. This low-contrast image then was copied onto a high-contrast print film (DuPont MRF-32) using a LogEtronics® model 4050 contact printer. Image contrast was reversed in the printing process. The principles of the LogEtronics® printer are illustrated in Figure 3.4.

With the LogEtronics® device, the exposure is accomplished with a scanning light beam that is generated by a cathode ray tube (CRT) and projected through a lens system onto the original chest radiograph and the print film. The intensity of the exposing light beam is modulated to compensate for different densities on the original image, with darker areas receiving more exposure and lighter areas receiving less. The size of the exposing light beam is about 2 cm at the printing stage; thus, the contrast of structures of this size or larger, eg, lungs versus mediastinum, is suppressed, whereas the contrast of smaller details, eg, pulmonary vasculature, is enhanced by the high-contrast print film. The result is a pleasing image with good density and contrast across the entire chest.

The LogEtronics® printer is essentially an automated system for achieving "photographic dodging," and, indeed, the usual application for

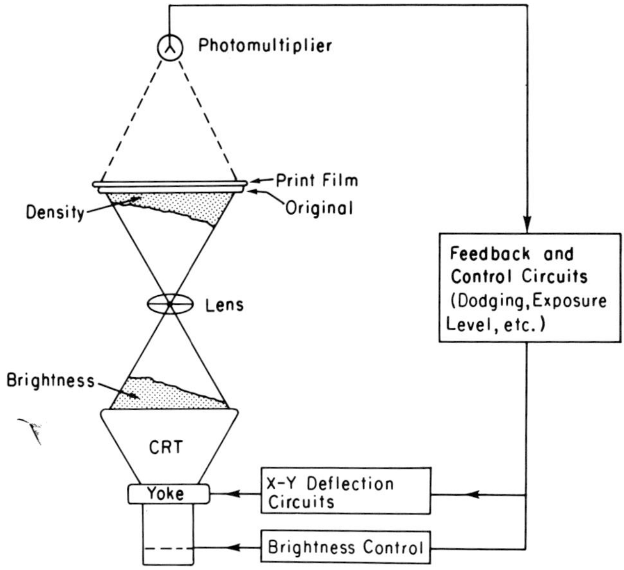

FIGURE 3.4. Principles of LogEtronics® printer, used to achieve unsharp-masking effect on chest radiograph in Figure 3.3. Exposure is obtained with a scanning light beam, the intensity of which is adjusted to accommodate varying densities in different areas of the original image. (Adapted, with permission, from Street et al, 1984.)

this device is for photographic printing. Unsharp masking and other forms of spatial filtering can also be performed by computer manipulation of digitized images [Johnson and Ravin, 1983]. Clinical studies have demonstrated that unsharp masking provides a significant improvement for the detection of nodules in the normally poorly exposed regions of the chest image, eg, behind the heart and below the diaphragm [Armstrong et al, 1983]. Additional research is needed to document the potential advantages of detail contrast enhancement in other areas of the chest image.

Another potentially useful image-processing method is the so-called "dual kilovoltage (kV)" technique. With this technique, images of the chest are recorded at two different x-ray tube kilovoltages (ie, a high- and low-energy image), and weighted subtractions are performed between the two images for selective suppression of bone or soft tissue [Sommer et al, 1982; Kruger et al, 1981; Lehmann et al, 1981]. The detector of a prototype dual-kV digital chest unit developed by Picker International is shown in Figure 3.5. This detector consists of two linear photodiode arrays each of 1,024 elements. Coupled to the front photodiode array is a low atomic number phosphor (Y_2O_2S) that absorbs the lower energy x-rays exiting the patient; thus, the front detector is used to create the low-energy image. The back detector has a high atomic number phosphor (Gd_2O_2S) that

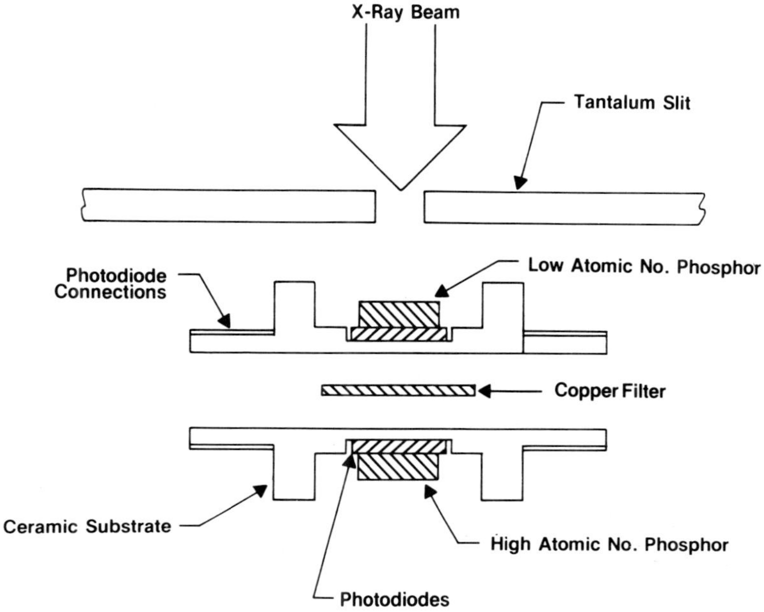

FIGURE 3.5. Cross-sectional view of the dual-kV detector demonstrating the front and back detectors, photodiodes, tantalum slit, and copper filter.

"sees" the hardened x-ray beam that has passed through the front detector and copper filter. The back detector is used to detect the high-energy x-rays. The low- and high-energy images (each 1,024 × 1,024 images) are then processed using the methodology of Lehmann et al [1981] to compute the bone (soft tissue cancelled) and soft-tissue (bone cancelled) images. This detector has a significant advantage over other dual-kV methods because both images are acquired simultaneously, eliminating patient motion artifacts between image acquisitions.

Figure 3.6A is a comparison of a conventional film-screen chest image and digital chest images obtained from the Picker dual-kV digital chest system. Dual-kV computed images of soft tissue and bone (Figures 3.6C, D) are shown along with a digital image created by addition of the high- and low-energy images to form a "conventional" digital chest image (Figure 3.6B). All of the digital images were obtained from a single exposure of a frozen, unembalmed human chest phantom.

The simulated soft-tissue nodules numbered 1 and 2 in Figure 3.6B are most clearly seen on the soft-tissue image demonstrating the advantage of removing the ribs. Nodule 3, which is calcified, is clearly demonstrated on the bone image. The bone image has significant clinical importance, since the presence of calcium in a pulmonary nodule, with rare exceptions, indicates the nodule is benign. There has been substantial interest in the

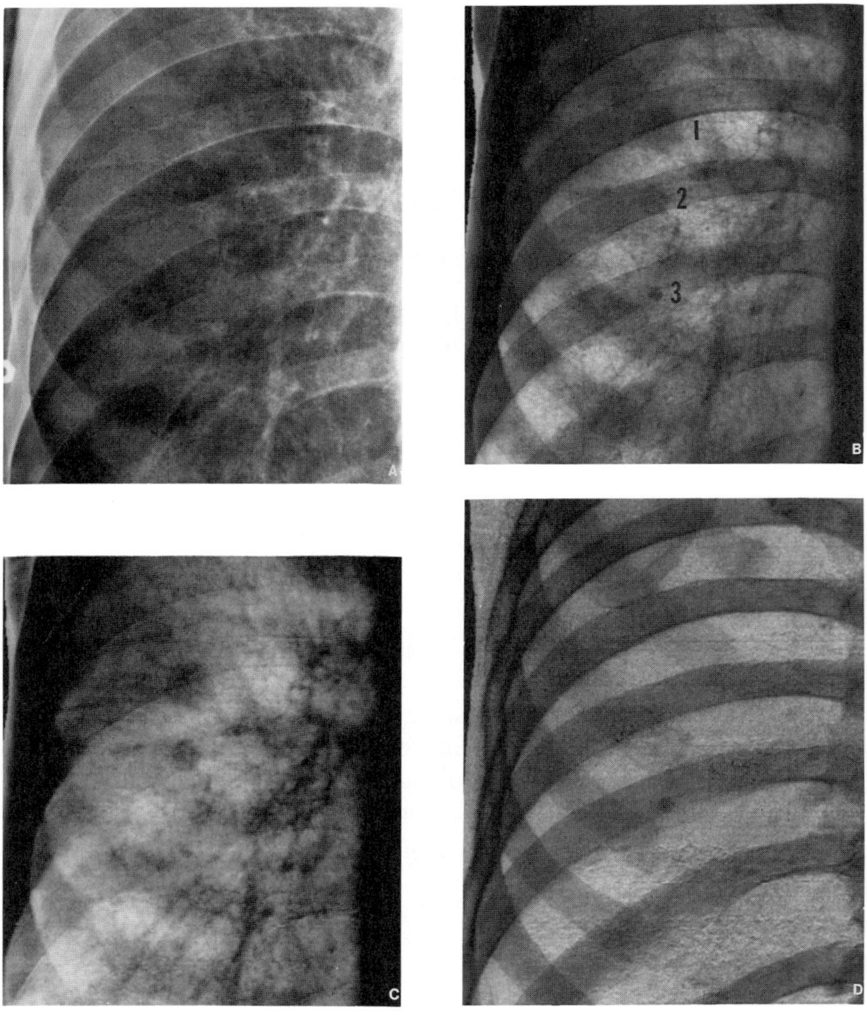

FIGURE 3.6. Conventional film-screen (**A**), conventional digital (**B**), soft-tissue (**C**), and bone (**D**) images of the right lung of a frozen human chest phantom. Simulated noncalcified nodule 1 and 2 (1.6 and 1.0 cm in diameter) and nodule 3, a calcified nodule (0.5-cm diam with 16 mg of calcium), are shown. Nodule positions are identical in both digital and conventional film-screen images. (Reproduced with permission from Fraser et al 1986.)

quantification of calcium in pulmonary nodules for the determination of a threshold value of calcium concentration above which a nodule may be considered benign [Siegelman et al, 1984]. Several studies have demonstrated dual-kV chest imaging is a simple and accurate method for the measurement of calcium concentration in pulmonary nodules [Hickey et al, 1987; Fraser et al, 1986]. Whether dual-kV chest imaging or CT is more accurate for the quantification of calcium has yet to be determined.

Another potential value of dual-kV imaging is illustrated by the data in Table 3.2, which shows nodule detection rates in a compartmentalized chest phantom from which various anatomic components could be removed. Removal of the ribs, which in clinical practice produces shadows that tend to obscure the shadows of nodules and other lung structure, resulted in substantial improvement in nodule detection rates in this study. A more recent study [Niklason et al, 1985] has demonstrated improved nodule detection using dual-kV images of a frozen human chest phantom recorded with the Picker dual-kV digital prototype. Both studies indicate significant improvement in nodule detection by dual-kV subtraction of rib shadows.

Efficient scatter rejection, in and of itself, usually is a worthwhile goal because of its potential for improvement of image contrast and SNR; however, it becomes even more important when image-processing techniques are employed. Such techniques frequently amplify the noise present in the original images. As discussed in the following section, scattered radiation decreases the SNR of the original images. When amplified by image-processing maneuvers, the presence of scattered radiation results in further noise degradation of the final image. Furthermore, some image-processing techniques, such as dual-kV bone subtraction, require quantitatively accurate recording of transmitted primary x-ray beam intensity. The presence of large and unpredictable levels of scattered radiation distorts this recording and can make the subtraction process inaccurate. Thus, effective suppression of scattered radiation is a desirable goal in any application involving image-processing maneuvers.

The remainder of this chapter is devoted to a discussion of scattered radiation in chest radiography, its effects on image quality, and current and potential means for highly efficient scatter rejection.

TABLE 3.2. Detection rates for simulated lung nodules in a modular chest phantom.*

	70 kVp	120 kVp
Whole phantom	31%	42%
Ribs removed	62%	65%
Lungs removed	74%	82%
Ribs and lungs removed	97%	95%

*From Christensen et al [1980].

Scatter Levels in Chest Radiography and Their Effects

Chest radiography sometimes is characterized as a "low-scatter" examination for which efficient scatter removal is unnecessary. This is true only for the thinnest tissue portions of the lung fields. In the mediastinum, subdiaphragmatic, and retrocardiac regions, and in other areas overlaid by thicker structures, eg, behind large blood vessels and ribs, scatter levels can approach and even exceed those encountered in abdominal radiography [Niklason et al, 1981].

The fraction of scattered radiation incident on the imaging detector typically is measured with a lead beam stop placed on the entrance surface of a phantom or patient. The lead beam stop used for scatter measurements must be thick enough to stop virtually all of the radiation incident upon it (typically > 3 mm of lead for diagnostic x-ray energies). The fraction of scattered radiation is obtained by measuring the film density in the beam stop shadow (scattered radiation only) and surrounding area (scattered and primary radiation). Film densities are converted to x-ray exposure using the characteristic curve (density versus log exposure) of the detector. The exposure values for scatter *(S)* and primary *(P)* radiation are used to calculate the scatter-to-primary *(S/P)* ratio.

Table 3.3 presents data obtained with a polystyrene chest phantom and illustrates the levels encountered with various scatter-rejection techniques. Figure 3.7 illustrates the cause of the very high levels encountered behind thicker portions of the chest, eg, the mediastinum. Scattered radiation exits the chest in a diffuse pattern that only coarsely resembles the distribution of primary radiation. Behind thick structures, primary radiation

TABLE 3.3. Scatter-to-primary ratios in different areas of the image of a polystyrene chest phantom for different scatter rejection methods at 120 kVp.*

	S/P Ratios**		
Technique	L	R	M
No grid, 3-cm air gap	1.93	6.20	13.30
No grid, 30-cm air gap	0.54	1.78	4.56
12:1 grid	0.54	1.07	1.60
Slit radiography with added 6:1 grid†	0.32	0.56	0.90
Crossed 10:1 tantalum air-interspace grid‡	0.21	0.31	0.63

*From Sorenson et al [1982].
**L = lung, R = retrocardiac, M = central mediastinum
†See Sorenson et al [1980]; also Figure 3.11.
‡See Sorenson et al [1982]; also Figures 3.9 and 3.10.

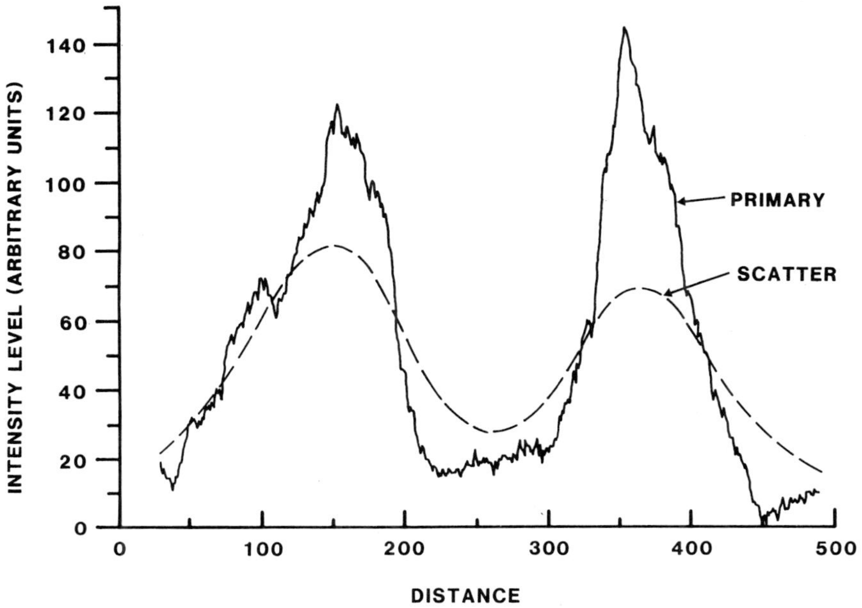

FIGURE 3.7. Scatter (S) and primary (P) radiation levels for a chest phantom (Humanoid Systems, Inc, Irvine, CA), illustrating high S/P ratios behind thick portions of the chest caused by cross-scatter from thinner areas. Scatter distribution was calculated from original image by a convolution algorithm, then subtracted from original image to estimate primary radiation intensity [Love et al, 1986]. Original image was recorded using a 35-cm-diam-image intensifier and conventional 8:1 antiscatter grid. Profiles were measured across the chest just above the heart shadow. (Courtesy of L. Alan Love.)

levels are more strongly attenuated than scattered radiation levels. Cross-scatter originating from thinner, better-penetrated portions of the chest may be quite intense; hence, relative *S/P* ratios behind thicker portions of anatomy may be quite large. Studies with phantoms indicate that the worst geometry for this "scatter enhancement" effect is approximately that of the human mediastinum [Niklason et al, 1981]; hence, this area of a chest radiography usually contains a very high *S/P* ratio.

The effects of scatter on image quality depend on the type of image recording employed. For direct recording on film, eg, using a film-screen cassette or an image-intensifier "spot-film" camera, the most noticeable effect is degradation of contrast. Defining contrast, C, as the difference in film density, ΔD, between a structure and its surroundings, one can show that for low-contrast objects [Sorenson and Nelson, 1976]:

$$C_s = C_p (1 - f_s) \quad (3.1)$$

where C_s = contrast with scatter, C_p = contrast without scatter (ie, "ideal" contrast), and f_s = fraction of exposure due to scattered radiation. If

S and P are the intensities of scattered and primary radiation recorded, respectively, then

$$f_s = (S/P)/[1 + (S/P)] \tag{3.2}$$

The ratio, C_s/C_p, is called the scatter degradation factor, SDF, and can be expressed as

$$\begin{aligned} \text{SDF} &= P/(P+S) \\ &= f_p \end{aligned} \tag{3.3}$$

where f_p is the fraction of total radiation exposure that is due to unscattered primary radiation. Figure 3.8 shows the effect of increasing S/P ratio on SDF and image contrast. For $S/P > 1$, the reduction of contrast exceeds 50%. Figure 3.9 demonstrates these effects visually for radiographs of a chest phantom recorded with some of the scatter-rejection techniques listed in Table 3.3.

Images recorded with electronic detection systems, or digitized from

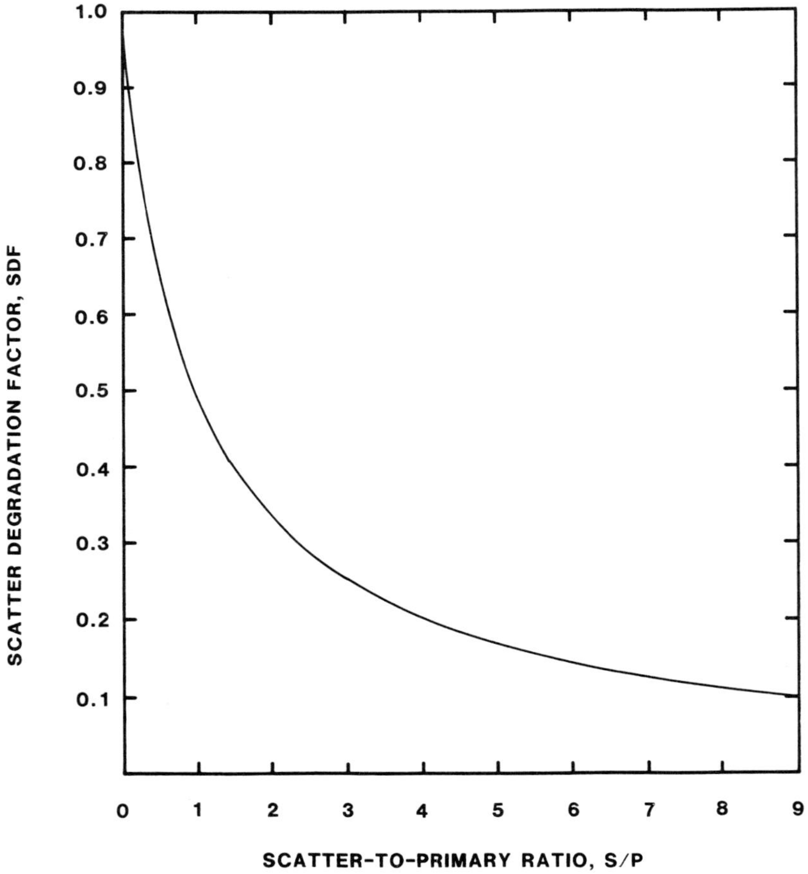

FIGURE 3.8. Scatter degradation factor *v* S/P ratio for radiographic imaging.

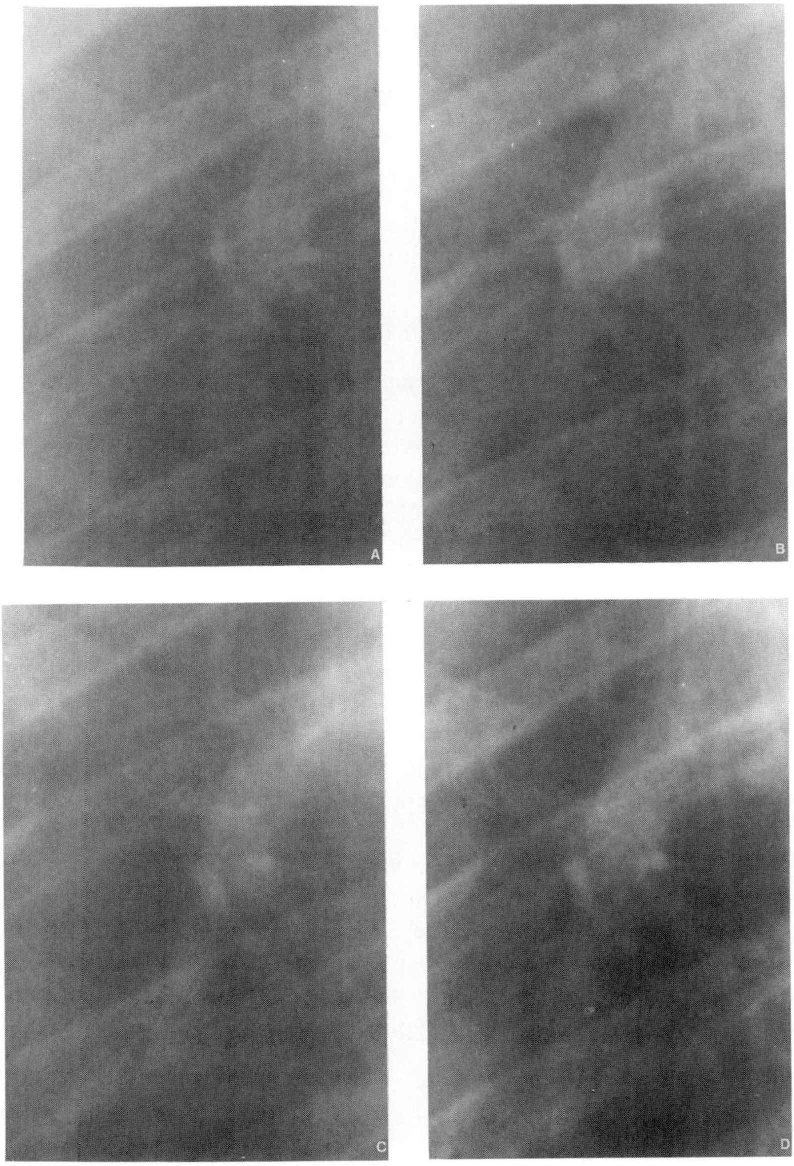

FIGURE 3.9. Portions of radiograph of a chest phantom illustrating effects of different levels of scatter rejection on contrast. **(A)** No scatter rejection; **(B)** 20-cm air gap; **(C)** conventional 12:1 antiscatter grid; **(D)** crossed 10:1 tantalum air-interspace grid. Compare with data in Table 3.3.

film for computer manipulations, can be enhanced to restore the contrast lost owing to scatter. In this circumstance, the dominant effect of scatter is degradation of SNR, as a result of the additional quantum mottle created by scattered x-ray photons. A commonly used parameter for characterizing image SNR is *detective quantum efficiency,* or DQE. The DQE of a system is the square of the ratio of its SNR versus that of an ideal system [Barrett & Swindell, 1981], ie,

$$\mathrm{DQE} = (\mathrm{SNR})^2 \text{ actual}/\ (\mathrm{SNR})^2 \text{ ideal} \tag{3.4}$$

The signal in this case might be the contrast of a low-contrast object, whereas noise might be due to quantum mottle, electronic noise, or other sources (excluding structured noise from overlying anatomic structures). This definition of DQE assumes equal radiation doses to the patient for the actual and ideal systems.

An ideal system is one for which neither contrast nor noise are corrupted by scattered radiation, fog, extraneous sources of quantum mottle, electronic noise, etc. For systems in which the predominating noise source is x-ray quantum mottle (as should be the case for any well-designed system), 1/DQE is the factor by which patient dose must be increased for the actual system to achieve the SNR that could be obtained with an ideal system. This relationship results from the fact that quantum noise, expressed as a percentage of the detected x-ray intensity level, is proportional to $\sqrt{D}$, where D is patient dose.

The presence of scattered radiation degrades SNR both by decreasing contrast (signal) and by increasing noise (additional quantum noise from the presence of scattered photons). In addition, the reduction of primary radiation intensity by a scatter-rejection device, eg, an antiscatter grid, also decreases the DQE. The effects can be analyzed as follows [Barnes, 1979].

Suppose that an ideal system provides contrast C_p and primary radiation intensity P at the image receptor and that an actual system provides primary and scatter radiation intensities, S' and P', respectively, at the image receptor for the same radiation dose to the patient. Here $P'/P = T_p$, the primary transmission factor for any scatter-rejection device used with the actual system (see Table 3.4 for typical values). For the ideal system,

$$\mathrm{SNR}_{\mathrm{ideal}} \propto C_p \sqrt{P} \tag{3.5}$$

whereas for the actual system

$$\mathrm{SNR}_{\mathrm{actual}} \propto \mathrm{SDF} \times C_p \sqrt{\mathrm{P}' + \mathrm{S}'} \tag{3.6}$$

where $\mathrm{SDF} = \mathrm{P}'/(\mathrm{P}' + \mathrm{S}')$ [Equation (3.3)]. Thus, DQE is given by

$$\mathrm{DQE} = [\mathrm{P}'/(\mathrm{P}' + \mathrm{S}')]^2\,(\mathrm{P}' + \mathrm{S}')/\mathrm{P} \tag{3.7}$$

$$= \mathrm{T}_p \times \mathrm{SDF} \tag{3.8}$$

TABLE 3.4. Performance characteristics for various scatter-rejection devices.

Method	T_p	T_s	Σ	K_c*	B_l*	D/D_o	DQE*	B_2
Reference conditions, no scatter rejection, 100-cm FFD	1.0	1.0	1.0	1.0	1.0	1.0	0.125	1.0
Conventional grids (0.05-mm Pb strips) 100 cm FFD								
8:1, 31.5 l/cm	0.67	0.112	6.0	3.69	5.51	5.51	0.31	1.0
12:1, 33.5 l/cm	0.62	0.085	7.3	4.08	6.58	6.58	0.32	1.0
Crossed 6:1, 33.5 l/cm	0.48	0.04	10.4	4.78	9.95	9.95	0.29	1.0
Conventional grid 12:1, 33.5 l/cm								
183 cm FFD	0.62	0.085	7.3	4.08	22.0	5.96	0.32	4.45
305 cm FFD	0.62	0.085	7.3	4.08	61.2	5.70	0.32	10.7
Crossed 10:1 tantalum air-interspace grid [Sorenson et al, 1982]	0.66	0.032	20.6	5.97	36.2	8.56	0.49	4.23
Rotating disk, slit-scanning device, +6:1 grid [Sorenson et al, 1980]	0.67	0.043	15.6	5.52	371	7.95	0.46	46.7
Digital chest unit [Tesic et al, 1983]	0.86	0.0004	215	7.75	4850	8.63	0.83	562
Raster-scanning system [Barnes and Brezovich, 1977]	1.0	0.167	6.0	3.69	1734	5.49	0.75	297
25-cm air gap technique (305-cm FFD)	1.0	0.234	4.27	3.03	28.2	3.13	0.38	9.0

*Assuming S/P = 7.0 for no scatter rejection.
FFD = focus-to-film distance.
From Sorenson and Floch [1985] and other references indicated.

Finally, as noted earlier, quantitative imaging techniques, such as dual-kV imaging for separation of bone versus soft tissue, require accurate measurement of primary x-ray beam transmission for the structures of interest in the image, eg, bones and calcified lung nodules [Sommer et al, 1982]. In the presence of scattered radiation, such measurements are distorted. ''Scatter corrections'' can be employed to minimize errors in quantitative measurements; however, in the chest, scatter levels are highly variable and difficult to predict on individual patients. It is also worth emphasizing that mathematical ''scatter corrections'' have no effect whatever toward improving the SNR and, indeed, are a potential source of additional noise if subtraction processes are involved. Thus, the preferred approach for reducing errors in quantitative studies is to reduce the amount of scatter recorded to its lowest practical level.

Methods for Removing Scattered Radiation

Conventional Methods

Conventional methods for scatter rejection in modern high-kilovoltage ($>$ 100 kVp, kilovolt peak) chest radiography include the use of antiscatter grids and air gaps (Figure 3.1). Typical recommendations vary, but air gaps of 20 to 30 cm, or ''medium-ratio'' linear (one-dimensional) focused grids (eg, R = 10:1) commonly are suggested for chest radiography. Here, R is the ratio of height to distance between the lead strips in the grid. Typical antiscatter grids are constructed using strips of lead 0.05 mm thick, with approximately 40 such strips per centimeter. Hence, the height and width of the interspaces for a 10:1 grid are about 2 mm and 0.2 mm, respectively. The interspace is filled with aluminum, plastic, or other low-absorbent material to provide structural stability. The grid oscillates during the x-ray exposure to minimize the visibility of grid lines, although for very short exposure times ($<$ 10 ms), this technique may be ineffective [Manninen et al, 1984].

Air gaps reduce the S/P ratio because there is a proportionately greater increase in distance from the image receptor to the scatter source (ie, the patient) than to the primary source (x-ray tube) when the air gap is introduced. A reasonable accurate approximation can be made for scatter rejection by an air gap by assuming that there exists an ''effective scatter source'' located 15 to 20 cm from the exit surface of the patient, ie, toward the radiation entrance side of the patient [Sorenson and Floch, 1985]. Air gaps have an inherent advantage in SNR over antiscatter grids because they cause no attenuation of primary radiation [T_p = 1 in Equation (3.8)]; however, they also are generally somewhat less effective for rejecting scattered radiation. Air gaps also must be used with relatively long focal-to-film distances, eg, 300 cm, to minimize blurring of the image because of magnification effects and the finite size of the x-ray tube focal spot. This can create complications where room size is a limiting factor.

Occasionally, recommendations are made that only a low-ratio grid or that no grid (or air gap) is needed for low peak kilovoltage chest radiography (eg, 60 to 70 kVp); however, experimental data suggest that there is not a significant reduction in scatter levels with decreasing peak kilovoltage [Niklason et al, 1981]. Presumably, the empirical advice suggested previously reflects the greater contrast of the primary image and relatively high efficiency of scatter cleanup by grids when lower kilovoltages are used, rather than intrinsically lower scatter levels.

Alternative antiscatter grid designs have been suggested for more efficient scatter removal. These include a "zig-zag" grid suggested by Barnes and Brezovich [1977], and a tantalum, air-interspaced crossed grid constructed by Sorenson et al [1982]. Both of these designs achieve more efficient scatter rejection by using the crossed-grid principle. A second set of scatter-attenuating grid strips are placed at right angles to those of a conventional linear grid to decrease the solid angle of acceptance of scattered radiation. The tantalum grid (Figure 3.10) has the additional advantage for scatter rejection of having thicker and more x-ray absorbent

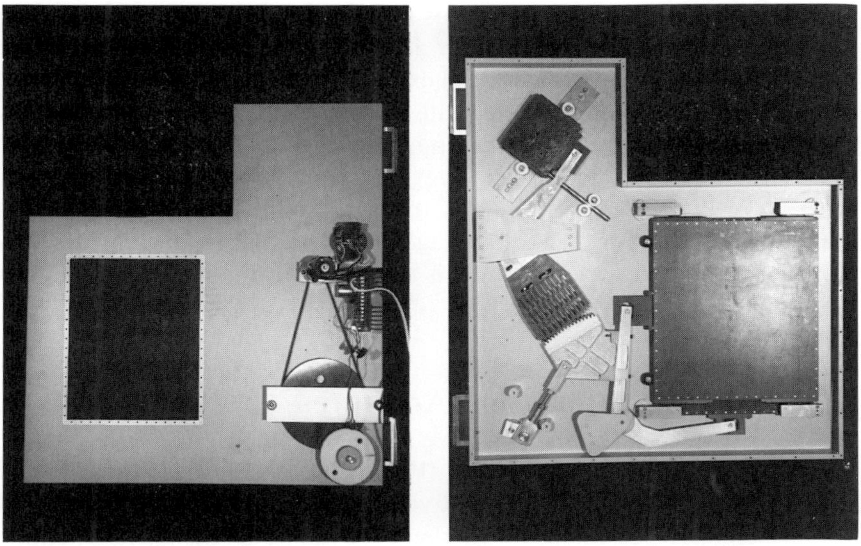

FIGURE 3.10. Tantalum air-interspace grid assembly, front (*left*) and back (*right*) views. Crossed grid consists of two perpendicular linear grids that reciprocate during the x-ray exposure to eliminate grid lines. Individual grids consist of 0.1-mm-thick tantalum strips, 1-mm apart, supported between thin graphite plates. Protective covers removed to illustrate components. (Reproduced, with permission, from Sorenson et al, 1982.)

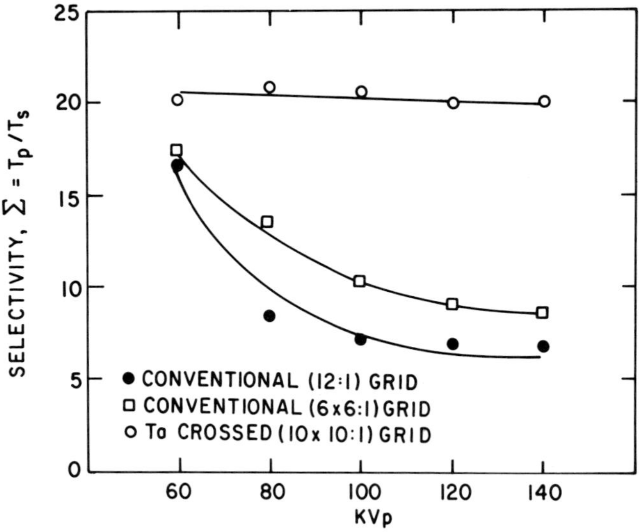

FIGURE 3.11. Selectivity, Σ, of tantalum air-interspace grid versus conventional linear (12:1) and crossed (6 × 6:1) grids at different kilovoltages. (Reproduced with permission, from Sorenson et al, 1982.)

grid strips (0.1-mm-thick Ta versus 0.05-mm-thick Pb in conventional grids), which is an important factor at the high peak kilovoltages (100 to 150 kVp) normally employed for chest imaging. The tantalum grid strips are supported between graphite plates, so that the interspace material is air, thus maximizing primary transmission. Figure 3.11 compares the effectiveness of the tantalum grids versus a conventional linear grid as a function of kilovoltage and demonstrates the relative ineffectiveness of conventional grids for high-kilovoltage chest imaging. This comparison is based on grid selectivity, Σ, which is the ratio of primary to scatter transmission factors (T_p/T_s).

Scanning Slits, Slots, and Apertures

Scanning slits, slots, and apertures (Figure 3.1) make use of the fact that small x-ray field sizes generate relatively low levels of scattered radiation. The principle of "scanning-slit radiography" is by no means new, dating back at least to a manuscript by O. Pasche in 1903, and a patent awarded to A. Dauvillier in 1921. For many years the development of scanning-slit radiography was impeded by a lack of x-ray tubes and generators having sufficient x-ray output intensities to provide the necessary film blackening within reasonable scanning and exposure times; however, recent advances in x-ray equipment and in x-ray image-recording technology have resulted in a renewed interest in this technique.

A scanning-slit device that was designed specifically for chest radiography and constructed at the University of Utah is shown diagramatically in Figure 3.12. It consisted of two concentric disks constructed from 1.6-mm-thick lead sandwiched between 3- and 12.5-mm-thick aluminum plates, with carefully aligned slit apertures. The tapered radial slits were 1 cm wide at the periphery of the large (back) disk and were separated, center-to-center, by five times their width. The disks rotated at high speed during the x-ray exposures (up to 10 rps) to blur the individual slit shadows. Adequate blurring was achieved for exposure times as short as 1/60 second. This slit scanner was used in combination with a conventional 6:1 linear grid. The grid lines were aligned at right angles to the slits for maximally efficient scatter rejection.

Table 3.3 includes scatter rejection capabilities of this device. Scatter cleanup was excellent; however, the device proved to be quite limited when used with conventional film-screen image recording. Because only 20% of available primary radiation was used, and additionally because of the long focal-to-film distance (305 cm) required by the geometry of the slit arrangement, relatively high tube loading and/or long exposure times were required. Very fast screens were used to compensate for this problem, but they produced correspondingly noisier and somewhat less sharp images. Clinical studies subsequently revealed that although measured

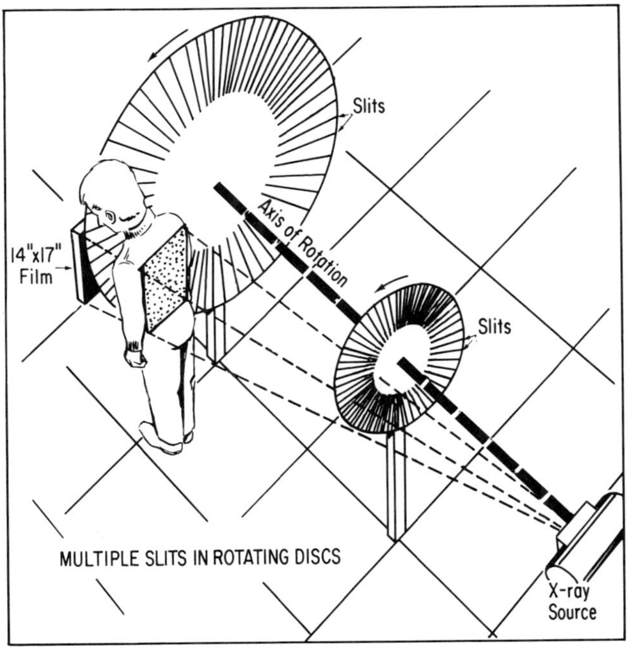

FIGURE 3.12. Rotating-disk, scanning-slit device for chest radiography. (Reproduced, with permission, from Sorenson et al, 1980.)

contrast was improved significantly with the slit-scanning device, detection accuracy for metastatic lung nodules was unaffected owing to the requirement for faster film-screen systems (which produced noisier and less sharp images) and long exposure times (which may have increased motion blurring) [Armstrong et al, 1983]. It was for these reasons that we developed the tantalum air-interspace grid previously described as an alternative in order to achieve highly efficient scatter rejection while maximizing the efficiency of utilization of x-ray output and permitting the use of film-screen systems of moderate speed.

The Picker dual-kV chest unit (Figure 3.13) is a scanning-slit device currently undergoing clinical evaluation [Tesic et al, 1983; Fraser et al, 1986; Fraser et al, 1983]. This device uses a fan beam of x-rays collimated by fore and aft slit apertures. The pixel dimensions of the 1,024 × 1,024 image matrix is 0.45 × 0.45 mm corresponding to a matrix resolution of 1.1 line pairs (lp)/mm. The image is digitized to 12 bits, and the dynamic range of the photodiode detector is 1000:1, which eliminates the exposure-latitude problem associated with film. Digital image-processing capabilities of the associated computer hardware for this device are used to manipulate and optimize the display of chest images.

In the Picker system only a single 1-mm-wide slit is used. The slit width is so narrow because it, in fact, determines the horizontal spatial resolution of the imaging system, ie, in the direction perpendicular to the slit aperture. Thus, the efficiency of the device for scatter rejection is very high, but its efficiency for utilization of x-ray output is correspondingly very low.

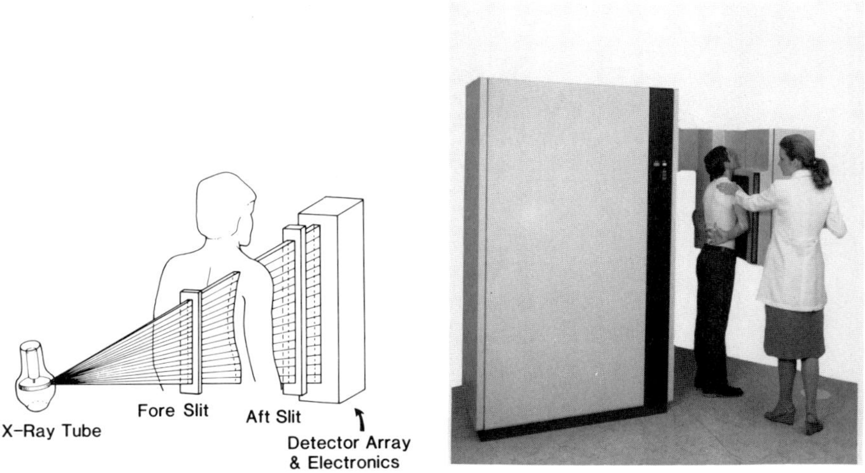

FIGURE 3.13. Scanning-slit digital chest imaging system. (*Left*) Scanning x-ray beam and detector. (*Right*) System in use. (Reproduced, with permission, from Tesic et al, 1983.)

Exposure (scan) time is about 5 seconds partly owing to mechanical scanning requirements, but primarily to achieve acceptable x-ray exposure levels at the detector.

The Picker chest system, with its combination of efficient scatter rejection and display manipulation capabilities, is capable of producing attractive chest images. Clinical testing to date suggests that it provides improved visibility of normal anatomy and may improve diagnostic accuracy for some tasks, particularly for disease in the mediastinum and other areas that normally are poorly visualized on conventional chest radiographs [Fraser et al, 1983]; however, its value as a total replacement for chest radiography remains to be demonstrated. Of particular concern is its capability for imaging structures having very fine detail, for which its 1,024 × 1,024 resolution capability (~ 1 lp/mm versus ~ 5 lp/mm for film-screen radiography) may be a limiting factor.

A final approach to efficient scatter rejection is a scanning raster-beam geometry. Plewes and Vogelstein [1983] developed a system in which the chest is scanned by an x-ray beam collimated to 3.5 × 3.5 cm^2 in area (Figure 3.14). The image is recorded on film. As the beam is scanned in a raster-like pattern over the subject, the intensity transmitted through the subject is monitored by a radiation detector mounted behind the film and consisting of a strip of fluorescent screen material coupled to a photomultiplier tube. The x-ray beam output intensity is modulated to compensate for differences in transmission across the chest. Output modulation is accomplished by varying the duration of x-ray exposure pulses. The exposure-modulation provides a form of unsharp masking during the im-

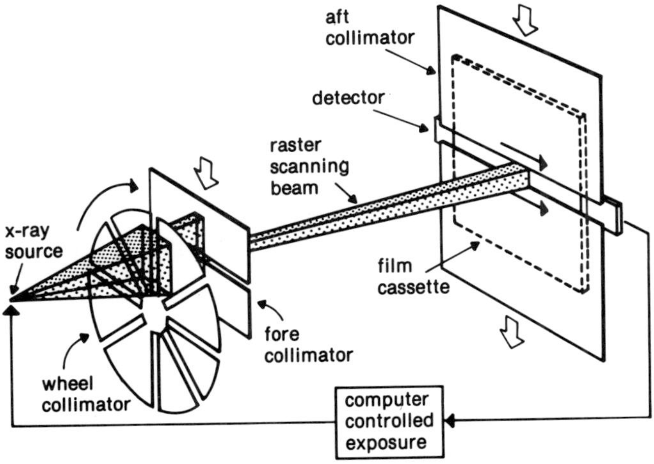

FIGURE 3.14. Raster-scanning chest radiography device. (Reproduced, with permission, from Plewes and Vogelstein, 1983.)

age-recording step; thus, the exposure-latitude limitation of the film is overcome.

The raster-scanning chest device of Plewes provides only moderately efficient scatter rejection, comparable to that of a conventional antiscatter grid, because of the relatively large area of the scanning beam. To achieve scatter rejection comparable to that of scanning slits or the tantalum grid described above, an aperture of only a few millimeters on a side would be needed. The principal advantage of Plewes' device is that it provides a means for compressing the exposure range that must be recorded, thus permitting the use of film for recording of chest radiographs without the usual exposure-latitude limitations. In comparison to the digital system described before, this "image-processing technique" also eliminates the effects of the finite size of digital detector elements on spatial resolution.

More efficient scatter rejection is accomplished with a pencil-beam scanning system, developed commercially by AS & E [Stein, 1975]. This device uses a slit collimator at the source, crossed by a second rotating slit aperture between the source and patient to create a scanning pencil beam (Figure 3.15). The detector is a single scintillation crystal monitored by photomultiplier tubes. Beam localization is achieved geometrically by the scanning aperture, rather than by position-sensitive response of the detector. The cross-sectional dimensions of the scanning beam at the detector for chest imaging are 1.1 × 1.1 mm.

The scatter-rejection capabilities of this device are very good because of the relatively small area of the scanning beam used. The scatter-rejection

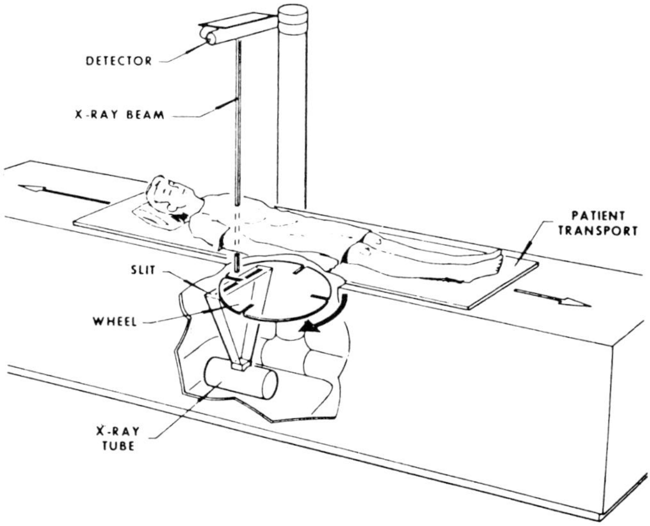

FIGURE 3.15. Pencil-beam x-ray imaging device. (Reproduced, with permission, from Stein, 1975.)

figure quoted in product literature is 99.4%. On the other hand, the system is limited by beam size and sampling intervals to 512 × 512 resolution, which is inadequate for general chest radiography [McLoud et al, 1984; MacMahon et al, 1984]. In common with other narrow-beam scanning devices, it also is very inefficient in utilization of available x-rays; thus, it requires relatively long imaging times for the chest. The system is claimed to be capable of providing x-ray images at very low radiation doses because of its highly efficient electronic detector system; however, this also is an inevitable consequence of the limited x-ray dose that can be delivered through a scanning 1 mm^2 aperture with conventional x-ray tubes. Thus, the "dose limitation" feature is not necessarily an advantage because the relatively few photons that are available result in higher levels of quantum noise than are usually present on chest radiographs.

Comparison of Scatter-Rejection Methods

A summary of comparative performance characteristics of the various scatter-rejection methods discussed in the previous section is presented in Table 3.4. Also included are some devices not designed specifically for chest radiography, but which presumably could be used for this application as well. In Table 3.4, T_p and T_s are primary and scatter transmission, respectively; $\Sigma = T_p/T_s$ is the selectivity of the device; K_c is the contrast improvement factor; B_1 is the bucky factor, ie, x-ray tube output must be increased to achieve the same level of x-ray intensity *at the image receptor* when the scatter-rejection device is used; and D/D_o is the relative radiation dose at midplane of the patient. Because of various geometric factors, $D/D_o \neq B_1$. DQE is the detective quantum efficiency as defined by Equation (3), and B_2 is the factor by which x-ray tube output must be increased to deliver *equal radiation dose* at patient midplane. Under these conditions (equal doses), SNR is related to DQE as given by Equations (4) to (8). Again, $B_2 \neq 1$, and $B_2 \neq B_1$, owing to various geometric factors.

Important considerations in the comparisons provided by Table 3.4 are as follows:

1. For film-screen recording, x-ray detection sensitivity and display contrast are relatively inflexible characteristics of the film-screen system. Thus, the most important comparative parameters are: (a) the contrast improvement factor, K_c, (b) the bucky factor, B_1, which is the increased tube loading required to achieve equal film density with versus without the scatter-rejection device, and (c) the relative patient dose delivered at midplane of the patient, D/D_o. Note that the bucky factor and relative dose factors given in the table take into account the different focal-to-film distances of the various devices listed.
2. For devices with electronic detection systems, and for which x-ray detection sensitivity and display contrast can be adjusted electronically,

the more important parameters are DQE, which reflects the relative SNR ratio for equal patient doses, and B_2, the factor by which x-ray tube output must be increased to deliver equal dose to the patient. The factors listed in the table again take into account the different focal-to-film distances of the devices listed.

Based on data presented in Table 3.4, it is apparent that efficient scatter rejection with film-screen recording probably will be limited to antiscatter grids and possibly multi-scanning systems because of large tube loading requirements (B_1) for narrow-beam scanning devices. Furthermore, given that efficient scatter rejection can be achieved, one still is faced with the problem of the exposure-latitude limitations of film. This could be overcome using the beam-intensity modulation technique of Plewes et al, or by using specially fabricated x-ray beam filters for individual patients [Peppler et al, 1982]; however, neither of these techniques has as yet been proven practical for routine clinical use. A further alternative would be to record the low-scatter image on a very-wide-latitude film, and then perform optical or computerized postprocessing for detail contrast enhancement and unsharp masking [Ishida et al, 1982; Sorenson and Mitchell, 1985] (see also Figure 3.3).

Electronic image-recording techniques can overcome x-ray detection sensitivity and display contrast/latitude limitations of film-screen recording. Thus, for these techniques, the most effective of the systems in Table 3.4 are those that maximize DQE. Slit- and raster-scanning devices accomplish this objective; however, tube-loading requirements (B_2) still are substantial. Thus, it is currently unclear that any narrow-beam slit- or raster-scanning device will become a practical alternative for general chest radiography, without dramatic breakthroughs in methods for producing x-rays, to achieve the levels of x-ray output required with these devices. For the near future, area-type detectors (digital arrays, film, or photostimulable phosphor plates [Sonoda et al, 1983], with digital storage and image-processing capabilities, used in conjunction with highly efficient grids or multislit-scanning devices, would appear to offer the best practical alternatives for achieving the benefits of highly efficient scatter rejection in chest radiography.

Acknowledgment. Chest radiography research at the University of Utah was supported by PHS grant number 5-R01-CA32854-09. The authors also thank Dr Charles Mistretta of the University of Wisconsin and Dr John Armstrong, Mercy Hospital, Des Moines, for helpful comments on the manuscript.

References

Armstrong JD, Sorenson JA, Nelson JA, et al: Clinical evaluation of unsharp masking and slit scanning techniques in chest radiography. Radiology 147: 351–356 (1983).

Barnes GT: Characteristics of scatter in Logan WW, Muntz EP (eds): Reduced Dose Mammography. New York, Masson, (1979) pp 223–242.

Barnes GT, Brezovich IA: A new type of grid. Med Phys 4: 451–453 (1977).

Barrett HH, Swindell W: *Radiological Imaging,* vol 1. New York, Grune and Stratton, (1981) p 194.

Brogdon BG, Kelsey CA, Moseley RD Jr: Factors affecting perception of pulmonary lesions. Radiol Clin North Am 21: 633–654 (1983).

Christensen EE, Dietz GW, Murry RC, et al: A modular chest phantom, in: *Optimization of Chest Radiography*. U.S. Dept. of Health and Human Services, HHS Publ. (FDA) 80-8124, (1980) pp 249–254.

Dauvillier A: Appariel pour la Realisation de la Radioscopic et de la Radiographic Integral. Brevet d'invention No. 521, 746. Office National de la Propiete Industrielle, Republique Francois. Issued March 14, 1921.

Fraser RG, Breatnach E, Barnes GT: Digital radiography of the chest: clinical experience with a prototype unit. Radiology 148: 1–5 (1983).

Fraser RG, Hickey NM, Niklason LT, et al: Calcification in pulmonary nodules: detection with dual-energy digital radiography. Radiology 160: 595–601 (1986).

Hickey NM, Niklason LT, Sabbagh E, et al: Dual-energy digital radiographic quantification of calcium in simulated pulmonary nodules. AJR 148: 19–24 (1987).

Ishida M, Kato H, Doi K, et al: Development of a new digital radiographic image processing system. SPIE 347: 42–48 (1982).

Johnson GA, Ravin CE: A survey of digital chest radiography. Radiol Clin North Am 21(4): 655–665 (1983).

Johnson JL, Abernathy DL: Diagnostic imaging procedure volume in the United States. Radiology 146: 851–853 (1983).

Kruger RA, Armstrong JD, Sorenson JA, et al: Dual-energy film subtraction technique for detecting calcification in solitary pulmonary nodules. Radiology 140: 213–219 (1981).

Kundel HL, Nodine CF, Carmody D: Visual scanning, pattern recognition and decision-making in pulmonary nodule detection. Invest Radiol 13: 175–181 (1978).

Lehmann LA, Alvarez RE, Macovski A, et al: Generalized image constructions in dual kVp digital radiography. Med Phys 5: 659–667 (1981).

Love LA, Kruger RA, Simons MA: Convolution filtering technique for estimating scatter distributions in radiographic images. SPIE 626(1): 275 (1986).

MacMahon H, Vyborny C, Powell G, et al: The effect of pixel size on detection rate of early pulmonary sarcoidosis in digital chest radiographic systems. SPIE 486: 14–20 (1984).

Manninen H, Terho EO, Wiljasalo M, et al: An evaluation of different imaging chains in clinical chest radiography. Br J Radiol 57: 991–995 (1984).

McLoud TC, Kushner DC, Dedrick CG, et al: Digital radiography in the assessment of pleural changes in an asbestos exposed population. Presented at 70th Scientific Assembly and Annual Meeting of RSNA/AAPM, November 1984, Scientific presentation No. 758.

Muhm JR, Miller WE, Fontana RS, et al: Lung cancer detected during a screening program using four-month chest radiographs. Radiology 148: 609–615 (1983).

Niklason LT, Hickey NM, Chakraborty DP, et al: Dual-energy digital vs. conventional chest radiography for the detection of lung nodules. Radiology 157(P): 92 (1985).

Niklason LT, Sorenson JA, Nelson JA: Scattered radiation in chest radiography. Med Phys 8: 677–681 (1981).

Pasche O: Uber eine neue Blendervorrichtung in der Rontgentechik. Deutsche Med Wochenschr 29: 267 (1903).

Peppler WW, Kudva B, Dobbins JT, et al: Digitally controlled beam attenuator. SPIE 347: 106–111 (1982).

Plewes DB, Vogelstein E: A scanning system for chest radiography with regional exposure control: practical implementation. Med Phys 10: 655–663 (1983).

Siegelman SS, Khouri NF, Scott WW, et al: Computed tomography of the solitary pulmonary nodule. Semin Roentgenol 19: 165–172 (1984).

Sommer FG, Brody WR, Macovski A, et al: Dual-energy scanned projection radiography. Appl Radiol; March/April, (1982) pp 59–66.

Sonoda M, Takano M, Miyahara J, Computed radiography utilizing scanning laser stimulated luminescence. Radiology 148: 833–838 (1983).

Sorenson JA, Armstrong JD, Niklason LT, et al: Letter to Editor: Enhanced unsharp masking technique. Invest Radiol (March/April, 1981), pp 59–66.

Sorenson JA, Floch J: Scattered rejection by air gaps; an empirical model. Med Phys 12(3): 308–316 (1985).

Sorenson JA, Mitchell CM: Optical unsharp masking and contrast enhancement of chest radiographs. Radiology 157(P): 240 (1985).

Sorenson JA, Nelson JA: Investigations of moving-slit radiography. Radiology 120: 705–711 (1976).

Sorenson JA, Nelson JA, Niklason LT, et al: Rotating disk device for slit radiography of the chest. Radiology 134: 227–231 (1980).

Sorenson JA, Niklason LT, Jacobsen SC, et al: Tantalum air-interspace crossed grid: design and performance characteristics. Radiology 145: 485–492 (1982).

Sorenson JA, Niklason LT, Nelson JA: Photographic unsharp masking in chest radiography. Invest Radiol 16: 281–288 (1981).

Stein JA: X-ray imaging with a scanning beam. Radiology 117: 713–716 (1975).

Street JN, McIntosh WL, Manack AW, et al: Optimizing photographic information transfer by CRT—a technology and applications review. SPIE 496: 162–172 (1984).

Tesic MM, Mattson RA, Barnes GT, et al: Digital radiography of the chest: design features and considerations for a prototype unit. Radiology 148: 259–264 (1983).

Tesic MM, Sones RA, Morgan DR: Single-slit digital radiography: some practical considerations. Radiology 142: 697–702 (1984).

4
Positron-Emission Tomography

JOEL S. KARP AND GERD MUEHLLEHNER

Introduction

The potential of positron imaging has been recognized for many years. In the mid-1970s two advances resulted in a dramatic increase in interest in this technique: (1) the development of positron-emission-computed tomographic devices [Phelps et al, 1975a; Hoffman, 1976a] and (2) the successful synthesis of [^{18}F]-fluorodeoxyglucose (^{18}F-FDG) [Ido, 1978; Reivich, 1977] and its application to the study of brain metabolism.

In the last few years the imaging properties of positron-emission tomographs (PET) have been improved significantly. These improvements resulted from a better understanding of design principles and trade-offs as well as from the introduction of novel concepts. Current efforts in positron instrumentation are directed toward accurate, quantitative reconstruction of three-dimensional distributions of radioactivity.

An example of state of the art in positron scanners is the ECAT III [Hoffman et al, 1986]. This is a multislice device designed for imaging both the brain and the heart. Figure 4.1 shows examples of a normal brain study and a gated cardiac study imaged with ^{18}F-FDG. The spatial resolution is about 5 mm. Although PET images have always provided metabolic information that is unavailable with other imaging modalities, the poor spatial resolution has been its major drawback. With high spatial resolution, such as that seen in Figure 4.1, not only is the detail of the structures much clearer, but the quantitative metabolic activity is more accurately measured [Hoffman and Ficke, 1985].

Rather than describing the instruments in detail, this chapter reviews the important parameters common to positron imaging. It summarizes options that groups have explored in designing tomographs and describes methods developed to overcome technical limitations.

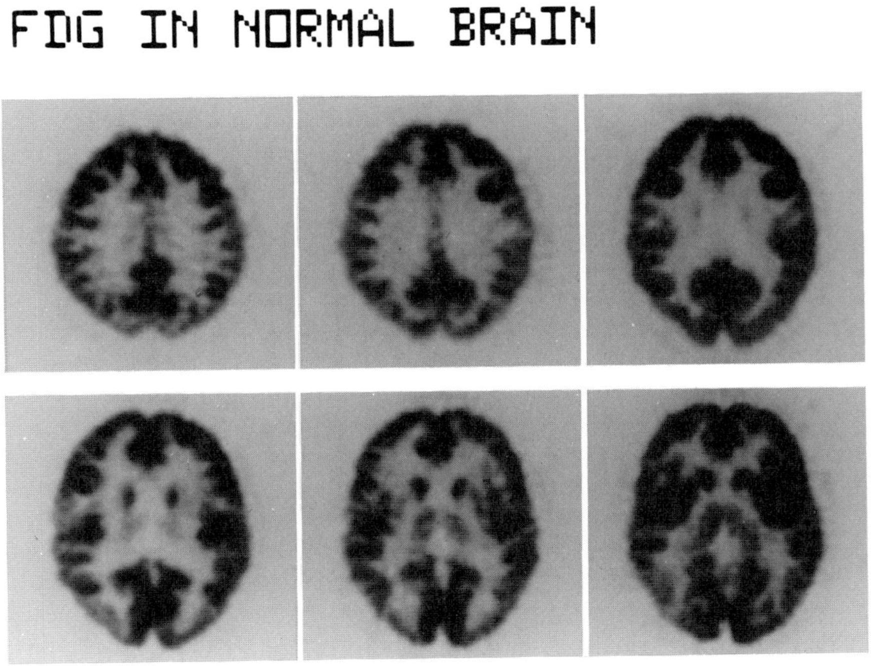

A

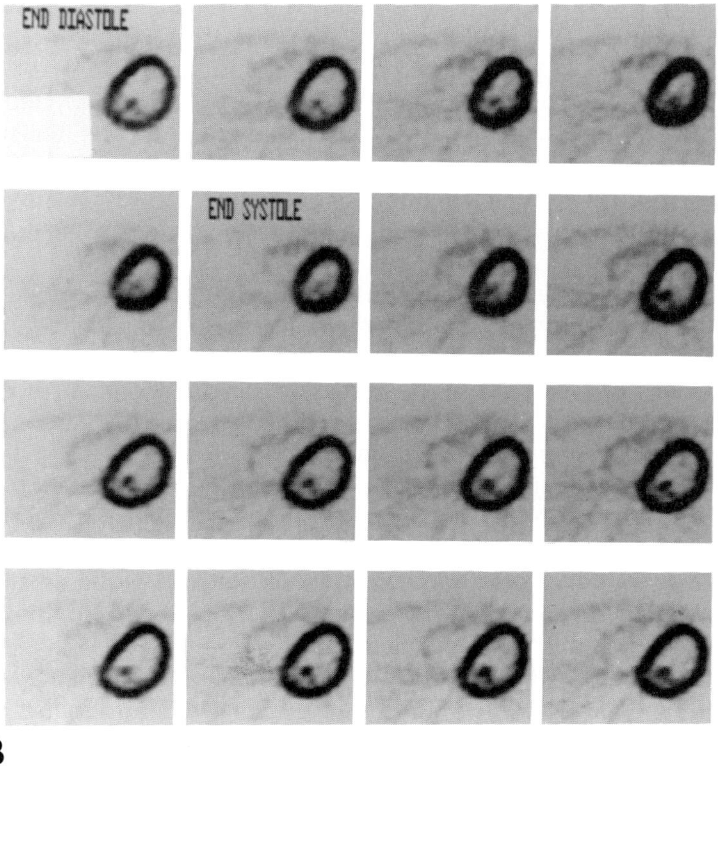

B

Design Considerations

Resolution

One of the main attractions of positron imaging compared with single-photon gamma-ray imaging is the basic localization mechanism. A positron emitted from a radioactive substance will travel a short distance, combine with an electron, and annihilate. The resulting energy is carried by two photons with an energy of 511 keV each, which travel in nearly opposite directions. Simultaneous detection of two photons of this energy thus signals the decay of a positron anywhere along a line connecting the two points of detection. Since localization is achieved by coincidence detection and not by mechanical collimation of the gamma rays, positron imaging can combine high spatial resolution with high detection efficiency given a large solid angle for detecting pairs of gamma rays.

Two fundamental processes limit the accuracy with which a source of radioactivity can be localized: (1) the finite range of positrons before annihilation and (2) the angular spread about 180° of the two gamma rays in the laboratory.

The effect on spatial resolution due to the finite range of the positron and its relative importance has been investigated extensively [Llacer and Graham, 1975; Cho et al, 1975; Phelps et al, 1975b; Derenzo, 1977; Muehllehner, 1976; Derenzo, 1979; Nohara, 1985]. The attenuation of positrons resulting from beta decay, which exhibits a continuous energy distribution, is nearly exponential and depends on the energy of decay. Since the spatial distribution is sharply peaked, it is difficult to quote a single meaningful number to characterize the width of the distribution. It was suggested [Hoffman, 1976b] that a width be chosen such that 75% of all events occur within that distance in the point spread function (PSF). This width would be analogous to the full-width-half-maximum (FWHM) of a Gaussian distribution since about 75% of the area under a Gaussian curve is between the half maxima. Table 4.1 shows that this width increases from a nominal 1.2 mm for fluorine 18, ^{18}F, to a substantial 12.4 mm for the radioactive isotope of rubidium, ^{82}Rb [Derenzo, 1985].

An equally fundamental source of resolution loss is the angular range of the two gamma rays. Because of the motion of the center of mass of an annihilation pair, the two 511 keV photons formed upon annihilation

FIGURE 4.1. **(A)** Images of a normal human brain taken with the ECAT III PET scanner. For each image, data were acquired for 15 minutes, starting 40 minutes after injection of 10 mCi ^{18}F-FDG, which resulted in 10 to 20 million events. Each image represents a different slice of the brain. **(B)** Gated heart ^{18}F-FDG study with 16 to 70-ms time frames. Each image represents the same slice of the heart at different time intervals.

TABLE 4.1. Positron range in water.*

Isotope	^{18}F	^{11}C	^{68}Ga	^{82}Rb
PSF: Range (75%)	1.2 mm	2.1 mm	5.4 mm	12.4 mm
PSF: FWHM	0.13 mm	0.13 mm	0.31 mm	0.42 mm

*The range is the diameter of a circle containing 75% of all counts. Data from Derenzo [1985]. ^{11}C = carbon-11, cyclotron-produced positron-emitting isotope; ^{68}Ga = radioactive isotope gallium.

exhibit an angular spread about 180° in the laboratory. The resulting distribution is roughly Gaussian with a FWHM of approximately 0.5° [De Benedetti, 1950]. Therefore, the loss of resolution depends on the distance between the detectors. Typically, tomographs for brain imaging have a ring diameter of 60 cm, and whole-body systems have a ring diameter of 100 cm. Table 4.2 summarizes the loss of resolution due to deviation from 180°, which is given by

$$\Delta x = 0.5D \cdot \tan 0.25° \qquad (4.1)$$

where D is the ring diameter and Δx is the loss of resolution in the center of the ring.

Spatial resolution in the object plane is related to the size of the detectors. Since two opposing detectors are in coincidence, the counting profile for a point source between them is given by the convolution of the two detector response functions. Thus, a rectangular detector of width W that has a rectangular response function (ignoring edge effects) leads to a detector pair response function at the midpoint between the detectors, which is triangular with a FWHM of $W/2$. Therefore, the response function in the center between the detectors is half as narrow as the detector response function. As a comparison, a continuous position-sensitive detector has a Gaussian response function with a FWHM, g, which leads to a Gaussian response function at the midpoint between two detectors with a FWHM of $g/\sqrt{2}$.

The image resolution is determined by combining the detector pair resolution with the uncertainties due to positron range and angular deviation. Table 4.3 shows this effect for a variety of radioisotopes and an 80-cm diameter ring [Nohara, 1985]. These values represent theoretical lower

TABLE 4.2. Loss of resolution due to angular range.

Ring diameter (cm)	Resolution loss (FWHM) (mm)
60	1.3
80	1.7
100	2.2

TABLE 4.3. Calculated system resolution including detector size, positron range, and angular range.*

	FWHM (mm)		
Detector pair resolution†	^{11}C	^{68}Ga	^{82}Rb
0.0 mm	2.3	3.1	3.3
2.0 mm	3.2	4.1	4.7
6.0 mm	6.6	7.4	9.2

*Data are for 80-cm ring diameter [Nohara, 1985].
†Detector resolution is calculated value midway between two detectors.

limits that can be achieved in PET imaging for a given detector resolution. Note that these calculated values, and indeed often quoted measured values, are at the center of the scanner. Penetration of the gamma rays through the crystal degrades the resolution at off-center locations, which is discussed in more detail later.

Sampling Requirements

In tomographs with arrays of discrete detectors, the radioisotope distribution is measured at points that are separated by a distance equal to the detector separation. As discussed above, the spatial resolution in the object plane is approximately half the size of the detector. Thus, the spacing of samples in a stationary system with discrete detectors exceeds the spatial resolution. Insufficient linear sampling has two effects: (1) contrast is reduced, especially at high spatial frequencies and (2) frequencies above the Nyquist frequency will show aliasing artifacts [Huang, 1980; Brooks, 1978]. The Nyquist frequency, f_n, is given by $f_n = (2d)^{-1}$, where d is the sample spacing. Although interpolation between measured points will reduce the intensity of the aliasing artifacts, the loss of contrast is severe enough that most systems with discrete detectors employ some form of scanning motion to increase linear or angular sampling or both. A sampling distance of one third the FWHM of the detector response function does not give noticeable artifacts [Huang, 1980].

The sampling requirement for angular samples is satisfied if the number of angles, w, is

$$w = \pi B/2\,d \tag{4.2}$$

where d is the linear sampling distance and B is the diameter of the object [Huesman, 1977; Snyder, 1977]. Under these conditions the distance between angular samples is equal to the linear sampling distance at the edge of the object. In practice, the *angular* sampling requirement can be relaxed significantly (a factor of 2 is not uncommon) without introducing noticeable artifacts, whereas the violation of the *linear* sampling requirement quickly leads to a loss of image quality.

Since the spatial resolution of two opposing detectors is less than the physical dimension of the detector, it is impossible to arrange discrete detectors closely enough to achieve a sampling interval that makes full use of the detector resolution without moving the detectors. Thus, most positron-imaging systems employ some kind of motion.

A hexagonal or octagonal array of detectors has been used in the PETT III (Phelps, 1975a) and the PETT IV (Ter-Pogossian, 1978a) as well as the ECAT [Phelps, 1978] and the neuro-ECAT scanners [Hoffman, 1981a]. This arrangement lends itself well to a linear translation of banks of detectors to increase the linear sampling and rotation of the whole detector array to improve angular sampling. It has the advantage that it is easy to achieve any desired sampling interval by controlling the amount of motion between data points.

Bohm [1978] and Ter-Pogossian [1978b] investigated another motion, wobbling, in which the orientation of the ring remains fixed while its center moves in a small circle (Figure 4.2A). This results in translation of each detector pair. Although it can substantially decrease the ray spacing, wobbling results in nonuniform sampling (Figure 4.2B). Some form of interpolation is thus required before the standard reconstruction algorithms can be employed.

Cho [1981] and Derenzo [1981] have each suggested splitting a ring of detectors into two halves and moving them relative to each other either in a "dichotomic" or a "clam-shell" motion. Alternatively, the detectors can be arranged at irregular intervals along the ring, and the ring can be rotated 360° [Nohara, 1980]. All these methods strive for a simple mechanical motion that can be quickly executed, that results in uniform sampling, and that does not unnecessarily increase the amount of data to be stored and sorted.

Problems associated with moving discrete detectors can be avoided by using continuous, position-sensitive detectors. In a continuous detector, the resolution can be potentially quite high, avoiding the use of very small detectors to improve resolution. Furthermore, linear sampling is continuous and can be digitized at a small fraction of the spatial resolution, obviating any need for motion [Muehllehner, 1983]. A scanner that is stationary during data acquisition is well suited to gated cardiac imaging as well, since it is difficult to synchronize the motion of the scanner with that of the beating heart.

Scattered Radiation

Scattered radiation must be excluded from the image since it degrades contrast and makes accurate quantitation of radioisotope uptake difficult. In single-photon imaging at low energies, the problem is simple: radiation

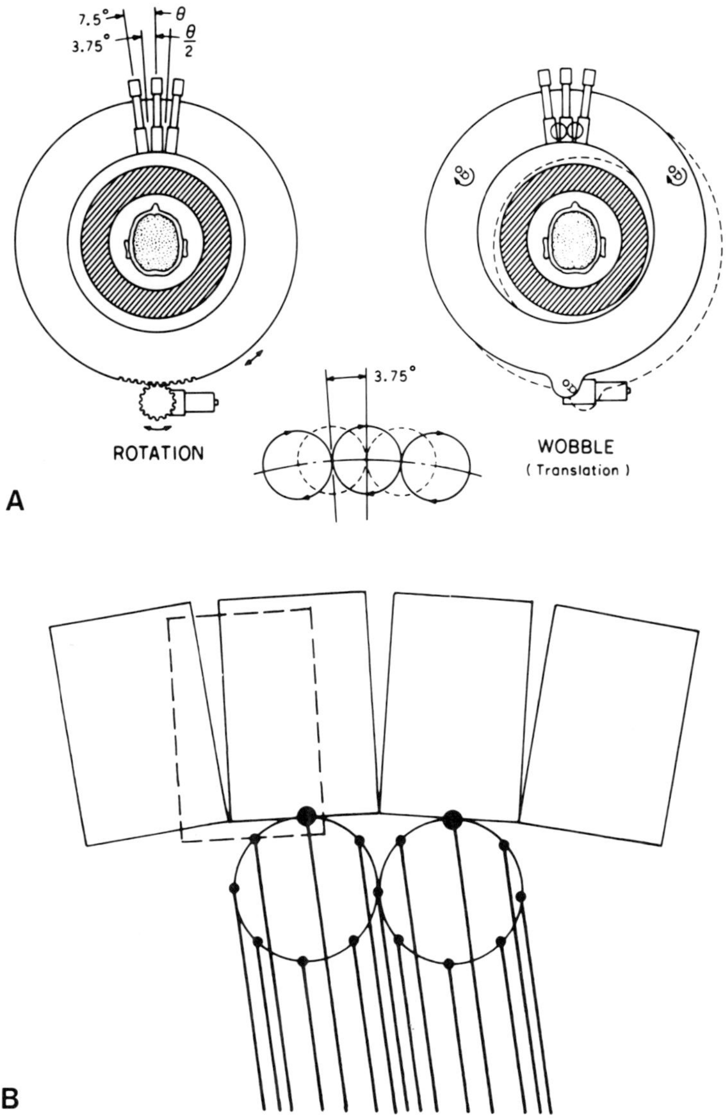

FIGURE 4.2. (**A**) Illustration of wobble motion and rotation to increase linear and angular sampling [Ter-Pogossian, 1978b]. (**B**) Illustration of irregular spacing of samples in a projection obtained with wobble motion with samples taken at fixed angular increments [Brooks, 1979].

reaching the detector without undergoing scattering in the patient has a high probability of being totally absorbed in the detector and will be detected as a photopeak event, while scattered radiation from the patient will have reduced energy. Thus, every interaction below the photopeak must be rejected. The problem is choosing the proper discriminator setting to minimize the amount of scattered radiation that is accepted without undue loss of sensitivity to the primary photopeak events. In positron imaging the two spectra—primary and scattered—overlap to a large extent. If an event is detected with an energy below the photopeak, either it can be due to an unscattered 511 keV gamma ray, which undergoes Compton scattering in the detector with subsequent escape of the scattered gamma ray, or it can be due to scattered radiation from the patient. Typical spectra for unscattered and scattered radiation are shown in Figure 4.3.

There are three methods of eliminating scattered radiation: (1) geometric configuration, (2) coincidence requirement, and (3) energy discrimination. Several investigators have examined the effect of the geometric configuration on system performance, specifically, on sensitivity, scattered coincidence fraction, and random coincidence fraction [Derenzo, 1980; Tanaka, 1982; Bergstrom, 1983; Atkins, 1978]. Derenzo [1980] and Tanaka [1982] derived formulas describing the behavior of these parameters as a function of shielding depth; however, the effect of energy discrimination, which also influences these parameters, cannot be readily included in these formulations. The difficulty in including the effect of energy discrimination is that energy spectra change dramatically as a function of detector material and object size. True, scatter and random rates vary as a function of energy

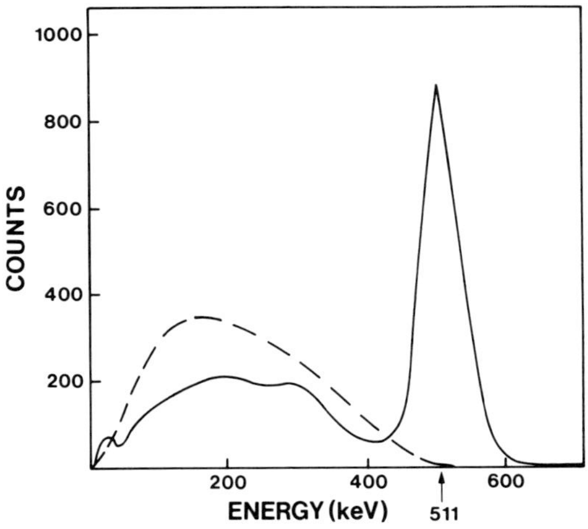

FIGURE 4.3. Energy spectrum of 511-keV primary radiation (solid line) and typical scattered secondary radiation spectrum from the patient (dashed line).

threshold, and for each shielding depth an optimum energy threshold can be determined [Mankoff, 1984].

Although the absolute amount of scattered radiation detected in coincidence is of some concern, the more important factor is the scatter fraction, which is defined as the ratio of scattered coincident events to unscattered true coincident events. This ratio is influenced by the geometric configuration because the true coincidence event rate is related to the patient port size and shielding depth as follows:

$$R_C \sim \frac{1}{X + P/2} \tag{4.3}$$

where $P/2$, the radius of the patient port, is the distance from the center to the lead shielding; and X, the shielding depth, is the length of lead shielding to the detector. The scattered coincidence event rate, R_{Sc}, on the other hand, is influenced by the geometric configuration as follows

$$R_{Sc} \sim \frac{1}{X(X + P/2)} \tag{4.4}$$

so that the scatter fraction, S, is given by

$$S = \frac{R_{sc}}{R_c} = f \cdot \frac{1}{X} \tag{4.5}$$

where f is a factor that is a function of the energy threshold.

Figure 4.4 shows measured and calculated values of scatter fraction for bismuth germanate (BGO) and thallium-doped sodium iodide [NaI(Tl)] crystals for a wide range of energy thresholds and shielding depths. Most positron-imaging devices use a lower threshold of 100 keV for NaI(Tl) detectors or 300 keV for BGO detectors. Although raising the energy threshold and increasing the shielding depth improves the contrast, and thus image quality, it does so at the expense of sensitivity. With NaI(Tl) detector systems, because of Compton scattering out of the crystal, raising the energy threshold above 100 keV may reduce system performance because of the loss of sensitivity. With BGO detector systems, however, where fewer events scatter out of the crystal, one can obtain equal data quality with low-energy thresholds with large shielding depths or high-energy thresholds with small shielding depths. At high data rates, however, it is necessary to raise the energy threshold to maintain data quality because of the higher random coincident fraction, which is discussed below.

Random Coincidences

Random coincidences occur when two gamma rays originating from two unrelated positron annihilations are detected within the coincidence resolving time. The probability of two random events falling within a given

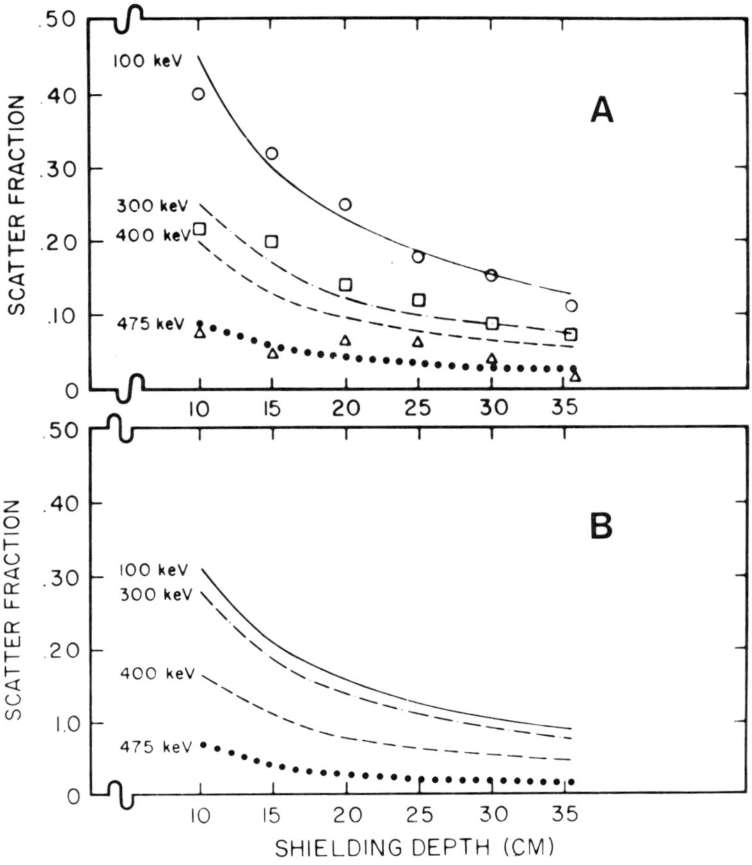

FIGURE 4.4. Scatter fraction for a ring system as a function of shielding depth for several energy thresholds for NaI(T1) detectors (**A**) and BGO detectors (**B**) [Mankoff, 1984].

time period in a detector pair is given by

$$R_r = 2t \cdot R_{S1} \cdot R_{S2} \tag{4.6}$$

where R_r is the random rate, R_{S1} and R_{S2} are the singles rates in the two detectors, and $2t$ is the coincidence resolving time. The coincidence resolving time is expressed as $2t$ since an event time, t, is associated with each detector, and an event is found in coincidence if there is any time overlap between the two events. Equation (4.6) is valid if the true coincidence rate is much smaller than the singles rate in each detector. If not, then the true coincidence rate must be subtracted from both R_{S1} and R_{S2}. Random coincidences generally add a uniform background to the projections, since the singles rates in all detectors vary slowly.

The true coincidence rate, R_C, is usually a small fraction of the average singles rates, R_S. If this fraction is denoted by k, then

$$R_C = k \cdot R_S \tag{4.7}$$

If we assume the singles rates in the two detectors to be equal ($R_S = R_{S1} = R_{S2}$), then the ratio of random coincidences to true coincidences is given by

$$\frac{R_r}{R_C} = \frac{2tR_S}{k} \tag{4.8}$$

It can be seen that the random fraction can be reduced by a smaller coincidence resolving time by counting at a lower singles rate or by increasing the fraction of counts found in coincidence through improved design. Even though the above expressions are only for a single pair of detectors, they can easily be expanded to a complete system. As more detectors are placed in coincidence with each other, the coincidence fraction, k, is improved (the singles rate per detector remains constant, but each new detector placed in coincidence increases the coincidence rate), but the total random rate also increases. Thus, for a uniform distribution, Equation (4.8) applies equally well to two banks of detectors.

Several methods have been used to subtract randoms from the image [Hoffman, 1981b]. The simplest but least accurate is to assume that the randoms vary smoothly over the field of view and to fit a curve based on the randoms outside the object. A second method is to measure the singles in each detector and determine the random rate from Equation (4.6). The third is to split one detector timing signal, delay it and measure the spatial distribution of the random coincidences as well as the usual true-plus-random coincidences so that the randoms can be subtracted either in real time or after data acquisition.

Signal-to-Noise Considerations

Nuclear medicine imaging has always had the advantage of high object contrast at the same time that it has suffered from a lack of sufficient count densities to achieve statistically smooth images. In emission computed tomography (ECT), these characteristics are emphasized. The uptake ratio in organs such as the brain, heart, or bone is often 5:1 (ie, an object contrast of 66%). In metabolic studies of the brain with ^{18}F-FDG, for example, the uptake ratio of gray matter to white matter is 4:1 (ie, an object contrast of 60%). In planar imaging, this object contrast is reduced, since the imaging process sums overlying and underlying activity, whereas in ECT the three-dimensional reconstruction preserves the high object contrast up to limitations imposed by spatial resolution, scatter, and attenuation.

Transverse section images require more counts than planar images to give statistically valid data. This supposition is supported by an analysis of statistical accuracy [Budinger, 1977], according to which more than 1 million counts are needed to achieve even 20% root-mean-square (RMS) noise in 1,000 "resolution cells." The requirement for a large number of counts per section translates into a need for high-sensitivity ECT imaging devices even at the cost of spatial resolution. According to Budinger [1977], as resolution is improved, more counts are needed in the image to maintain the same RMS noise. The argument seems to contradict formulations of lesion detectability and modulation transfer [Jaszczak, 1981; Phelps, 1982; Lim, 1982], according to which improved spatial resolution increases image contrast and thereby *reduces* the number of counts necessary to detect the lesion.

These two points of view are expressed in the two basic approaches to PET scanner design. One group of researchers selects very small scintillation crystals, typically BGO, to achieve the best possible spatial resolution, thereby aiming to preserve the object contrast in the imaging process, while another group of researchers uses time-of-flight (TOF) information to reduce the noise in the image and thereby effect an equivalent improvement in sensitivity. To understand the interplay of the various design and performance parameters such as detector resolution, TOF resolution, sensitivity, and reconstruction filter, it is important to formulate a concept relating image quality to these parameters.

Although a complete analysis must include the object, the imaging device, and the observer, we will restrict ourselves to a specific issue. Assuming the primary aim is to generate an image of a given radioisotope distribution that will be analyzed visually or quantitatively, we can restrict our analysis to a formulation of signal-to-noise (SNR) ratio. Such an analysis specifically excludes the effect of structured noise on lesion detectability and other visual factors influencing detectability and stresses instrument performance instead [Muehllehner, 1985].

To calculate the RMS noise in an ECT image, it is necessary to derive the error propagation in the reconstruction process. If the time of arrival of the two photons is determined to within a few nanoseconds, then this information is sufficient to ascertain that both events are probably due to a positron annihilation within the patient port, thereby excluding most random coincidences. If, however, the time of arrival can be determined to a fraction of a nanosecond, then the location along the path connecting the points of detection can be determined approximately from the TOF information. While TOF-assisted scanners employ reconstruction algorithms that are quite different in their specific implementations from conventional ECT reconstruction algorithms, the calculation of the variance can be formulated in general terms applicable to both situations [Tomitani, 1981]. Conceptually, the essential difference between the two reconstruction methods is the method of back-projection. If the timing information is insufficient to determine the origin of the annihilation within the object, then a constant value that is proportional to the measured intensity in that particular projection angle (θ) and location (r) (Figure 4.5A) is back-projected into the image space. If the time of arrival is known accurately enough (Figure 4.5B), then the back-projected values are "confidence-weighted" [Snyder, 1981; Wong, 1983]; that is, the back-projected intensity is replaced by a one-dimensional Gaussian function along the projection

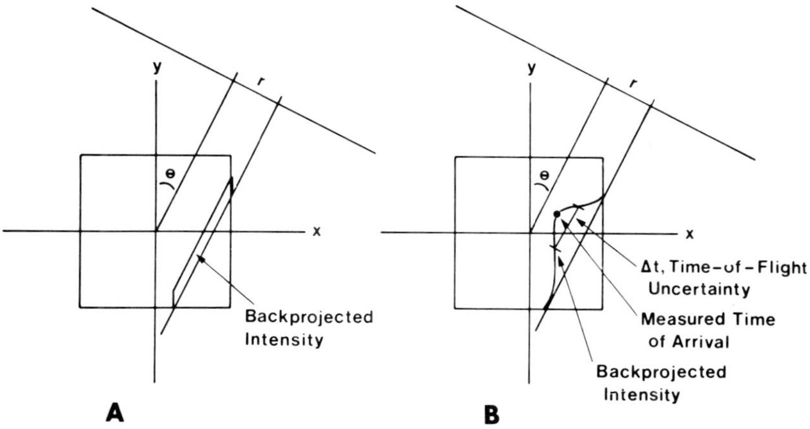

FIGURE 4.5. In conventional back-projection a constant intensity is projected along a line or strip (**A**); in back-projection using TOF information, the back-projected intensity reflects the confidence about the true location of the event (**B**).

ray with the mean at the measured point of annihilation and with a standard deviation appropriate for the accuracy of the coincidence timing. In calculating the variance [Tomitani, 1981] for a given point in a large uniform distribution of radioactivity, the noise contribution from each measured value is added, taking into account the appropriate weight given by the reconstruction filter. A complete derivation goes well beyond the scope of this chapter and the interested reader is referred to Tomitani [1981] and Wong [1983]. For systems utilizing the TOF information, the variance, V_{TOF}, is given by

$$V_{TOF} = \frac{0.69\, I_o \cdot S \cdot T \cdot t_{TOF}}{R^3} \tag{4.9}$$

where I_o is the positron-emission rate, S is the detection efficiency, T is the total measurement time, t_{tof} is the FWHM time resolution in spatial units, and R is the FWHM resolution in the reconstructed image due to the smoothing effect of the reconstruction filter.

If the TOF information is not utilized, then the variance is given by

$$V = \frac{0.43\, I_o \cdot S \cdot T \cdot d}{R^3} \tag{4.10}$$

for a uniform disk source of diameter, d. An equivalent diameter, d_{eq}, can be defined for TOF, which results in the same amount of variance as conventional PET without TOF. Comparing Equations (4.9) and (4.10), $d_{eq} = 1.6\, t_{TOF}$ and d_{eq}/d measures the variance reduction using TOF information. Thus, if the timing resolution is 300 ps, $d_{eq} = 14$ cm. The variance is therefore reduced with TOF information for objects larger than 14 cm. The factors, I_o, S, and T determine the total counts in the image

N, so that for a given object the variance is proportional to N and the RMS noise is proportional to $N^{1/2}$.

$$\mathrm{SNR} = \frac{C_{\mathrm{image}}}{\text{RMS noise}} = \frac{C_{\mathrm{image}}}{V^{1/2}} \tag{4.11}$$

where C_{image} is given by

$$C_{\mathrm{image}} = \frac{C_{\max} - C_{\min}}{C_{\max} + C_{\min}} \tag{4.12}$$

$C_{\max}$ and $C_{\min}$ are the counts per pixel in the target (ie, signal) area and the background area, respectively. While other definitions of contrast have been used (such as using only $C_{\min}$ in the denominator instead of $C_{\max} + C_{\min}$) this particular definition has the advantage that it allows the background to go to zero.

The image contrast is a function of both the object contrast and the spatial resolution of the imaging device. It can be calculated directly as a function of spatial resolution for simple objects such as those with sinusoidally varying intensities [Beck, 1968] or spheres in a uniform background [Whitehead, 1977], but generally not for more complicated objects. Figure 4.6 shows the measured contrast for the Derenzo phantom with uniformly spaced cylindrical structures of size and spacing as described in the figure caption. As the spatial resolution of an imaging device is improved, image contrast increases rapidly at a rate dependent on the size of the object until nearly 100% contrast is achieved.

From this analysis of SNR it is clear that the benefit of either high spatial resolution or TOF information is object-dependent. If the imaging task is to visualize a large, low-contrast target in a background of activity, then utilizing TOF information, and thereby reducing the noise, is more important than improving the spatial resolution, since little contrast improvement is achieved for large targets, although sharpening the edges improves visualization somewhat. If, on the other hand, the object is small and contains fine structures and the contrast is high, then TOF information is of little benefit, whereas an improvement in spatial resolution quickly improves the image contrast, thereby improving the SNR.

Detectors

Several different types of scintillation crystals are used in PET, depending on the design of the system. High-resolution systems with small discrete detectors generally use BGO because of its very high stopping power [Cho, 1977]. Compared with NaI(Tl), the crystal replaced by BGO in most PET scanners, BGO is about 50% more efficient and is not hygroscopic; but it has a much lower light output, thus poorer energy resolution, and somewhat poorer timing characteristics. Table 4.4 compares these scintillators with those used for TOF systems, which have fast decay times.

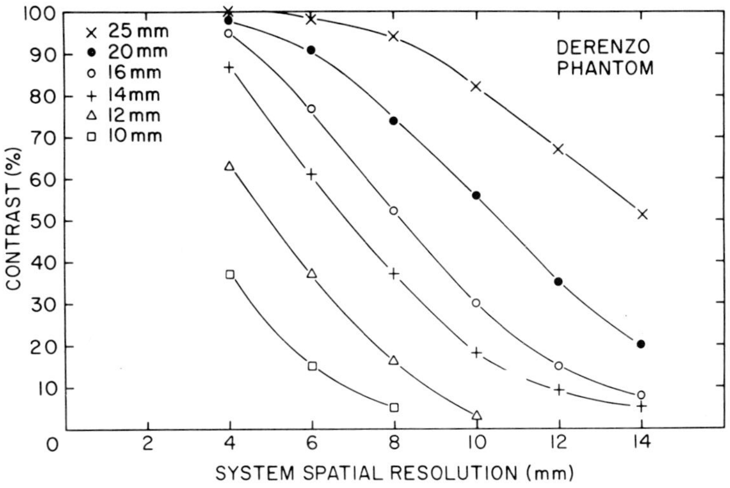

FIGURE 4.6. Image contrast measured in noise-free reconstructions plotted as a function of spatial resolution of the imaging device for all six sectors of the Derenzo phantom. Object contrast was 100%. Disks in Derenzo phantom are 6.25 mm in diameter spaced 25 mm center-to-center (x); 5-mm diameter spaced 20 mm (•); 4-mm diameter spaced 16 mm (○); 3.5-mm diameter spaced 14 mm (+); 3-mm diameter spaced 12 mm (△); and 2-mm spaced 10 mm (□) [Muehllehner, 1985].

Early efforts to use TOF information in medical imaging were not successful because the available scintillators had either insufficient timing accuracy [Brownell, 1969] or insufficient stopping power [Nickles, 1978]. Allemand et al [1980] investigated the use of cesium fluoride (CsF) and subsequently barium fluoride (BaF_2) [Laval, 1983], both of which have high stopping power and, owing to their fast decay time, good timing characteristics. Barium fluoride, like BGO, has the advantage that it is not hygroscopic, and therefore a high packing fraction can be achieved, re-

TABLE 4.4. Properties of scintillators used in PET.

	NaI(Tl)	BGO	CsF	BaF_2
Attenuation coefficient at 511 keV, cm^{-1}	0.34	0.92	0.44	0.47
Detection efficiency for 3-cm thick crystal	64%	94%	73%	76%
Decay time, ns	230	300	2.5	0.8 620
Resolving time* (FWHM), ns	1.0	2.5	0.4	0.3
Light output (relative)	100	15	6	4 (fast) 20 (slow)
Energy resolution (FWHM)	7%	15%	25%	13%

*Resolving time between two identical crystals in coincidence.

sulting in improved sensitivity compared with CsF. In addition, the fast component of BaF_2 is emitted at 220 nm and requires photomultipliers with a front entrance window and photocathode suitable for detection of ultraviolet light. Generally, non-TOF systems using BGO or NaI(Tl) have coincidence resolving times 5 to 10 times those quoted in Table 4.4. A timing window of at least 3 ns is needed to accept the gamma-ray transit-time differences in a typical whole-body scanner. In addition, the resolving times quoted in Table 4.4 can be achieved only if good timing photomultipliers are used, but these tend to be expensive and relatively large. Furthermore, the timing resolution deteriorates as scintillation crystals are made smaller. It is therefore difficult to build a PET tomograph that combines very high spatial resolution, generally achieved through the use of small, closely packed crystals, with good TOF characteristics.

In comparing positron scanners with different scintillation crystals, it is necessary to consider many factors. Among these are sensitivity, scatter fraction, random fraction, and the deleterious effect on spatial resolution of penetration through or scatter from one crystal to an adjacent crystal. This last effect, which is discussed later, is more pronounced for NaI(Tl) and BaF_2 than for BGO because of their lower stopping powers. For this reason, BGO has replaced NaI(Tl) in multicrystal scanners (eg, Donner Ring; see Derenzo, 1981) and is considered superior to BaF_2 as well in non-TOF scanners, despite the better timing characteristics and thus better randoms rejection of BaF_2 [Wong, 1985a]. If we assume that we are limited to a dose of activity such that the random fraction does not exceed unity [see Equation (4.8)], then the maximum true coincidence rate is given by

$$R_C = k^2/2t \tag{4.13}$$

The higher coincidence sensitivity of BGO, about 50% higher than BaF_2 and 100% higher than Na(Tl), offsets its longer coincidence time window. For fast dynamic studies with radioisotopes that have short decay times, one needs to be able to handle high doses and count at high rates at the onset of data collection in order to collect a large number of counts in the image. With BGO, one can achieve a given true coincidence rate, and thus total counts, with a lower dose of activity than with other scintillators. At the same time, the scatter fraction will be lower for BGO because a higher energy threshold can be used to eliminate scatter from the patient without eliminating true coincident events (see, for example, Figure 4.4).

High-Resolution Systems

Coding Schemes

The design of high-resolution positron cameras has led to the development of scintillation detectors that can localize a gamma-ray interaction to within a few millimeters with high efficiency. Multiwire proportional gas chambers

[Townsend, 1987] can detect charged particles and low-energy x-rays with very good spatial resolution in two dimensions (2 to 3 mm), but these counters are not very efficient for detecting 511 keV gamma rays, making their application to PET difficult. The recent trend in positron instrumentation has been to design multiple rings of densely packed detector arrays with many crystals that localize the gamma-ray interaction within the width of each discrete crystal. As mentioned earlier, the geometrical response in the center of two rectangular crystals each of width W is a triangular function with a FWHM of $W/2$. Therefore, to achieve an image resolution of 3 mm, at which point the effect of positron range and gamma-ray noncolinearity become important, the crystal width must be less than 4 mm (see Table 4.3). The straightforward solution of coupling one photomultiplier to each crystal to determine which crystal is "hit" is neither technologically feasible at the present time for such small crystals nor desirable in terms of complexity or cost. However, many ingenious solutions have been proposed and implemented so that systems are now being designed with crystals as narrow as 2 mm. Rather than mention every solution, we will discuss several general schemes and some unique approaches.

Generally, the width of the crystals being used today is less than 6 mm, since this dimension defines the in-plane resolution. The height, which defines the slice thickness and axial resolution, is usually at least 10 mm. The smallest photomultipliers available today are still larger than the crystals of interest. For example, Hamamatsu Corporation offers a 10-mm circular diameter photomultiplier and a 6 × 24-mm rectangular photomultiplier. To preserve the one-to-one coupling and the close packing of crystals, several methods have been devised to couple a small crystal to a large photomultiplier. The Donner ring [Derenzo, 1981] does this with a special set of light guides that place the photomultipliers at five different angles with respect to the image plane (Figure 4.7A). This ring uses 9.5 × 32 × 32-mm BGO crystals with 38-mm diameter photomultipliers. A newer design from this group [Derenzo, 1987] partially couples 14-mm diameter photomultipliers directly to three sides of 3 × 10 × 30-mm BGO crystals (Figure 4.7B). Both configurations, however, do not allow expansion to a multislice system, since the photomultipliers point out of the image plane. The ECAT III [Hoffman, 1986], a multislice device, uses larger BGO crystals, 5.6 × 29 × 30 mm, and partially couples photomultiplier tubes (PMTs) only to the back face in a staggered arrangement (Figure 4.7C). All partial coupling schemes, however, compromise both the energy and the timing resolution somewhat, since only part of the scintillation light reaches the photomultiplier.

Several specialized photomultipliers have been developed that can differentiate among several small crystals coupled to them. Photomultipliers with grid wires under each crystal are being used in PET [Murayama, 1984] to identify 4-mm-wide crystals (Figure 4.7D). Several photomultipliers with multiple anodes have been built, notably a 24 × 24-mm

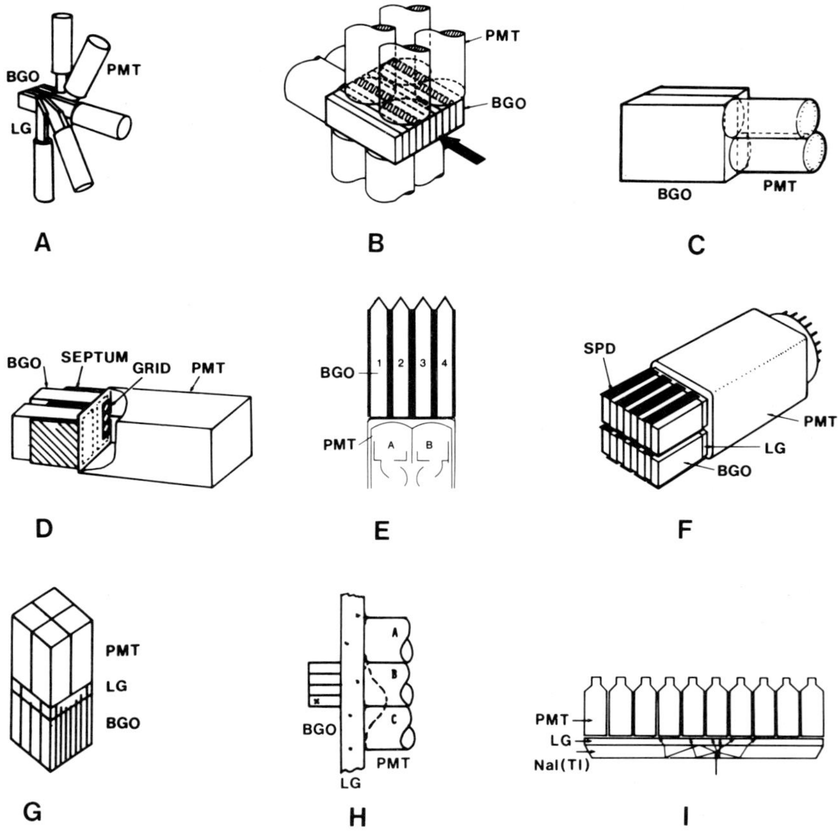

FIGURE 4.7. Representative types of coding schemes. I. *One-to-one coupling schemes:* **(A)** Donner ring [Derenzo, 1981] uses bismuth germanate (BGO) crystals coupled to photomultiplier tubes (PMT) through uniquely shaped light guides (LG). **(B)** Derenzo et al [1987] use partial coupling of PMTs to three faces of BGO crystals. **(C)** ECAT III partially couples PMTs in staggered arrangement to BGO crystals [Hoffman, 1986]. II. *Special photodetectors:* **(D)** Murayama et al [1984] use PMT with grid wire to identify two BGO crystals. **(E)** Roney et al [1984] use dual-anode PMT to identify four BGO crystals by comparing crosstalk between anodes. **(F)** Derenzo et al [1984] use silicon photodiodes (SPD) to identify each BGO crystal and a large PMT for coincident timing information. III. *Light distribution:* **(G)** Nutt et al [1985] use notched light guide to channel light in different ratios to four PMTs to identify 32 BGO crystals. **(H)** Burnham et al [1984] employ an analog-logic scheme to determine BGO crystal nearest the center of light distribution. **(I)** Muehllehner et al [1986] use a position-sensitive NaI(Tl) crystal and Anger position logic to localize event.

dual-anode photomultiplier [Yamashita, 1982]. Although a multiple-anode photomultiplier is likely to be more expensive than a conventional photomultiplier with photocathodes of similar size, it permits closer packing because of the reduction of the dead space of the thick glass walls. An interesting application of the dual-anode PMT [Roney, 1984] takes advantage of the intrinsic optical crosstalk between the two segments. This allows identification of four crystals, each 4.5 mm wide, since the center crystals exhibit more crosstalk to the opposite segment than the outside crystals do (Figure 4.7E).

It is possible to eliminate the photomultiplier altogether for crystal identification. Derenzo [1984] proposes using silicon photodiodes coupled to the top or bottom surface of 3-mm-wide BGO crystals (Figure 4.7F). These solid-state devices have high quantum efficiency and good energy resolution when cooled. While silicon photodiodes can provide photopeak energy discrimination as well as crystal identification, the timing characteristics are relatively poor. Derenzo suggests coupling a conventional photomultiplier to the back of a group of 16 crystals for coincidence time information. Since silicon photodiodes are thin, a multislice system can be formed by stacking several layers of crystals.

With conventional photomultipliers, several methods of crystal identification rely upon the characteristics of the crystal and its scintillation light distribution. Eriksson et al [1985] uses two different crystals with dissimilar timing characteristics coupled to a single photomultiplier. Bismuth germanate and cesium-doped gadolinium silicate (GSO) have similar stopping powers, but GSO has a much shorter decay constant (60 ns) than BGO (300 ns). The identification of the crystal is based upon rise-time discrimination. The feasibility of this scheme has been investigated using 6-mm-wide crystals and 14-mm-diameter photomultipliers.

The relative amount of light received by a photomultiplier is the principle behind the quad BGO detector [Murayama, 1982] used in Positologica II. The quad BGO detector consists of four 15-mm-wide crystals coupled to two 29-mm-diameter photomultipliers. By optically shielding between crystals 1 and 2 and between 3 and 4 and optically coupling between crystals 2 and 3, the "hit" crystal can be identified by the ratio of the light measured by the two photomultipliers.

An approach that carries this further is that used by Nutt et al [1985] with a block of 32 BGO crystals (8×4) coupled through a light guide to four photomultipliers (Figure 4.7G). By putting notches of varying lengths in the light guide between each crystal boundary in the transverse direction, the light from each crystal is channeled into each photomultiplier in different proportions.

A more continuous light sampling is used in the PCR-1 [Burnham, 1984] (Figure 4.7H). In this single-ring tomograph, 360 BGO crystals 4 mm wide are coupled through a light guide to 90 photomultipliers. The light distribution spreads in the light guide, and an analog encoding scheme determines the crystal closest to the center of brightness.

A truly continuous detector has been designed for use in PET [Muehllehner, 1986] whereby the position is determined from the centroid of the light distribution in the crystal (Figure 4.7I). This is the principle first suggested by Anger [1958]. Using 50-mm-diameter photomultipliers, the position is determined to 4 mm, or better than one twelfth the diameter of the photomultiplier. Since the spatial resolution depends on the statistical fluctuations of the photomultiplier signals [Karp, 1985], a scintillator with a high light output is essential. For this reason NaI(Tl) is used, since it has about seven times the light output of BGO, although only two-thirds the stopping power. This approach has been extended to a multislice device using two-dimensional position-sensitive detectors [Muehllehner, 1988].

Although many of these schemes seem different in application, there are two basic approaches. The first is to couple each crystal individually to a photodetector. This approach becomes increasingly difficult and costly as crystals get smaller. The second approach, which is becoming more popular, is to differentiate among as many crystals (or resolution elements, in the case of a continuous detector) as possible with as few photomultipliers as possible. The major drawback to most of these schemes is their inability to identify events that deposit energy in more than one crystal, leading to some loss of resolution. The effects of crystal penetration and interdetector spillage on spatial resolution are discussed in detail in the next section.

Depth of Interaction

As crystals are made narrower for high-resolution PET, penetration of the gamma ray through the crystal becomes more probable. In particular, oblique gamma rays may penetrate the first crystal and then interact in an adjacent crystal. Also, a gamma ray that undergoes Compton scattering in the first crystal produces a secondary gamma ray, which may pass through the crystal and interact in an adjacent crystal. Both cases result in mispositioning of the event. Even with BGO, as the crystal width is reduced to several millimeters, crystal penetration and interdetector spillage become significant. Cho et al [1984] have calculated the probability of normally or obliquely incident 511-keV gamma rays having any interaction in crystals of varying widths. For an 8 × 20 × 40-mm BGO crystal, the efficiency of detecting the gamma ray in the primary crystal is 73% at normal incidence and 35% at 30° from normal incidence, an angle common in a whole-body tomograph. For a 4-mm-wide BGO crystal, the primary detection efficiency drops to 20% at 30° incidence. Therefore, many of the obliquely incident gamma rays interact in the adjacent-to-the-primary crystal or even the next-to-adjacent crystal, causing position uncertainty.

Most current PET rings are circular with the crystals radially directed, as shown in Figure 4.8A. A high probability of crystal penetration and interdetector spillage causes the spatial resolution to worsen, particularly

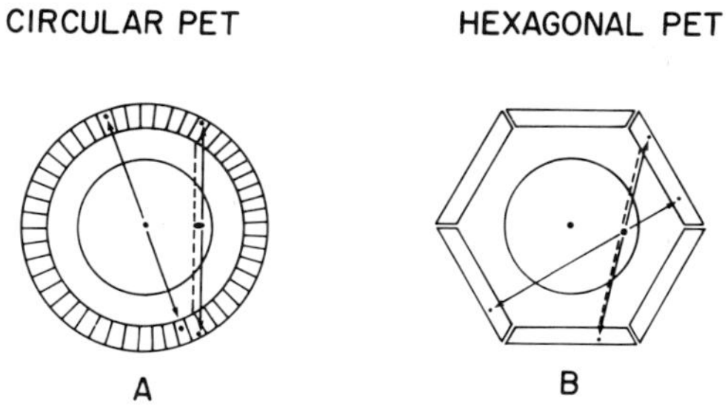

FIGURE 4.8. Spatial resolution is degraded in circular systems as a function of radial position owing to penetration of radiation into adjacent crystals (**A**). In hexagonal systems spatial resolution is degraded uniformly throughout the field of view owing to uncertainty regarding the depth of interaction in the crystal (**B**).

in the radial direction, as the distance from the center of the ring increases. Figure 4.9 shows this effect for BGO and NaI(Tl) crystals of similar width [Derenzo, 1981]. The radial elongation results from obliquely incident gamma rays that may either interact near the front of the primary crystal, or pass through and interact near the rear of an adjacent crystal. Since the depth of interaction is not known, the gamma ray is assumed to interact at the surface of the crystal, introducing an error in its trajectory and thus the line along which the event is backprojected.

In addition to using BGO, the densest scintillator available, there are several ways of reducing crystal penetration and interdetector spillage. Interdetector tungsten septa have been used [Erikson, 1985]; however, this reduces the packing fraction and thus the sensitivity. A more recent idea [Cho, 1984; Holmes, 1985; Roney, 1984] is to shape the front of the crystal to form a wedge to create a longer path for obliquely incident gamma rays. Cho et al [1984] calculated that with a 20° wedge, the primary detection efficiency increases by 33% relative to a conventional rectangular shape for an 8-mm wide crystal and by 40% for a 4-mm-wide crystal. Another idea to reduce the effect of interdetector spillage is to use two energy discriminators to reject those events that interact in two adjacent crystals through Compton scattering [Derenzo, 1983].

Optimally, one would determine the depth of interaction to compensate for the uncertain path length of the gamma rays. As a first step, Wong [1985b] has proposed a stratified ring with two layers of 15-mm-deep BGO crystals. The back layer is staggered relative to the front layer. This allows determination of the depth to within 15 mm and doubles the linear sampling but does not permit a tomograph of more than two rings. Wong predicts that this would reduce radial elongation at distances from the center of the ring to the same extent as 2-mm tungsten septa without the loss of sensitivity.

Derenzo [1984] mentions the possibility of using position-sensitive silicon

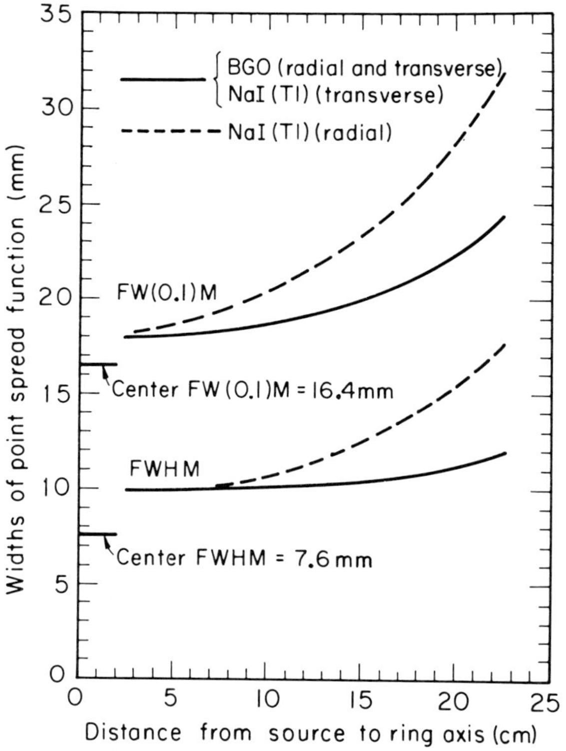

FIGURE 4.9. Variations of response as a function of position and direction in the Donner ring with a detector diameter of 92 cm [Derenzo, 1981].

photocathodes. Coupled along the top or bottom of a narrow crystal, the silicon photocathode could then identify the crystal and provide depth of interaction information.

With a continuous crystal such as that used in the PENN-PET [Muehllehner, 1986], it is possible to reduce the parallax error by determining the depth of interaction [Cook, 1985]. Compared to a circular scanner, the spatial resolution of a hexagonal scanner does not worsen as significantly as the distance from the center increases; rather, every point in the image has contributions from gamma rays entering the crystals obliquely (Figure 4.8B). Still, the uncertainty of the depth of interaction causes an average degradation to the spatial resolution of 1.5 mm. Preliminary experiments have shown that the depth of interaction may be determined by measuring the width of the light response distribution in the crystal [Karp, 1986; Rogers, 1986] or by measuring the scintillation decay time with a temperature gradient across the crystal [Karp, 1987].

The uncertainty about the depth of interaction can cause the resolution to worsen at the edge of the field of view of a circular tomograph by as much as 50%. Although the object being imaged does not usually extend out to the edge of the field of view, the quality and versatility of a scanner are compromised without the ability to produce distortion-free images anywhere inside the field of view. Particularly with image resolutions reaching 3 to 5 mm, the uncertainty of the depth of interaction becomes the most important limiting factor in spatial resolution.

Time-of-Flight Systems

In conventional positron tomography two gamma rays are detected in coincidence defining a line through the object to be imaged, the endpoints of which are given by the location of the detectors. The origin of the two gamma rays, ie, the point of annihilation of the positron, can be anywhere along that line, since typically the coincidence resolving time is too long. In TOF-assisted PET systems, the time difference between the two annihilation photons arriving at the detectors is measured to obtain additional information about the point of annihilation within the object. The time difference between the arrival of the two photons, Δt, is related to the location of the annihilation:

$$\Delta t = \frac{2\Delta x}{c}$$

where Δx is the distance from the source to center of the scanner and c is the velocity of light. Since $\Delta x = 15$ cm results in a time difference of only 1 ns, very good coincidence resolving times must be achieved to gain meaningful additional information. Good timing information can be achieved only through careful choice of scintillation crystals and photomultipliers. Using the values of timing resolution in Table 4.4 for CsF or BaF_2, one can expect position information with an error of about 5 cm FWHM. The main advantage of TOF PET imaging is an improvement in the SNR through a reduction in the noise (as discussed previously), but the technique also results in instruments with high countrate capability and good rejection of random coincidences, making TOF PET systems particularly useful for fast dynamic studies.

Countrate Capability and Randoms Reduction

The short decay time of fast scintillators allows a PET system to handle very high countrates without interference between consecutive pulses, ie, it results in low deadtime and negligible pulse pileup. Alternatively, high countrate capability can be achieved by distributing the counts over more independent channels by using many small crystals. Generally, neither TOF systems nor systems using hundreds of BGO crystals per ring are limited by the scintillation decay time in the maximum achievable countrate but rather by deadtime effects in some associated electronics or data-handling system.

One of the primary advantages of TOF-assisted PET systems is their increased ability to reject random coincidence events. While in equilibrium studies such as ^{18}F-FDG brain metabolism measurements or neuroreceptor studies, the random coincidence fraction is negligible or relatively low; dynamic studies often involve the injection of up to 100 mCi of short-lived tracers such as oxygen-15, ^{15}O ($t_{1/2}$ = 120 seconds), or ^{82}Rb ($t_{1/2}$ = 75 seconds), resulting in a high random fraction. Particularly in body-imaging systems, the overall coincidence window must be at least 3 ns if one wishes to accept most of the true coincidences. In TOF systems, however, the random coincidences that fall within the coincidence window are back-projected over a relatively small volume, thereby reducing the noise contribution from the random coincidences in the same way that the noise contribution from true coincidences is reduced [Holmes, 1984].

Discussion

The dramatic improvements in image quality of PET systems in recent years has been achieved by improving the SNR with the use of two methods: (1) improving the spatial resolution to achieve better image contrast and therefore a higher signal, and (2) using TOF information to reduce the image noise. Obviously, if the object contains mostly large, low-contrast structures, then an improvement in spatial resolution will not improve the image quality significantly. Similarly, if the object has a small diameter and high contrast, then TOF measurements will add little useful information. The type of study thus determines the selection of the PET scanner. If the primary use of the PET scanner is to image equilibrium distributions of radiopharmaceuticals in small high-contrast objects such as the brain, then a high-resolution scanner would be selected. If, on the other hand, one wishes to study dynamic processes in large-volume distributions of activity, then TOF-assisted tomography gives the best results. The ideal PET scanner would combine the virtues of both techniques, ie, it would combine the use of small crystals to achieve good spatial resolution with the ability to reduce the noise through good coincidence resolving time. Technical difficulties in the proper crystal-photomultiplier configuration make this goal elusive at the present time.

As the performance of PET scanners has improved, their complexity has also increased. State-of-the-art systems often use more than 500 scintillation crystals, and some groups are contemplating multislice PET scanners with thousands of crystals [Burnham, in press, 1988]. Such systems are designed so that each crystal does not require a separate photomultiplier. At the same time, the task of processing the data to obtain reconstructed images has grown in complexity, particularly in TOF-assisted scanners. As the computing power of associated minicomputers increases rapidly through advanced integrated circuit technology, the use of special processors for operations such as back-projection is also growing.

The relationship between the spatial resolution in the transverse and axial directions needs to be studied in more detail. Thus far, we have discussed only the transverse resolution. The axial resolution is typically two to three times the transverse resolution. Limited axial resolution blurs objects in the axial direction and reduces the contrast of small objects. While many groups have reduced the crystal width in the transverse plane to 6 mm or less, crystal height ranges from 10 to 30 mm. Many crystal-PMT coupling schemes allow fine spatial sampling in only one direction.

If a slice is thick relative to the anatomical structures, then these structures will often be partly within a slice and partly outside, resulting in partial volume effects. With an unknown radioisotope distribution, it is possible to underestimate the activity by up to a factor of two, depending on the location of the structure relative to the center of the slice. One way to reduce this problem is to increase the axial sampling by moving either the patient or the scanner in the axial direction to interleave slices; however, this partial remedy increases the imaging time. As slices are made thinner, the sensitivity decreases if the axial angle of acceptance is also decreased. This is the current practice where only straight-across coincidences and coincidences between adjacent rings, "cross slices," are used.

These considerations point to the advantages of a scanner with equal spatial resolution and fine spatial sampling in all three dimensions, which would allow the object to be reconstructed as a true volume distribution. The ability to reslice the imaged object in any direction after data acquisition, as is currently done in single-photon ECT, would be very desirable. Such a scanner would require three-dimensional volume reconstruction and display techniques, which have been investigated for other imaging modalities and can be applied, in part, to PET.

In conclusion, there have been major strides during the past few years in PET instrumentation. Many systems now being designed are approaching the fundamental limits of spatial resolution, although determining the depth of interaction in a thick crystal is still a limiting factor. Many ingenious crystal-PMT encoding schemes, which simplify the system without sacrificing performance, are of great interest, as lower costs will mean PET can be moved from university research to the clinical setting.

References

Allemand R, Gresset C, Vacher J: Potential advantages of a cesium fluoride scintillator for a time-of-flight positron camera. J Nucl Med 21: 153–155 (1980).

Anger HO: Scintillation camera. Rev Sci Instrum 29–33: 27 (1958).

Atkins F: Monte Carlo analysis of photonscattering in radionuclide imaging. Doctoral thesis, Univ. of Chicago, Chicago, IL, 1978.

Beck RN, Harper PV: Criteria for Evaluating Radioisotope Imaging Systems. Gottschalk A, Bek RN (eds). Springfield, IL, Charles C Thomas (1968) pp 348–382.

Bergstrom M, Eriksson L, Bohm C, et al: Correction for scattered radiation in a ring detector positron camera by integral transformation of the projections. J Comput Assist Tomogr 7:42–50 (1983).

Bohm C, Eriksson L, Bergstrom M, et al: A computer assisted ringdetector positron camera system for reconstruction tomography of the brain. IEEE Trans Nucl Sci NS-25: 624–637 (1978).

Brooks RA, Sank VJ, Talbert AJ: Sampling requirements and detector motion for positron emission tomography. IEEE Trans Nucl Sci NS-26: 2760–2763 (1979).

Brooks RA, Weiss GH, Talbert AJ: A new approach to interpolation in computed tomography. J Comput Assist Tomogr 2:577–585 (1978).

Brownell GL, Burnham CA, Wilensky S, et al: New developments in positron scintigraphy and the application of cyclotron-produced positron emitters, in: *Medical Radioisotope Scintigraphy,* vol 1, IAEA, Vienna (1969) p 466.

Budinger TF, Derenzo SE, Gullberg GT, et al: Emission computer assisted tomography with single-photon J Comput Assist Tomogr 1:131–145 (1977).

Burnham CA, Bradshaw J, Kaufman D, et al: A stationary positron emission ring tomograph using BGO detector and analog readout. IEEE Trans Nucl Sci NS-31: 632–636 (1984).

Burnham CA, Kaufman DE, Chesler DA, et al: Cylindrical PET detector design. IEEE Trans Nucl Sci NS-35 (1988) (in press).

Cho ZH, Chan JK, Eriksson L, et al: Positron ranges obtained from biomedically important positron-emitting radionuclides. J Nucl Med 16: 1174–1176 (1975).

Cho ZH, Farukhi MR: Bismuth germanate as a potential scintillation detector in positron camera. J Nucl Med 18: 840–844 (1977).

Cho ZH, Hong KS, Ra JB, et al: A new sampling scheme for the ring positron camera: dichotomic ring sampling. IEEE Trans Nucl Sci NS-28: 94–98 (1981).

Cho ZH, Lee HS, Hong KS: Wedge-shaped BGO scintillation crystal for positron emission tomography: Concise communication. J Nucl Med 25: 901–904 (1984).

Cook WR, Finger M, Price TA: A thick Anger camera for gamma-ray astronomy. IEEE Trans Nucl Sci NS-32: 129–133 (1985).

De Benedetti S, Cowan CE, Konneker WR, et al: On the angular distribution of two-photon annihilation radiation. Phys Rev 77: 205–212 (1950).

Derenzo SE: Precision measurement of annihilation point spread distributions for medically important positron emitters, in Proceedings of the Fifth International Conference on Positron Annihilation, Japan, (1979), pp 819–823.

Derenzo SE: Method for optimizing shielding in positron-emission tomographs and for comparing detectors materials. J Nucl Med 21: 971–977 (1980).

Derenzo SE: Initial characterization of a BGO-photodiode detector for high resolution positron emission tomography. IEEE Trans Nucl Sci NS-31: 620–626 (1984).

Derenzo SE, Budinger TF: Resolution limit for positron imaging devices. J Nucl Med 18:491 (1977).

Derenzo SE, Budinger TF, Huesman RH: Imaging properties of a positron tomograph with 280 BGO crystals. IEEE Trans Nucl Sci NS-28: 81–89 (1981).

Derenzo SE, Budinger TF, Huesman RH: Detectors for high resolution dynamic positron emission tomography, in Greitz T, Ingvar DH, Widen L (eds): *The Metabolism of the Human Brain Studied with Positron Emission Tomography*. New York, Raven Press (1985) pp 21–31.

Derenzo SE, Budinger TS, Vuletich T: High resolution positron-emission tomography using small bismuth germanate crystals and individual photo sensors. IEEE Trans Nucl Sci NS 30: 665–670 (1983).

Derenzo SE, Huesman RH, Cahson JL, et al: Initial results from the Donner 600 Crystal Positron Tomograph, IEEE Trans Nucl Sci NS-34: 321-325 (1987).

Eriksson L, Bohm C, Kesselberg M, et al: A high resolution positron camera, in Greitz T (ed): *The Metabolism of the Human Brain Studied with Positron Emission Tomography*. New York, Raven Press (1985) pp 33–46.

Hoffman EJ, Huang S, Phelps ME, et al: Quantitation in positron emission computed tomography: 4. Effect of accidental coincidences. J Comput Assist Tomogr 5: 391–400 (1981b).

Hoffman EJ, Phelps ME: An analysis of some of the physical aspects of positron transaxial tomography. Comput Biol Med 6:345–360 (1976b).

Hoffman EJ, Phelps ME, Huang SC, et al: A new tomograph for quantitative positron emission computed tomography of the brain. IEEE Trans Nucl Sci NS-28: 99–103 (1981a).

Hoffman EJ, Phelps ME, Huang SC, et al: Dynamic gated and high resolution imaging with the ECAT 3. IEEE Trans Nucl Sci NS-33: 452–455 (1986).

Hoffman EJ, Phelps ME, Mullani NA, et al: Design and performance characteristics of a whole-body positron transaxial tomograph. J Nucl Med 17:493–502 (1976a).

Hoffman EJ, van der Stee M, Ricci AR, et al: System design features that are determinate in precision and accuracy in positron emission tomography, in Greitz T, Ingvar DH, Widen L (eds): *The Metabolism of the Human Brain Studied with Positron Emission Tomography*. New York, Raven Press (1985) pp 69–83.

Holmes TJ, Ficke DC: Analysis of positron-emission tomography scintillation-detectors with wedge faces and inter-crystal septa. IEEE Trans Nucl Sci NS-32: 826–830 (1985).

Holmes TJ, Snyder DL, Ficke DC: The effect of accidental coincidences in time-of-flight positron emission tomography. IEEE Trans Med Imaging MI-3: 68–79 (1984).

Huang SC, Hoffman EJ, Phelps ME, et al: Quantitation in positron emission computed tomography: 3. Effect of sampling. J Comput Assist Tomogr 4: 819–826 (1980).

Huesman RH: The effects of a finite number of projection angles and finite lateral sampling of projections on the propagation of statistical errors in transverse section reconstruction. Phys Med Biol 22: 511–521 (1977).

Ido T, Wan CN, Casella V: Labeled 2-deoxy-D-glucose analogs. F-18 labeled 2-deoxy-2-fluoro-D-glucose, 2-deoxy-2-fluoro-D-mannose and C-14 2-deoxy-12-fluoro-D-glucose. J Label Compds Radiopharm 24: 174–183 (1978).

Jaszczak RJ, Coleman RE, Whitehead FR: Physical factors affecting quantitative measurements using camera-based single photon emission computed tomography (SPECT). IEEE Trans Nucl Sci NS-28: 69–80 (1981).

Karp JS, Daube-Witherspoon ME: Determination of depth-of-interaction in scintillation crystals using a temperature gradient. Nucl Inst and Meth A260: 509–517 (1987).

Karp JS, Muehllehner G: Performance of a position-sensitive scintillation detector. Phys Med Biol 30: 643–655 (1985).

Karp JS, Muehllehner G, Beerbohm D, et al: Event localization in a continuous scintillation detector using digital processing. IEEE Trans Nucl Sci NS-33: 550–555 (1986).

Laval M, Moszynski M, Allemand R, Cormoreche E, Odru R, Vacher J: Barium fluoride: inorganic scintillator for subnanosecond timing. Nucl Instr Meth 206: 169–176 (1983).

Lim CB, Han KS, Hawman EG, et al: Image noise, resolution, and lesion detectability in single photon emission CT. IEEE Trans Nucl Sci NS-29: 500–505 (1982).

Llacer J, Graham LS: The effect of improving energy resolution on gamma camera performance: a quantitative analysis. IEEE Trans Nucl Sci NS-22: 309–320 (1975).

Mankoff D, Muehllehner G: Performance of positron imaging systems as a function of energy threshold and shielding depth. IEEE Trans Med Imaging MI-3: 18–24 (1984).

Muehllehner G: Resolution limit of positron cameras. J Nucl Med 17: 757 (1976).

Muehllehner G: Effect of resolution improvement on required count density in ECT imaging: a computer simulation. Phys Med Biol 30: 163–173 (1985).

Muehllehner G, Colsher, JG, Lewitt RM: A hexagonal bar positron camera: problems and solutions. IEEE Trans Nucl Sci NS-30: 652–660 (1983).

Muehllehner G, Karp JS: A positron camera using position-sensitive detectors: PENN-PET. J Nucl Med 27: 90–98 (1986).

Muehllehner G, Karp JS, Mankoff DA, et al: Design and performance of a new positron tomograph. IEEE Trans Nucl Sci NS-35 (1988) (in press).

Murayama H, Nohara N, Tanaka E, et al: A quad BGO detector and its timing and positioning discrimination for positron computed tomography. Nucl Instr Meth 192: 501–511 (1982).

Murayama H, Tanaka E, Nohara N, et al: Twin BGO detectors for high resolution positron emission tomography. Nucl Instr Meth 221: 633–640 (1984).

Nickles RJ, Meyer HO: Design of a three-dimensional positron camera for nuclear medicine. Phys Med Biol 23: 686–695 (1978).

Nohara N, Tanaka E, Tomitani T, et al: Positologica: a positron ECT device with a continuously rotating detector ring. IEEE Trans Nucl Sci NS-27: 1128–1136 (1980).

Nohara N, Tanaka E, Tomitani T, et al: Analytical study of performance of high-resolution positron emission computed tomographs for animal study. IEEE NS-32: 818–821 (1985).

Nutt R, Casey M, Carroll LR: A new multi-crystal two dimensional detector block for PET. J Nucl Med 26: P28 (1985).

Phelps ME, Hoffman EJ, Huang SC, et al: Effect of positron range on spatial resolution. J Nucl Med 16: 649–652 (1975b).

Phelps ME, Hoffman EJ, Huang SC, et al: ECAT: a new computerized tomographic imaging system for positron-emitting radiopharmaceuticals. J Nucl Med 19: 635–647 (1978).

Phelps ME, Hoffman EJ, Mullani NA, et al: Application of annihilation coincidence detection to transaxial reconstruction tomography. J Nucl Med 16: 210–224 (1975a).

Phelps ME, Huang SC, Hoffman EJ, et al: An analysis of signal amplification using small detectors in positron emission tomography. J Comput Assist Tomogr 6: 551–565 (1982).

Reivich M, Kuhl D, Wolf A: Measurement of local cerebral glucose metabolism in man with F-18 2-fluoro-2-deoxy-D-glucose. Acta Neurol Scand 56 [suppl 64]: 192–193 (1977).

Rogers JG, Saylor DP, Harrop R, et al: Design of an efficient position sensitive gamma ray detector for nuclear medicine. Phys Med Biol 31: 1061–1090 (1986).

Roney JM, Thompson CJ: Detector identification with four BGO crystals on a dual PMT. IEEE Trans Nucl Sci NS-31: 1022–1027 (1984).

Snyder DL, Cox JR: An overview of reconstructive tomography and limitations imposed by a finite number of projections, in *Reconstruction Tomography in Diagnostic Radiology and Nuclear Medicine*. Baltimore, MD, University Park Press. (1977) pp 3–32.

Snyder DL, Thomas LJ, Ter-Pogossian MM: A mathematical model for positron-emission tomography systems having time-of-flight measurements. IEEE Trans Nucl Sci NS-28: 3575–3583 (1981).

Tanaka E, Nohara N, Tomitani T, et al: Analytical study of the performance of a multilayer positron computed tomography scanner. J Comput Assist Tomogr 6: 350–364 (1982).

Ter-Pogossian MM, Mullani NA, Hood J, et al: A multislice positron emission computed tomograph (PETT IV) yielding transverse and longitudinal images. Radiology 128: 477–484 (1978a).

Ter-Pogossian MM, Mullani NA, Hood J, et al: Design considerations for a positron emission transverse tomograph (PETT V) for imaging of the brain. J Comput Assist Tomogr 2: 539–544 (1978b).

Tomitani T: Image reconstruction and noise evaluation in photon time-of-flight assisted positron emission tomography. IEEE Trans Nucl Sci NS-28: 4582–4589 (1981).

Townsend D, Frey P, Jeavons A, et al: High Density Avalanche Chamber (HIDAC) Positron Camera. J Nucl Med 28: 1554–1562 (1987).

Whitehead FR: Quantitative analysis of minimum detectable lesion-to-background uptake ratios for nuclear medicine imaging systems, in *Medical Radionuclide Imaging*. Vienna, IAEA (1977) pp 409–434.

Wong WH: An analytical comparison between bismuth germanate and barium fluoride scintillators for non-time-of-flight positron emission tomography cameras. J Nucl Med 26: P100 (1985a).

Wong WH: A new stratified detection system for positron emission tomography cameras. J Nucl Med 26: P28 (1985b).

Wong WH, Mullani NA, Phillippe EA, et al: Image improvement and design optimization of the time-of-flight PET. J Nucl Med 24: 52–60 (1983).

Yamashita T, Ito M, Hayashi T: New dual rectangular photomultiplier tube for positron CT. Proceedings of the International Workshop on Physics and Engineering in Medical Imaging. Asilomar, CA, March 15–18, 1982.

5
Introduction to Ultrasonic Pseudo-Random Code Systems

VERNON L. NEWHOUSE, D. CATHIGNOL, AND J.Y. CHAPELON

Objectives and Scope

In the last decade many successful and advantageous sonic systems for imaging, flow measurement, and medium characterization have been published, which use pseudo-random coding to provide performance greatly superior to that of classic pulsed-echo techniques. However, no review of such pseudo-random code systems has been published, nor is the subject treated in any textbooks. Chapters 5 through 9 are intended to fill this gap by providing a tutorial introduction to the design of pseudo-random code systems and by reviewing the published work in this field as well as the work that either has not been published before or exists only in conference proceedings or theses. Since the review is dedicated to the use of pseudo-random code in ultrasound applications, it does not cover their use in spread spectrum communications or in spectroscopy. These two subjects are respectively covered by Dixon [1979] and Harwit and Sloan [1979].

Historically speaking, noise imaging and Doppler systems [Newhouse et al, 1974; Newhouse and Bendick, 1974] preceded their pseudo-random code counterparts and served as a point of departure for them. Hence, in this review we start by describing the properties of noise imaging and Doppler systems as a pedagogical device before studying the somewhat more complex properties of pseudo-random code systems.

The presentation is divided into chapters on one-dimensional imaging and system characterization,* on flow measurement, on the characterization of time-changing media, and on multimode imaging (usually two-dimensional). Chapter 6 on one-dimensional imaging also serves to introduce important properties of noise and pseudo-random codes, relating to the correlation functions of noise, of maximum length codes, and of Golay codes. Chapter 7 on flow measurement also introduces the subject of noise

*The published work on system characterization does not involve ultrasound and is not therefore strictly appropriate for this review. It is included because of its close conceptual relationship to the use of pseudo-random code for A-scan imaging.

and code ambiguity functions. These functions are of importance both for that chapter as well as for Chapter 8, which describes methods used to measure the time-changing properties of the ocean, considered as an acoustic transmission medium.

The last part of the review (Chapter 9) on multimode operation describes work on the simultaneous transmission of coded signals in different directions from a phased array. This work has not yet come to fruition, but promises to have important applications to systems for imaging moving targets. The remainder of this chapter describes the problems addressed in the remainder of the review.

Noise Correlation A-Scan Imaging

To introduce the subject of our review we start by describing the noise correlation A-scan receiver and by summarizing those of its properties that distinguish it from the better-known pulse-echo systems. As previously mentioned, the noise correlation receiver forms the basis of most of the systems described later.

The basic ultrasound noise correlation system [Newhouse et al, 1974; Furgason et al, 1975] transmits a pulsed or continuous broadband noise signal $x(t)$ via a conventional piezoelectric ultrasound transducer (Figure 5.1). With a single plane reflector at a distance corresponding to a time of flight τ_e, the echo can be written as $x(t - \tau_e)$ if the bandwidth limiting effects of the transmit and receive transducers are neglected. The echo is amplified and then multiplied with the so-called reference signal, after which it is integrated over a period T. The reference signal $x(t - \tau_r)$ consists of a copy of the transmitted signal, delayed by τ_r seconds by means of a delay line, which in the earliest ultrasound noise correlation system consisted of two transducers identical with those used for transmission and reception, separated by an adjustable water gap. The output of the correlation system can be written as

$$y(t) = \int_{t-T}^{t} x(t - \tau_e)\, x(t - \tau_r)\, dt \equiv R_x^T(\tau_e - \tau_r) \tag{5.1}$$

where $R_x^T(\tau)$ is defined as the time average of $x(t)x(t - \tau)$ (assumed stationary) over time T.

Provided that the integration time T is sufficiently large with respect to the decorrelation time θ of the transmitted noise $x(t)$, $R_x^T(\tau_e - \tau_r)$ approaches the autocorrelation function of $x(t)$, which will be written as $\Gamma_x(\tau_e - \tau_r)$. The decorrelation time θ is defined as the shortest time interval for which two successive samples of the noise signal $x(t)$ are no longer correlated. It can be shown [see, for instance, Papoulis, 1965] that

$$\theta = \frac{1}{B} \tag{5.2}$$

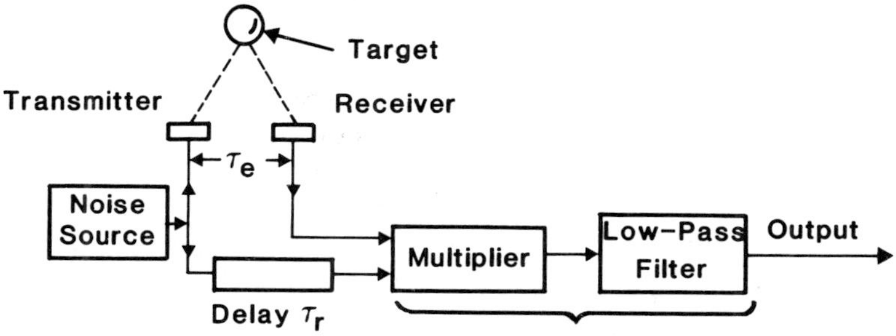

FIGURE 5.1. Basic ultrasound random signal correlation system (Newhouse et al. [1974].

where B is the bandwidth of $x(t)$. It should also be recalled that $R_x(\tau)$, the autocorrelation function of the signal $x(t)$, is the Fourier transform of the spectral density $S(\omega)$ of $x(t)$, ie,

$$R_x(\tau) \leftrightarrow S(\omega) \tag{5.3}$$

Thus if, for example, the transmitted noise spectrum is bell-shaped of form

$$S(\omega) = \frac{B/2\pi}{(B/2)^2 + (f - f_o)^2}$$

as shown in Figure 5.2a, the correlator output will approach the form shown in Figure 5.2b:

$$R_x(\tau_e - \tau_r) = \exp\left[-\pi B \,|\, \tau_r - \tau_e \,|\, \right] \cos 2\pi f_o(\tau_r - \tau_e) \tag{5.4}$$

If the reference signal delay in Figure 5.1 is changed slowly enough with respect to time, then in the presence of a single target having range delay τ_e, the correlation receiver output signal, which can be displayed on the y axis of a pen recorder for instance, traces out the function $R_x^T(\tau)$ of the transmitted signal, provided that the x-axis excursion of the pen recorder is controlled by the time delay τ_r of the reference signal. The peak of $R_x(\tau)$ is at $\tau_r = \tau_e$, corresponding to the location of the target. For the type of spectrum illustrated, the "1/e power" width of the envelope of $R_x(\tau)$ is seen to be $2/\pi B = \theta$ seconds. Hence the spatial resolution of the system may be written as

$$\Delta R = \frac{2c}{\pi B} \tag{5.5}$$

where c is the ultrasound velocity. This is close to the resolution $c/2B$ obtained with a pulsed radio frequency (RF) system emitting pulses of the same bandwidth as the noise signal, ie, having length $t = 1/B$.

Advantages of Noise Correlation Systems

Peak-to-Average Power Ratio

The result derived immediately above shows that the great advantage of noise systems, which is shared by pseudo-random code systems, is that high spatial resolution can be obtained with signals of any length, whereas with the classic pulsed RF system, high bandwidth, and thus high resolution, can only be obtained with a correspondingly short transmitted signal. Thus when observing isolated targets, a pseudo-random code system can use signals of arbitrary length without sacrificing resolution and thus requires a much smaller peak transmitted power than does an RF system of equal resolution transmitting the same average power. This can be important in ultrasonic imaging of thick specimens because the maximum transmitted power for pulsed RF systems under these circumstances is usually limited by the peak voltage that can be applied to the transducer. With a noise or pseudo-random code system the power can be transmitted with a larger on-off ratio, so that the limit on transmitted power is determined by the average power-carrying capacity of the transducer. For this reason noise systems can usually transmit more average power than pulsed RF systems.

For medical imaging it is also desirable to distribute the radiated power as uniformly as possible, so as to minimize the peak power to which the tissues are exposed. This again favors a noise or pseudo-random code correlation system over a pulsed RF system.

High Signal-to-Noise Ratio Enhancement

One of the remarkable properties of the noise correlation receiver is its ability to enhance a poor input signal to noise ratio. If, for example, the target has a reflectivity a, then the amplifier input signal can be written as

$$s_{\text{in}}(t) = a\, x(t - \tau_e) + n(t)$$

where $n(t)$ is the thermal noise voltage of the amplifier referred to the input. The input power signal-to-noise ratio may then be written as

$$\text{SNR}\,|_{\text{in}} = \frac{a^2 < x^2(t - \tau_e) >}{< n^2(t) >}$$

where $<>$ indicates averaged with respect to time. The output signal may be written as

$$s_{\text{out}}(\tau_e - \tau_r) = \frac{1}{T} \int_0^T x(t - \tau_r) \left[ax(t - \tau_e) + n(t) \right] \mathrm{dt}$$

$$= aR_x^T(\tau_e - \tau_r) + \frac{1}{T} \int_0^T x(t - \tau_e)\, n(t)\, \mathrm{dt}$$

It is found that the time averaged noise term of the output signal (1/T) $\int_0^T x(t - \tau_e)\, n(t)\, dt$, is much smaller than the input signal noise $n(t)$. The corresponding enhancement in signal-to-noise power ratio produced by the systems is found to be

$$\text{SNRE} = \frac{\text{SNR}|_{\text{out}}}{\text{SNR}_{\text{in}}} \simeq \eta\, BT = \eta \frac{T}{\theta} \tag{5.6}$$

where B is the bandwidth of the transmitted noise signal assumed equal to that of the amplifier thermal noise, T is the correlator integration time, η is the on-off ratio of the transmitted signal, and θ is the decorrelation time as defined in Equation (5.2). (If continuous noise is transmitted, $\eta = 1$.)

Equation 5.6 shows that the signal-to-noise power ratio enhancement equals the number of independent (noise) samples of the amplifier thermal noise, which are averaged during the integration time T. This result is reminiscent of the fact that the variance of a random signal $x(t)$, which is equal to its power, can be reduced by a factor n by averaging over n independent samples $x(t)$, $x(t_2) \cdots x(t_n)$.

A similar enhancement can also be produced by adding the echoes of conventional pulse-echo systems in phase. However, to retain good range resolution the transmitted signal must be short, and for adequate range the repetition frequency must be sufficiently low. Hence, for a pulsed-echo system,

$$\eta \not> \frac{\Delta R}{R}$$

where ΔR and R are the lengths of the range cell and the maximum range, respectively. $\Delta R/R$ is usually less than 10^{-3}, which shows that the on-off ratio of a noise-imaging system, which can be of order unity, can be 1,000 times larger than that of a conventional pulse-echo system. Hence, for a given peak transmitted power and given integration time, the signal-to-noise ratio enhancement of a noise correlation system can usually be three orders of magnitude larger than that of a pulse-echo system of equal resolution.

This advantage is lost, however, when one attempts to image a weak target that is close enough to a stronger target so that with a long transmitted signal, echoes from the two arrive at the same time. In that case it is preferable to use short pulses of noise, as is demonstrated below.

Ease of Digitization

Most current imaging systems use some form of digital processing, which means that the echo signal must be digitized at some point. For pulsed RF systems this causes a difficulty owing to the speed of the analog to digital converter required. Thus, if the incoming echo has a maximum frequency of 30 MHz, which is often desired, the analog-to-digital (A/D) converter must operate at minimum of 150 MHz, which is at the limit of current technology. With an analog correlation system of the type described here, we need only digitize the correlator output signal that contains no high frequencies.

Noise Correlation Doppler

Just as we will find it useful to precede our discussion of pseudo-random code imaging with a review of the noise correlation receiver, we find it useful, as is done in this section, to discuss the noise Doppler system [Newhouse and Bendick, 1974] before analyzing the more complex pseudo-random code Dopplers.

We recall that if a conventional pulsed Doppler system transmits a short-wave train of sine waves of frequency f_o, which is reflected from a target moving with velocity v away from the transducer, the reflected echo will be of frequency $f_o - f_d$. The so-called Doppler shift frequency f_d is given by the relation

$$f_d = \frac{c - v}{c + v} f_o \sim -\frac{2v}{c} f_o \qquad (5.7)$$

$$\text{for } v << c.$$

If the noise correlation system of Figure 5.1 illuminates a moving target having velocity v at range delay τ_e, with a random signal $x(t)$, then the echo will be of form $x(\alpha t - \tau_e)$, where

$$\alpha = 1 - \frac{2v}{c}$$

The correlator output will now be

$$y(t) = \int_{t-T}^{t} x(\alpha t - \tau_e)\, x(t - \tau_r)\, dt$$

$$\sim R_x^T \left(\left[\alpha - 1 \right] t + \tau_r - \tau_e \right) \qquad (5.8)$$

provided that the integration time T is sufficiently small, as determined below.

Combining Equations (5.4) and (5.8), we find that the output signal of the correlation system in the presence of a moving target becomes

$$R_x^T(t) = \exp\left\{ -\pi B \left[\tau_r - \tau_e - \frac{2v}{c} t \right] \right\} \cos 2\pi f_o \left[\tau_r - \tau_e - \frac{2v}{c} t \right] \qquad (5.9)$$

Equation (5.9) shows that a single scatterer passing through the range cell produces a train of sine waves of frequency $(2v/c)f_o$, amplitude modulated by the function $\exp\left\{ -\pi B \left[\tau_r - \tau_e - \frac{2v}{c} t \right] \right\}$.

The upper limit on the integration time T is now seen to reduce to the condition that T be much smaller than the period of the correlation system output frequency $(2v/c)f_o$, ie,

$$T << (\frac{2v}{c} f_o)^{-1} \tag{5.10}$$

Comparison of Equation (5.9) with Equation (5.7) shows that the noise correlation system output signal has the same Doppler shift frequency as that produced by conventional pulsed RF Doppler systems transmitting a spectrum centered at f_o. Hence, pulsed or continuous noise correlation systems can be used for ultrasonic Doppler flow measurement.

Noise Dopplers have both advantages and disadvantages with respect to pulsed RF Dopplers. One advantage is that noise Dopplers do not have the velocity-range product maximum that limits the performance of pulsed RF Dopplers. This limitation arises from the fact that pulsed Dopplers emit pulses at intervals T_R, whose lower limit is determined by the time required for echoes to return from the furthest range, ie,

$$T_R \not< \frac{2R_{max}}{c} \tag{5.11}$$

Since the Doppler signal produced after demodulation is constructed from echo samples with repetition frequency $f_r = 1/T$, the maximum unambiguous Doppler frequency that can be measured is

$$f_{dmax} \not> 2f_r = 2/T_R \tag{5.12}$$

Combining these two equations with Equation (5.7), which relates the Doppler frequency with target velocity, gives us the well-known range-velocity product limit in the form

$$v_{max}R_{max} \not> \frac{c^2}{8f_o} \tag{5.13}$$

This limit is low enough to be of practical significance in most ultrasound or acoustic applications of Doppler systems, and it provided the original motivation for the development of both the noise Doppler and the multicode Doppler. Chapter 7 describes the two very different ways in which these two systems overcome the range-velocity product limitation.

The second advantage of noise Doppler with respect to pulsed RF Doppler is that noise Doppler, just like noise-imaging systems, can use a lower peak power without sacrificing amplifier input signal-to-noise ratio.

The major disadvantage of noise Doppler with respect to pulsed RF Dopplers relates to the fact that in both kinds of systems, echoes from the flow being measured often arrive at the same time as interfering echoes from other stationary reflectors. In pulsed RF Dopplers, after being multiplied by the transmitted frequency, these interfering echoes produce a DC output that can be easily filtered out from the desired Doppler signal.

However, in noise Dopplers these interfering echoes are incoherent with the reference signal and thus appear as noise at the output. This noise can only be removed by increasing the integration time. This, however, has an upper limit given by Equation (5.10).

The disadvantages of noise Dopplers can be removed without losing the advantages by substituting pseudo-random codes. This is summarized in the section that follows and is discussed at greater length in Chapter 7.

Advantages of Pseudo-Random Codes

It is known [see, for example, Papoulis, 1965] that, as illustrated in Figure 5.2, for $\tau >> \theta = 1/B$, the function $R_x^T(\tau)$ of white noise varies randomly with respect to its argument, with standard deviation (SD)

$$\text{SD } (R_x^T(\tau))_{\tau>>\theta} = \frac{R_x^T(0)}{\sqrt{n}} = \frac{R_x^T(0)}{\sqrt{BT}} \tag{5.14}$$

where $n = T/\theta$ can be interpreted as the number of independent samples of $x(t)\ x(t - \tau)$ that have been summed during the period T.

If $R_x^T(\tau)$ is interpreted as a measure of the SD of the variable

$$\frac{1}{T} \sum_{t-T}^{T} x(t)\ x(t - \tau)\text{dt},$$

then Equation (5.14) states the familiar result that the SD of the sum of n independent terms of the Gaussian random variable $x(t)\ x(t - \tau)$ is reduced by a factor $\sqrt{n}$ with respect to the SD of $x(t)\ x(t - \tau)$, ie,

$$\text{SD}\left[\text{of sum of } n \text{ samples of } x(t)\ x(t - \tau)\right] = \frac{1}{\sqrt{n}} \text{SD}\left[x(t)\ x(t - \tau)\right]$$

The random peaks of $R_x^T(\tau)$ for $\tau > \theta$ are known as the "time side lobes" of this function. When receiving a signal from a target at $\tau = 0$, the existence of these time side lobes permits targets outside the range cell, ie, with $\tau > \theta$ to introduce incoherent interference signals that reach the receiver at the same time as the echoes from the region $\tau < \theta$. This type of interference, which reduces the input signal-to-noise power ratio, is known as "clutter" and limits the performance of noise correlation receivers when viewing targets whose separation in terms of time of flight differences τ are less than the length of the emitted signal.

This problem, which is discussed at length in Chapters 6 and 7, can be reduced by either emitting the noise signal in the form of pulses whose length is smaller than the minimum target spacing, or replacing the noise with pseudo-random codes having lower time side lobes.

These codes, discussed further in Chapters 6 and 7, consist of trains of binary pulses, which give the appearance of being random but which, in

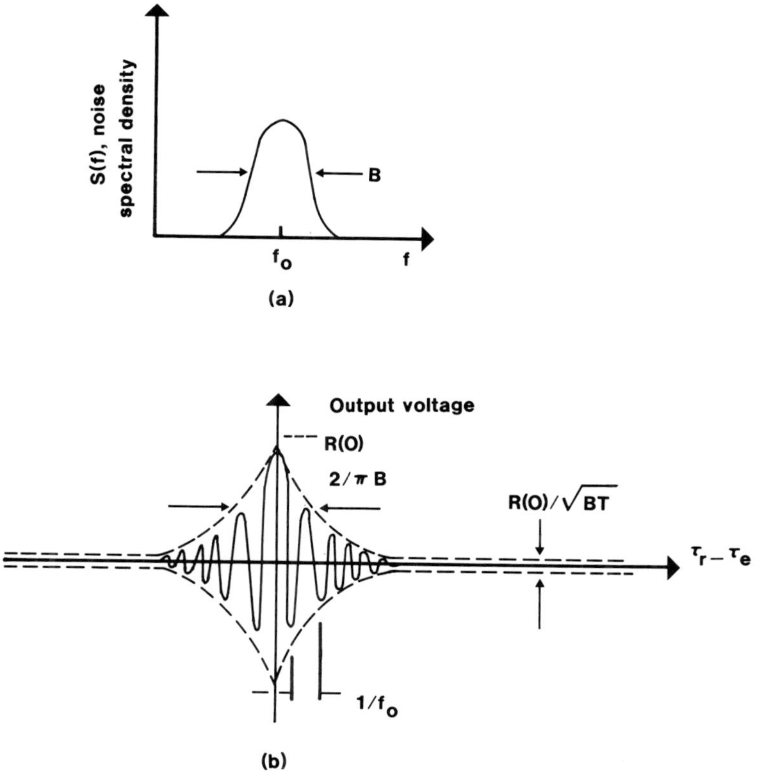

FIGURE 5.2. **(a)** Transmitted noise spectrum for random-signal correlation system. **(b)** Noise correlation system output for single target.

fact, are periodic. Figure 6.2a shows a type of pseudo-random code known as a maximum length sequence code ('m code'). This consists of a continuous train of pulses each of length θ, which repeats after a period $T = N\theta$ containing N pulses. The autocorrelation function of such a code, shown in Figure 6.2c, can be shown to have a half-width of $\Delta\tau = \theta$ [see, for example, Golomb, 1967], which shows that the binary signal whose spectrum is the Fourier transform of the autocorrelation function must have a bandwidth of $B = 1/\theta$.

The striking fact revealed by comparing Figures 6.2b and 6.2c is that whereas the time side lobes of noise are of height $R^T(0)/\sqrt{BT}$, those of the m-code are only of magnitude $R^T(0)/BT$. An even more remarkable code is discussed in Chapter 6, for which the time side lobes are completely eliminated. Since time side lobes limit the performance of correlation receivers in the presence of distributed targets, a pseudo-random code transmitted signal will usually be preferable to the use of noise in imaging stationary distributed targets.

In flow measurement, codes have an equally strong advantage over noise when flow is being measured in the proximity of stationary targets. In that case the echo from the flow, which is usually very weak, arrives at the receiver simultaneously with the usually much stronger echo from a stationary target. With noise as the transmitted signal, the stationary target echo is incoherent with respect to the flow echo and can therefore, in principle, be removed by time averaging; however the time required may be impractically long, particularly in measuring time-changing flows. Using a code of period T as the transmitted signal, however, the interfering echo also has period T and if sampled at intervals T, is changed into a DC signal that can easily be filtered out from the desired Doppler signal. (The same technique is used in classical pulsed RF systems, which also sample at the repetition period of the periodic transmitted signal.) This advantage of pseudo-random code over noise as well as other specifics of these codes are further discussed in Chapter 7.

Characterization of Systems and Time-Changing Media

The applications discussed so far may be said to use codes because of their ability to identify the different portions of a long signal. The work to be described in Chapter 8 uses codes because of their ability to measure range and velocity simultaneously with the maximum possible accuracy. This feature of codes is discussed in connection with the task of characterizing the sound propagation characteristics of an undersea channel, if necessary, when the transmitter and/or receiver is in motion. This characterization, more fully described in Chapter 8, consists of measuring the velocity spectra of distributed scatterers at various ranges. The "classic" way of accomplishing this, which forms the unacknowledged basis of all the work described up to now, is to demodulate the echo corresponding to each range and to use 10 to 100 of these samples to build up a section of the Doppler frequency signal, whose frequency spectrum is then measured.

The frequency resolution of this procedure is analyzed in Chapter 8 for perhaps the first time and is found to be very high. However, the price paid for this high resolution is twofold. In the first place the process is relatively time-consuming and takes many periods of the transmitted waveform. In the second place, for the process to work it is necessary that the scatterers at a given range retain their relative positions with respect to one another for the time necessary to build up the Doppler waveform. In other words, the spatial coherence time of the scatterers must be 10 to 100 times larger than the time of flight between the target and the transmitter/receiver. In the ocean, where propagation times become large, this condition is often not met. Yet a third problem, which makes its appearance when Doppler frequencies are measured with samples, is the appearance of a range-velocity product limit given by Equation (5.13). However, as explained in Chapter 7, the range-velocity product limit is not fundamental, since it can be overcome by using multiple codes for the transmitted signal.

The three problems listed above, namely excessive measurement time, target decoherence, and range-velocity product limit, can be overcome by analyzing the Doppler frequency shifts of each set of echoes in a manner independent of subsequent or previous echoes. Economical methods of doing this are described in Chapter 8. It is shown there that to maximize spatial and frequency resolution simultaneously, the transmitted waveform must have an ambiguity function with the narrowest possible peak and the lowest possible time-frequency plateau. The manner in which various types of noise and pseudo-random code waveforms approach this ideal is discussed in Chapters 7 and 8.

Simultaneous Multimode Operation

The applications of noise and codes described so far refer to essentially one-dimensional imaging and flow measurement. Chapter 9, the last of this review, describes results that suggest that noise and/or codes may also have advantages in two-dimensional imaging of moving targets. In such systems it is necessary to emit beams along the largest possible number of directions to conserve lateral resolution and in the shortest possible time, so as to conserve temporal resolution. In current ultrasound-imaging systems these beams are emitted sequentially by phased arrays or rotated transducers. For a range R the emission time T between successive beams must exceed $2R/c$. Hence if the frame time is F, the number of beams per frame has an upper limit of

$$m = \frac{F}{T} \ngtr \frac{Fc}{2R}$$

Clearly m, the number of beams per frame, could be improved if several beams could be emitted simultaneously. This can be achieved if these beams can be modulated or coded so that their echoes are distinguishable. Although no such system appears to have been published as yet, the concept has been demonstrated using noise. Chapter 9 describes these published experimental and theoretical results and discusses various system design aspects.

References

Dixon RC: *Spread Spectrum Systems*. New York, Wiley (1979).

Furgason ES, Newhouse VL, Bilgutay NM, Cooper GR: Application of random signal correlation techniques to ultrasonic flaw detection. Ultrasonics: 11–17 (January 1975).

Golomb SW: *Shift Register Sequences*. San Francisco, Holden-Day (1967).
Harwit M, Sloane JN: *Hadamard Transform Optics*. New York, Academic Press (1979).
Newhouse VL, Bendick PJ: An ultrasonic random signal flow measurement system. Jour Acoust Soc of America 56: 860–865 (September 1974).
Newhouse VL, Furgason ES, Bilgutay NM: Random signal flaw detection, in Proceedings of the IEEE Ultrasonics Conference, November 1974.
Papoulis A: *Probability Random Variables and Statistic Processes*. New York, McGraw-Hill (1965).

6
Pseudo-Random Correlation Imaging and System Characterization

J.Y. CHAPELON

Correlation Function in Imaging

We have seen earlier that in all conventional systems the average signal power is very weak. At a signal-to-noise ratio of unity, a significant increase in the energy received can be obtained by increasing the length of the transmitted signal, or by increasing its repetition frequency. In a conventional ultrasound system the length of the transmitted signal is known to be limited by the range resolution, while the repetition frequency of the signal is related to problems of range ambiguity. These restrictions can be avoided by using transmitted signals of long duration especially chosen to be time compressible in the receiver by means of the correlation technique.

Definition of the Echo

Assume that an ultrasound emitter transmits a signal $x(t)$ that is sent back by a fixed reflector situated at a distance r_o. If the only change produced in the signal by this reflection is a time delay τ_e, the echo of the target is of the form

$$x_r(t) = ax(t - \tau_e) \tag{6.1}$$

where a is the reflecting power of the target and τ_e is the time of flight equal to

$$\tau_e = 2r_o/c \tag{6.2}$$

and c is the velocity of sound in the medium. In terms of the analytic signal one defines

$$x(t) = R_e\{X(t)\} \tag{6.3}$$

where $X(t)$ is the analytic signal associated with $x(t)$:

$$X(t) = A(t)\exp 2\pi i\nu t \tag{6.4}$$

The analytic signal of the echo is

$$X_r(t) = aA(t - \tau_e)\exp[2\pi i\nu(t - \tau_e)] \qquad (6.5)$$

This representation has particular interest in the case of narrow band signals because in this case the instantaneous frequency ν becomes the same as the carrier frequency f_o.

Relation Between Resolution and Correlation Function

We recall that we are interested in the system range resolution. When the receptor has to conjointly treat two echoes whose analytic signals are $X_{r_1}(t)$ and $X_{r_2}(t)$, reflected by two targets with range delay τ_1 and $\tau_2 = \tau_1 + \tau$, the possibility of resolving the two echoes depends on the function

$$D(X_{r_1}, X_{r_2}) = |X_{r_1} - X_{r_2}|^2 = |X_{r_1}|^2 + |X_{r_2}|^2 - 2R_e < X_{r_2}, X_{r_1} > \qquad (6.6)$$

Only the scalar product $< X_{r_2}, X_{r_1} >$ reflects the "distance" between the echoes because the magnitudes $|X_{r_1}|^2$ and $|X_{r_2}|^2$ are independent of the range delay. In the case of narrow band signals,

$$< X_{r_2}, X_{r_1} > = a_1 a_2 \exp(2\pi i f_o \tau) \int A(t - \tau_1)A^*(t - \tau_l - \tau)dt \quad (6.7)$$

In this expression $\int A(t - \tau_1)\, A^*(t - \tau_1 - \tau)dt$ represents the envelope of $< X_{r_2}, X_{r_1} >$ and $\exp(2\pi i f_o \tau)$ the high-frequency modulation at frequency f_o, which does not have to be taken into account in considering the resolution. It is therefore unnecessary to retain the information given by this factor. The ability of the detection system to distinguish between two fixed targets is thus evaluated simply by the quantity

$$R(\tau) = \int A(t)A^*(t - \tau)dt \qquad (6.8a)$$

which is the correlation function of $A(t)$. $R(\tau)$, therefore, depends uniquely on the time dependence $A(t)$ of the transmitted signal. Because $R(0) = |A|^2$, the signal that will best separate two echoes due to two neighboring targets is that in which the value of the correlation function $R(\tau)$ always stays very small with respect to $|A|^2$ whenever $\tau \neq 0$ and in particular in the neighborhood of $\tau = 0$. The correlation function of such a signal is similar to that of a Dirac delta function.

Correlation Functions for a Random Signal of the Gaussian Type

Continuous Emission with Integration Time T

One of the best known types of usable signals for which the correlation function approaches a Dirac delta function is continuous random noise. This signal has been used in radar and has found applications in the field of ultrasound, notably in nondestructive testing [Newhouse et al, 1974; Furgason et al, 1975].

$$X(t) = A(t) \exp(2\pi i f_o t) \tag{6.8b}$$

where $A(t)$ is a random function of time.

The spectral density is

$$S_x(f) = \frac{B/2\pi}{(B/2)^2 + (f - f_o)^2} \tag{6.9}$$

where f_o is the center frequency and B the bandwidth. If the integration time T approaches infinity, the output of the correlator approximates the Fourier transform of $S_x(f)$ and can then be written

$$R_x(\tau) = \exp - \pi \mathrm{B}\tau \cdot \cos(2\pi f_o \tau) \tag{6.10}$$

The effective width $\Delta\tau$ of the correlation function that characterizes the resolution is in this case equal to $2/\pi B$. Consequently, the range resolution of this system is approximately

$$\Delta R \simeq c/\pi B \tag{6.11}$$

The result is to be compared with that obtained with conventional ultrasound signals where the range resolution is

$$\Delta R = c/2B \tag{6.12}$$

The slight difference between the two results is essentially due to the form of the signal and the definition of the range resolution.

Outside of the correlation peak at $\tau = 0$, the correlation function tends toward the value $R(0)/\sqrt{BT}$ (Fig. 5.2b). That is to say the two targets can only be distinguished if the ratio of their scattering cross-sections is no greater than $\sqrt{BT}$. Within these limits the absence of time side lobes signifies that the use of this signal in an ultrasound detection system avoids the customary problems of range ambiguity.

Random Pulsed Signal with On/Off Ratio η and Integration Time T

The correlation function is identical with the preceding one (Figure 6.1) except for the values of τ contained between the values ηT_R and $(1 - \eta)T_R$ for which the correlation function is zero. For this type of pulsed signal, two different kinds of range resolution are distinguishable. The first corresponds to adjacent targets (cross-hatched zone) for which the permissible ratio of scattering cross-sections is less than $\eta\sqrt{BT}$, as in the preceding case where $\eta = 1$. The second case corresponds to two targets whose separation lies between $\eta T_R c/2$ and $(1 - \eta)T_R c/2$, for which the permissible ratio is limited only by the signal-to-noise ratio at the output of the correlator.

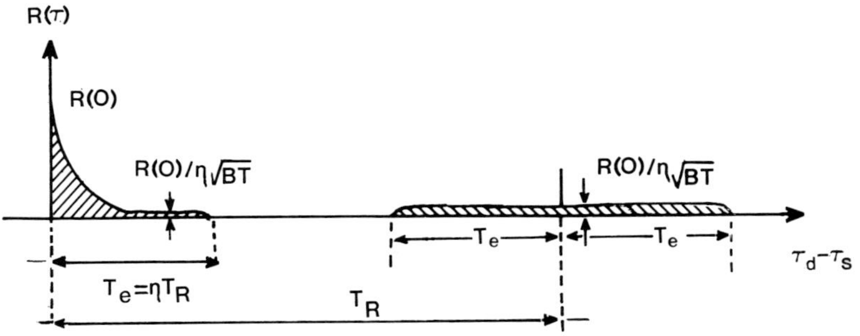

FIGURE 6.1. Estimation of the correlation functions of a noise train of length T_e at repetition frequency T_R. The main peak $R(0)$ is at $\tau = 0$. Each noise burst is independent of the preceding one. Time side lobes in the form of a plateau exist around $\tau = 0$ and $\tau = T_R$ of average height: $\dfrac{R(0)}{\eta\sqrt{BT}}$.

Correlation Function for Codes

We are looking for a transmitted signal in which the peak-to-average power ratio approaches unity and in which the correlation function approaches the Dirac delta function. A criterion in the choice of transmitted signal can be the fashion in which one can delay it with respect to time for the purpose of realizing the reference signal used in the calculation of the correlation function. For these reasons it is advantageous to use binary signals whose characteristics approach those of noise in this application. In fact, a number of binary signals possess the characteristic of having a correlation function similar to a Dirac delta function.

Binary signals that possess this property may or may not be synchronized with a clock. The characteristics of the second type of binary signal are so close to that of Gaussian noise that they present hardly any advantages over it. We recall that this type of code can be obtained by passing white noise through a Schmitt trigger [Newhouse et al, 1974]. In this review we mainly consider signals of the synchronous type, specifically circular and noncircular M-sequences and Golay codes.

M-SEQUENCES

The M-sequences (maximum length binary sequences) have been discussed by Golomb [1976]. They are, by definition, codes of maximum length that can be generated by a shift register using a particular type of feedback. An M-sequence is composed of $N = 2^n - 1$ states, where n is the number of stages in the shift register. If $(a_0, a_1 \ldots a_{N-1})$ is an M-sequence with $a_i = \pm 1$ and $N = 2^n - 1$, then $R(k)$ the correlation function of this code is defined by

$$R(k) = \sum_{i=0}^{N-1} a_i \cdot a_{i-k} \text{ for } k \in [-\frac{N-1}{2}, \frac{N+1}{2}] \tag{6.13}$$

CIRCULAR M-SEQUENCES

An important property of circular M-sequences is that

$$R(k) = \begin{cases} N \text{ for } k = 0, N, 2N, \ldots \\ -1 \text{ otherwise} \end{cases} \tag{6.14}$$

Since this sequence is generated by a clock of period θ (Figure 6.2a), the correlation function of the resulting signal $X(t)$ is given by

$$\mathrm{R}_x(\tau) = \begin{cases} N\theta - (N+1)|\tau| \text{ for } |\tau| \leq \theta \\ -\theta \text{ for } \theta < |\tau| \leq \dfrac{N\theta}{2} \end{cases} \tag{6.15}$$

This function is periodic, of period $\mathrm{T} = \mathrm{N}\theta$ (Figure 6.2c). It can be considered as the sum of a triangle function and a rectangular function. Meanwhile, for large values of N, the passband of this signal is determined principally by the triangular portion in such a way that the spectral energy density of the signal may be written

$$S_x(\nu) \simeq (N\theta)^2[\sin\pi\nu\theta/\pi\nu\theta]^2 \tag{6.16}$$

This function is shown in Figure 6.2b. The equivalent bandwidth is $B = 1/\theta$.

The correlation function of a circular M-sequence is therefore particularly interesting, because outside its main peak, it has a constant level whose amplitude is $1/N$ times the amplitude of this peak (Figure 6.2c) equal to $R(0)/N = R(0)\theta/T = R(0)/BT$. It is easier to separate two targets with M-codes than with noise, since the allowable ratio of their reflectivity is BT instead of $\sqrt{BT}$.

NONCIRCULAR M-SEQUENCES

In the applications envisioned in the pseudo-random code correlation system, the M-sequences are not used in a continuous cyclic mode. In effect, for technical reasons it is necessary to split the sequence into many small separated sequences. It has been shown that as a result, the correlation function of the entire sequence deteriorates. The level outside the main peak is now no longer constant, and time side lobes appear of which the maximum level is $1/\sqrt{N}$ times the amplitude of this peak (Figure 6.3b). This case is the equivalent of the pulsed noise system treated earlier.

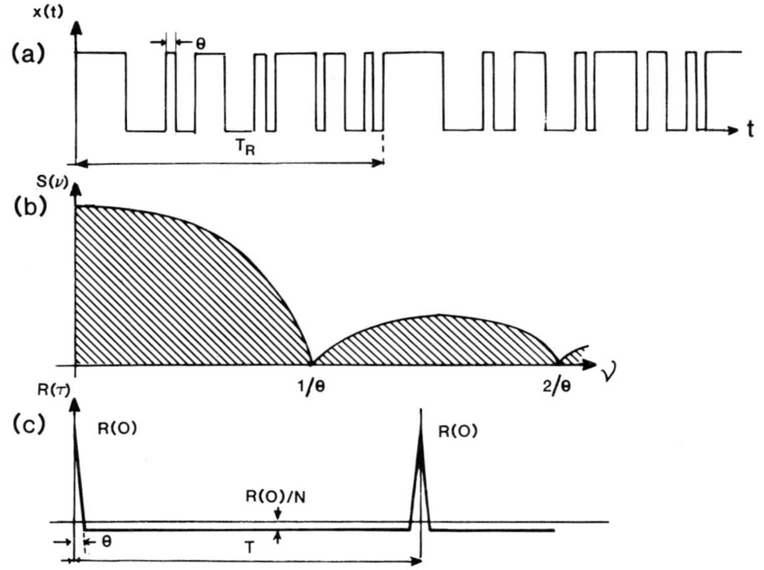

FIGURE 6.2. Circular M-sequence. (**a**) Form of the signals—*θ:* clock period, T_R: M-sequence repetition period $T_R = N\theta = (2^n - 1)\theta$. (**b**) Energy spectral density—envelope: $S(\nu) = (N\theta)^2 (\sin \pi\nu\theta/\pi\nu\theta)^2$, line spectrum at frequency intervals $1/T_R$. (**c**) Correlation function—*T:* integration time, $T = T_R$.

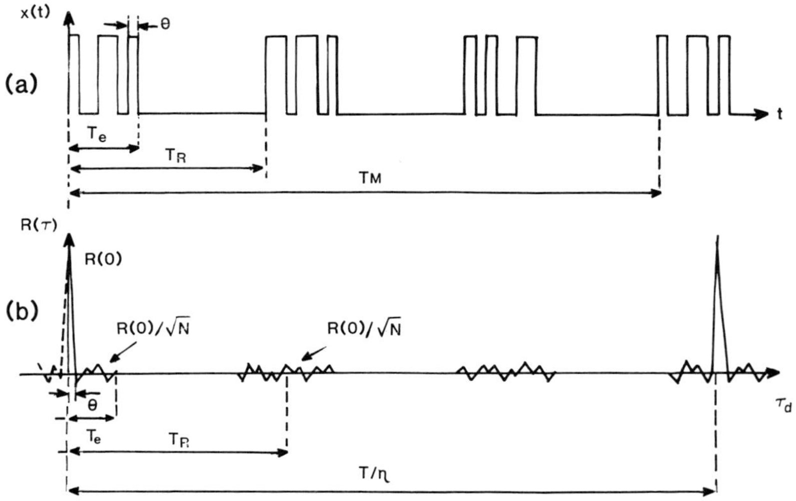

FIGURE 6.3. Broken M-sequence. (**a**) Form of the signal—*θ:* clock period; T_R: burst repetition period; T_e: duration of one burst $T_e = \eta T_R$, T_M: repetition period of the M-sequence. (**b**) Correlation function—*T:* integration time. The function is periodic with period T/η.

Golay Codes

There exists a type of code that has a particularly interesting correlation function. These are the codes named after Golay [1963]. He showed that it is always possible to associate a specific code A with a code B so that the sum of the two correlation functions corresponding to each of the codes is zero outside of the main peak. This peak has a triangular shape of height $2N$ and base width 2θ, where N is the number of states of each of the codes and θ is the clock period used to generate these codes (Figure 6.4). This characteristic is remarkable in that its use permits us to distinguish two targets, whatever their reflectivities, provided that the output signal-to-noise ratio is adequate and providing that the range separation between the two targets is greater than $\Delta R = \theta c/2$.

Effect of Modulation

Figure 6.2b shows that the spectral density of an M-sequence becomes small beyond the clock frequency $1/\theta$. If this signal is used directly for transmission, only the frequencies contained in the passband of the transducer are transmitted. This signal corresponds to the convolution of the transducer input signal with the transducer impulse response. Thus, it is necessary to match the transmitted spectrum to the passband of the transducer. To do this, one can use a carrier f_o modulated in frequency or in phase by the signal to be transmitted. The resulting signal is of the form

$$X_m(t) = X(t) \sin[2\pi f_o t + \phi(t)] \tag{6.17}$$

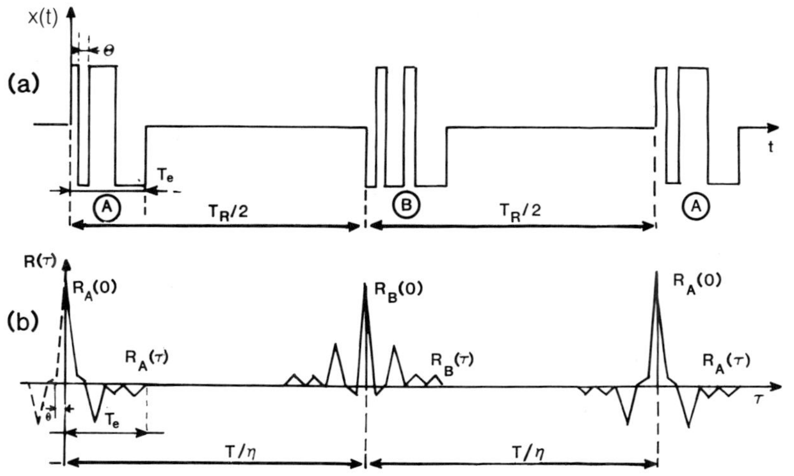

Figure 6.4. Golay code. (**a**) Form of the signal—θ: clock period; T_R: repetition period of trains A as well as of trains B; T_e: duration of signals A and B. $T_e = \eta T_R/2$. (**b**) Correlation function—$R_A(\tau)$: correlation function of code A; $R_B(\tau)$: correlation function of code B. Integration time, $T = T_e$.

where $X(t)$ is the envelope of the signal to be transmitted, $\phi(t)$ the instantaneous phase of the signal, and $X_m(t)$ the modulated transmitted signal.

In the case where f_o is a multiple of the clock frequency $1/\theta$ of the pseudo-random sequence $\{a_k\}$ utilized in the phase modulation, one can show that the correlator output signal is written

$$R_{x_m}(\tau) = \frac{1}{2} R_x(\tau) \cdot \cos \frac{2\pi pt}{\theta} \qquad (6.18)$$

with an associated spectral density

$$S_{x_m}(\nu) \simeq \frac{1}{4} [S_x(\nu + \frac{p}{\theta}) + S_x(\nu - \frac{p}{\theta})] \qquad (6.19)$$

Here $X_m(t)$ is the modulated signal corresponding to $X(t)$, and p is such that

$$p/\theta = f_o \qquad (6.20)$$

It is clear that the spectrum of the modulated signal can now be matched in a more precise way to the transfer function of the transducer provided that

$$1/\theta = B \qquad (6.21)$$

where B is the bandwidth of the transducer. Under these conditions the shape of the correlation function of the transmitted signal can be represented by an isosceles triangle whose base width is $\Delta\tau = 2\theta$. The corresponding range resolution is given by the equation

$$\Delta R = c\theta/2 \qquad (6.22)$$

Noise Correlation Systems

Continuous Noise Systems

The first ultrasound systems based on correlation techniques used a continuous random signal. These systems were developed for use in the field of nondestructive evaluation by Newhouse et al [1974]. The block diagram of such a correlation system is shown in Figure 6.5. The noise source produces electric signals that are converted into ultrasound and transmitted into the sample by the piezoelectric transmitting transducer. The echoes reflected by the different interfaces are reconverted into electrical signals by the receiving transducer. The received signal after amplification is correlated with a copy of the transmitted signal delayed by a time τ_d. The delay was originally obtained by means of an acoustic transmission line composed of two transducers separated by water. In a more recent version [Lee and Furgason, 1982] the analogue delay line has been replaced by a

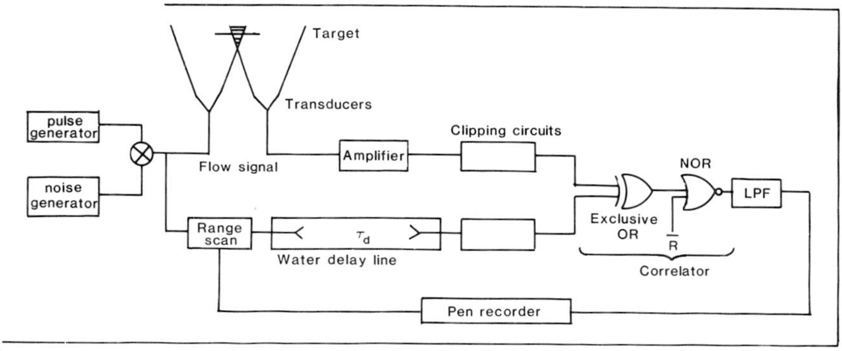

FIGURE 6.5. Experimental random signal flaw-detection system employing signal clipping and polarity coincidence correlator (Furgason et al, 1975).

digital delay line in which the noise signal is sampled and stored before being read out with delay τ_d.

The correlator consists of a multiplier followed by an integrator in the form of a low-pass filter. With the single target, the correlator output is maximum when the delay τ_d is approximately equal to the signal time of flight τ_s. If the length of the delay line changes slowly, τ_d varies and gives a correlator output signal proportional to the correlation function of the transmitted signal. In a practical flaw-detection system, correlator output is displayed on the Y axis of a pen recorder in which the X deflection is proportional to the delay τ_d.

Analysis of Signal-to-Noise Ratio Effect of Clipping

The improvement in signal-to-noise ratio produced by a correlation receiver can easily be deduced in the following way. Consider that the correlation system illustrated in Figure 5.1 of Chapter 5 transmits a series of sinusoidal waves of a duration T_e with an on/off ratio η. If the delay line is adjusted so that $\tau_d = \tau_s$, the output of the multiplier is a rectified sinusoidal signal of duration T_e and on/off ratio η. The output of the low-pass filter is an averaged signal, and the enhancement in signal-to-noise ratio, SNR, may be written

$$\frac{\mathrm{SNR}_{\mathrm{out}}}{\mathrm{SNR}_{\mathrm{in}}} = \frac{nT}{T_e} = \eta BT \tag{6.23}$$

where B is the bandwidth of the transmitted signal $B = 1/T_e$, and T is the equivalent integration time of the low-pass filter. In the practical flaw-detection system using continuous noise emission, the ratio η is equal to unity and the output signal-to-noise ratio is maximum.

A simplified correlation procedure for a random signal-flaw detection receiver consists of passing the amplified echo signal and the signal transmitted through the delay line through Schmitt triggers functioning as clipping circuits. It is shown by Newhouse et al [1974] that this method of transforming a noise signal of many levels into a signal of only two levels reduces the signal-to-noise ratio by a factor $2/\pi$, provided the input signal-to-noise ratio is weak. Clipping the echo and reference signals makes possible the advantages of digital processing in more elaborate versions of the system. On the other hand it has the disadvantage of losing the information contained in the amplitude of the signal. In the system considered here, the clipping action makes it possible to provide the multiplication required for the correlation by using a simple "exclusive-or" gate as shown in Figure 6.5, for which experimental results have shown an enhancement of the signal-to-noise ratio corresponding to the equation

$$\frac{\mathrm{SNR}_{\mathrm{out}}}{\mathrm{SNR}_{\mathrm{in}}} \simeq \left(\frac{2\eta}{\pi}\right)BT \tag{6.24}$$

A very important property of a random signal system of detection is that its sensitivity, in the presence of a single target, can augment indefinitely by simply increasing the time of integration T of the correlator.

Meanwhile, if the transmitted signal simultaneously covers many targets, their echoes are received simultaneously at the transducer. The echoes of targets external to the target zone visualized are uncorrelated and lead to a nonzero residue of which the amplitude corresponds to the effective output noise of the correlator. This residue, also known as the clutter signal, is in this case $\sqrt{BT}$ times smaller than the maximum amplitude of the targets.

The signal-to-clutter ratio, SCR, at the correlator output is defined as the ratio between the maximum amplitude of the echo signal of a target and the residue that accompanies it owing to noncorrelated targets. It is thus defined as

$$\mathrm{SCR} = a_o\sqrt{BT/\Sigma a_i} \tag{6.25}$$

where the a_i are the scattering cross-sections of the clutter and a_o is that of the target considered. It is this relationship that limits the dynamic range of echoes in a correlation system using continuous noise emission.

A more advanced system [see Bilgutay et al, 1976] that uses pulsed noise for the transmitted signal has been constructed. Transmission and reception are made using the same transducer in order to simplify the examination of samples. Besides this advantage, it is noteworthy that the signal-to-clutter ratio can be minimized by limiting the length of the transmitted signal so that it does not exceed the time of flight between two targets, while making sure that the repetition frequency does not lead to the range ambiguities of conventional pulse echo systems. However, in the case of targets very close to one another, this procedure completely does away with the advantage of noise systems in lowering the required peak-to-average power ratio.

Pseudo-Random Code Correlation System

Advantage of Codes*

One of the principal properties of these codes is the periodic character of the sequence. Because of this periodicity, the time side lobes that appear outside the principal peak have a well-determined value that never varies for two successive correlations. This characteristic permits, in certain applications and with certain types of codes, the elimination of residual clutter that is found outside the main lobe. Meanwhile, this periodicity creates a range ambiguity similar to that of pulse-echo systems. The ambiguity is not important, however, since in ultrasound applications the range of visualization is relatively small, so that it is always possible to lengthen the code sufficiently to eliminate any range ambiguity.

The binary character of the signal eases the processing with respect to storage, delay, and generation, etc. Also in view of the periodic nature of the sequence, it is easier to generate a second code of which the starting time is different, rather than to delay the transmitted code by means of a delay line in which the number of stages is proportional not to the length of the code, but to the maximum range.

Note, too, that it is possible to transpose all the spectral energy of the binary sequence around the characteristic central frequency of the transducers by a phase or frequency modulation that only requires a multiplication by $+1$ or -1.

Finally, as mentioned above, there are many well-known classes of pseudo-random binary signals such as M-sequences, Barker codes, Golay codes, etc, and within each class it is possible to construct a great number of pseudo-random codes of varying degrees of orthogonality whose applications are considered in this review.

Each of these pseudo-random codes is characterized by its correlation function, which is particularly interesting for ultrasound applications in medical imaging [Gerson et al, 1971; Chapelon et al, 1979] or in nondestructive testing [Elias, 1980].

Basic M-Sequence System, Basic Principle

The block diagram of a pseudo-random correlation system functioning by means of a carrier frequency modulated in phase by a sequence of maximum length is represented in Figure 6.6. The principle of operation is analogous to that of the random noise systems developed earlier, except

*Only codes synchronized with a clock are considered in this review.

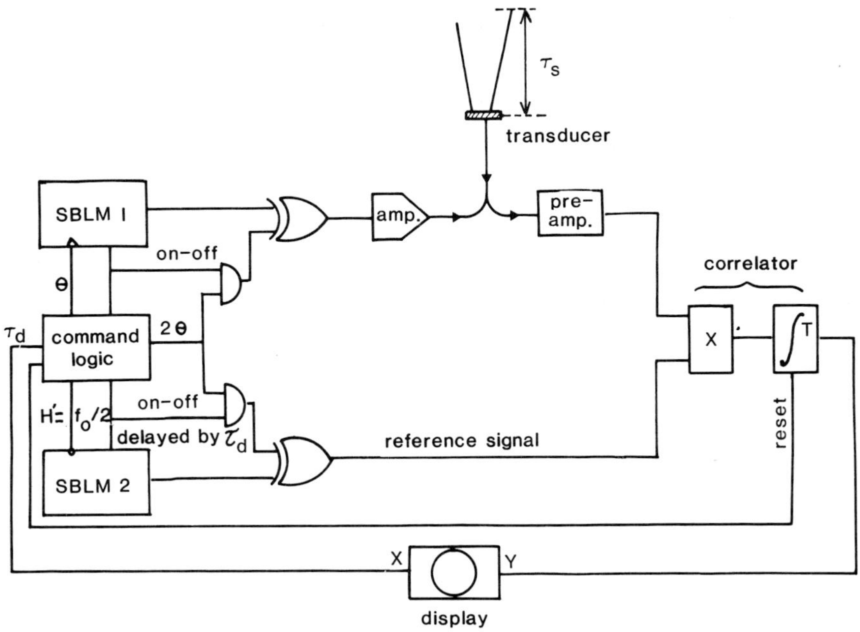

FIGURE 6.6. Basic block diagram of an M-sequence system. θ: code generation clock period; T: duration of integration $T = (2^n - 1)\theta$; τ_d: incremental delay; SBLM: maximal length binary sequence created by n-stage shift registers using modulo 2 feedback.

that the delay line disappears, considerably simplifying the system. The system uses two binary pseudo-random generators, SBLM1 and SBLM2, for generating the trains of required M-sequences. The phase modulation can be obtained simply by an "exclusive-or" between the codes and the main clock [Chapelon et al, 1979]. The reference signal, a delayed replica of the transmitted signal, is obtained by means of a second shift register, identical to the first, in which the initial state is delayed by τ_d with respect to the initial state of the first register. A control logic permits the synchronization of the SBLM1 and SBLM2 registers between each pulse train so that the M-sequence in its entirety can be transmitted at each correlation. The correlator is an analogue type. The correlation can be made directly either between the amplified echo signal and the delayed modulated binary sequence coming from a second shift register [Chapelon et al, 1979] or between unmodulated code produced by a second shift register and the amplified echo signal from which the carrier has been eliminated [Elias, 1980].

The second solution is justified in narrow band systems such as radar or sonar. In ultrasound systems the constraints related to resolution, to attenuation in the medium, and to transducer fabrication lead to a code spectral density bandwidth equal to at least 50% of the carrier.

The integration time is $T = (2^n - 1)\,\theta$, where n is the number of bits of the shift register. For a complete visualization of a part of the ultrasonic field, the delay τ_d is incremented by a quantity $\Delta\tau_d$ after each correlation. The output of the correlator is then the impulse response of the medium sampled at $1/\Delta\tau_d$.

As in the preceding systems, the correlator output is displayed on the Y axis of a recorder (which can be an oscilloscope) in which the X deflection is proportional to the length of the delay τ_d.

Analysis of the Signal-to-Noise Ratio

In the presence of a target the system produces an output signal proportional to the correlation function of the transmitted signal. The integration time is the time T necessary to generate the total M-sequence. One can see that

$$T = q(\mathrm{m}\theta) = N\theta \tag{6.26}$$

where q is the number of integrated pulse trains and m is the number of bits forming each transmitted train. In the absence of clutter one can define, as previously, the gain in signal-to-noise ratio as being given by time-bandwidth product BT of the receptor multiplied by the ratio η. Since in this particular case $T = \eta T_R$, where T_R is equal to the repetition period of the entire M-sequence

$$\frac{\mathrm{SNR}_{\mathrm{out}}}{\mathrm{SNR}_{\mathrm{in}}} \simeq \eta B T_R \tag{6.27}$$

The gain is the same as with a system of pulsed random noise.

Analysis of the Clutter Signal

For an M-sequence, the clutter signal, created by a target situated in a different position from the target under observation, can be defined precisely.

The clutter, of which a part of the signal is received simultaneously with the signal of the target being observed, creates at the correlator output a signal proportional to a time side lobe of its correlation function, whereas the observed target creates a signal proportional to the peak of the correlation function. Since the two correlation functions are stationary and only depend on the scattering cross-sections of the targets, one can define the signal-to-clutter ratio at the correlator output in the least favorable case by

$$\mathrm{SCR} = \frac{a_o}{\Sigma a_i}\sqrt{N} = \frac{a_o}{\Sigma a_i}\sqrt{\eta B T_R} \tag{6.28}$$

where a_i is the scattering cross-sections of ith clutter target, a_o that of the main target, and N the total number of bits forming the pseudo-random binary sequence. The performance of pulsed M-code systems has been compared with that of pulsed noise and pulsed radiofrequency (RF) systems by Nahamoo and Kak [1981], as well as by Lee and Furgason [1982].

Remark. A better signal-to-clutter ratio can be obtained at the correlator output by the use of an M-sequence that is transmitted continously in a cyclic mode. In this case the signal-to-noise ratio reduces to

$$\mathrm{SCR} = \frac{a_o}{\Sigma a_i} BT_R \tag{6.29}$$

However, for this solution one needs to employ a probe with two transducers, which is awkward for ultrasound nondestructive evaluation and medical-imaging systems, where it is customary to use only one single transducer probes.

Description of a System with an Increased On/Off Ratio η

As the use of only one transducer is a desirable option, the ratio η between emission and nonemission is necessarily less than 1:2. In addition, the transmitted pulse trains may not be longer than the echo time of flight of the nearest target to be visualized. Furthermore the nonemission time cannot be less than the time of flight of the most distant echoes to be visualized. Thus, if the area of visualization is sufficiently large, the ratio η is small and one is again confronted by a conventional system.

In order to visualize, in real time, a substantial exploration zone while conserving the ratio η large, it has been shown by Chapelon et al [1979] that it is possible to visualize the different parts of this zone successively. When the series of transmitted bursts corresponding to the whole of the M-sequence is reflected from a part of the studied zone so that its correlation is completed, new parameters are chosen for the immediately adjacent portion. These parameters are the length of the transmitted burst (which is determined by the transducer-target distance) and the number of transmitted bursts, during the time of integration T, which is held constant (Figure 6.7). Thus by varying the length of the transmitted burst after each integration, it is possible to optimize the ratio η to a value near 1:2.

An on/off ratio of 1:3 has made it possible [Chapelon et al, 1979] to visualize in A mode a tissue zone 5-cm thick in real time, with the utilization of correlators at 16 range points simultaneously. The basic transmitted code of this system is shown in Figure 6.7. The number of states N of the sequences SBLM1 and SBLM2 is determined either by the amelioration of the noise-to-signal ratio given by Equation (6.27), or by the ratio of signal to clutter shown in Equation (6.28). In the system of Figure 6.8, $N = 1{,}023$, which presents a good compromise for the specific application envisaged.

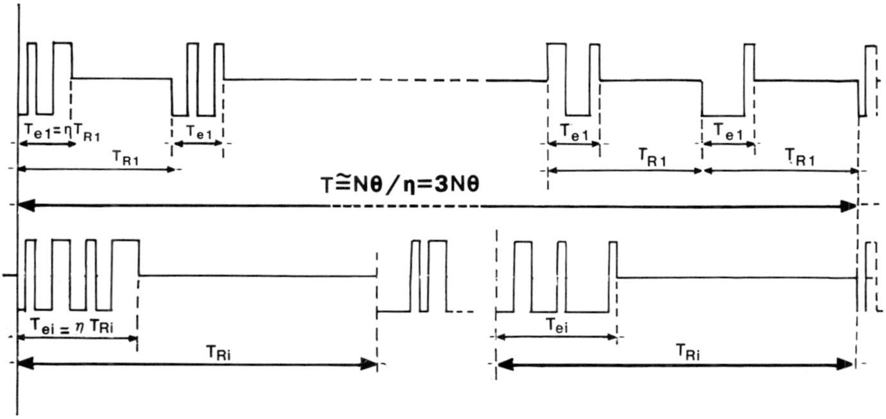

FIGURE 6.7. Nonmodulated sequences used in a system with on/off ratio $\eta = 1{:}3$. $N = 2^n - 1$ is the number of bits of each M-sequence. The system starts by emitting bursts of duration T_{e1} formed of just enough bits to visualize the points in the near zone. This first region is visualized after a time T corresponding to the integration time. Then the system increments the length of the burst to create a new burst of duration T_{e2}, with a corresponding repetition period T_{R2} chosen to maintain the on/off ratio $\eta = 1{:}3$, so that the integration time T remains constant. The system increments the bursts in this manner until the furthest region of the visualization zone is reached.

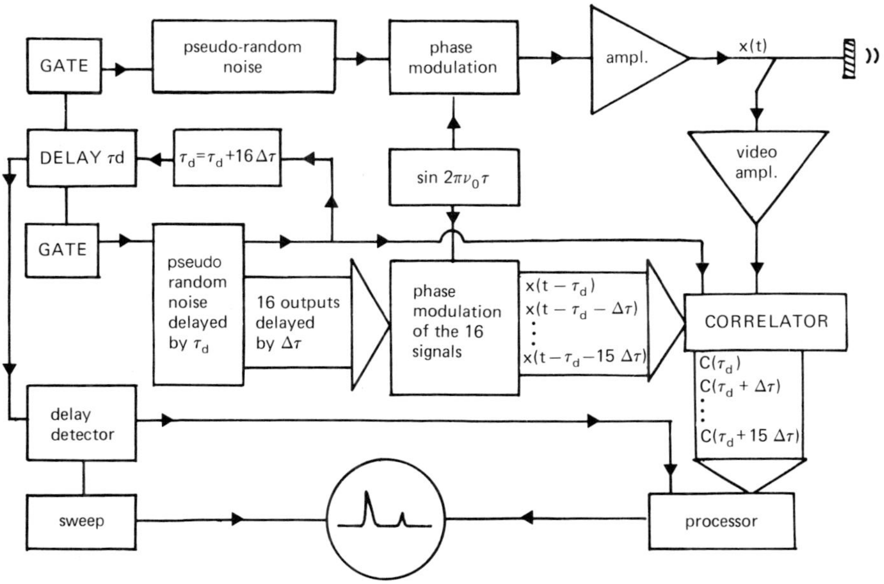

FIGURE 6.8. Block diagram of the pulsed multipoint high-resolution ultrasonic M-code system.

Conforming to the bandpass of the transducers used, the value of θ was chosen equal to $2/f_o$ [Equation (6.20)].

Even though the systems mentioned above permit an enhancement in signal-to-noise ratio, they are limited to specific applications where the number of reflecting targets and the dynamic range between the different echoes of these targets are both small.

Golay Code Correlation System

Limit of the Previous Systems with Respect to Medical Imaging

The two principal characteristics of the medium encountered in medical ultrasound are strong attenuation of the ultrasound with depth and the multitude of targets. These create a very large echo dynamic range of the order of 100 to 110 dB at the input of the system (taking account of the noise of the first stage of the amplifier). For correcting the attenuation effects, time-varying gain is used to eliminate the dynamic range between near and distant echoes of targets with identical reflectivity. Only the dynamic range due to the fluctuations in target reflectivity is retained. This is of the order of 30 to 40 dB, which is compatible with most of the visualization systems currently used.

In the practical systems cited in the first two sections, this correction of gain with depth cannot be directly realized during reception because the echoes of the targets situated at different ranges arrive simultaneously at the transducer. Furthermore, it can not be applied after the correlation, since the signal-to-clutter ratio is too small (30 dB in the system described by Chapelon).

Another constraint connected with these systems is in obtaining real-time B scans. In effect, this would necessitate the use of multipoint correlators, which would lead, with current techniques, to gigantic size systems.

Solution

To satisfy the exigencies cited above, the signal-to-clutter ratio must be enhanced. With the use of a cyclic M-sequence, this could be done by augmenting the number of states N until the desired signal-to-clutter ratio is reached at the correlator output. This solution is, unfortunately, not feasible in real time because of the excessive code length required to reach 120 dB dynamic range. A more feasible scheme is to use Golay codes for which the correlation function has theoretically no time side lobes, as Takeuchi [1979] has proposed.

The principle of such a system is represented in Figure 6.9. The emitter transmits, alternatively, code A and code B while the correlator analyzes

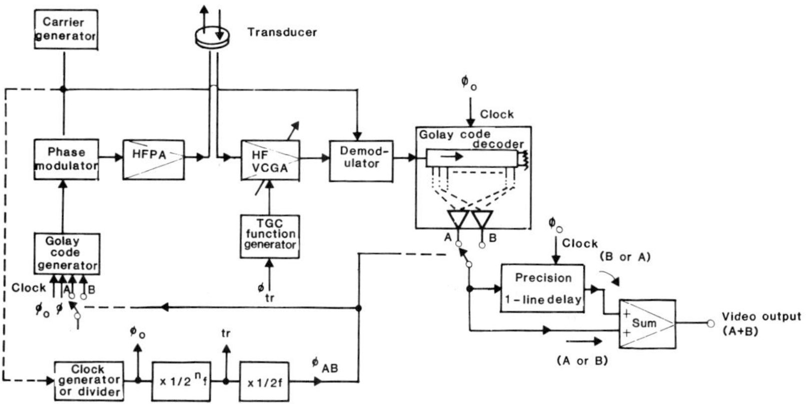

FIGURE 6.9. Basic schematic view of time serial Golay code pulse-echo system (Takeuchi, 1979).

the received signal with each one of these codes. The correlated signal corresponding to the first output is stored in a memory before being summed with the correlated signal corresponding to the emission of the second code.

Even though it is possible to generate Golay codes with a large number of bits, this solution nevertheless necessitates codes whose length is not too long to permit time gain compensation at the receiver input. The correlator can work with an input dynamic range of the order of 40 dB and permits real-time operation. The gain in signal-to-noise ratio of such a system is of the order of 20 dB compared with a pulsed system that uses the same peak intensity [see also Lee and Furgason, 1981].

System Characterization

One of the major advantages of noise or pseudo-random code transmitted signals for A-scan imaging over conventional pulse-echo techniques is that their use makes it possible to lengthen the transmitted signal without impairing resolution, thus increasing the average transmitted power (and hence the receiver input signal-to-noise ratio) without increasing the peak transmitted power. A related situation in which pseudo-random codes have been used to increase signal-to-noise ratio is in the study of neurological-evoked responses.

In the classical method of studying evoked responses, a light flash or audio click is administered to the subject, and the subsequent neurological-evoked response $h(t)$ is recorded from electrodes glued to the scalp.

In conventional evoked-response measurement, the stimuli used are separated by a time interval long enough for $h(t)$ to decay. An improvement in received signal-to-noise ratio could be obtained if responses could be evoked at a higher repetition rate. A technique of doing this using an M-sequence code has been reported by Eysholdt and Schreiner [1981].

If an M-sequence $a_0, a_1 \ldots . a_{N-1}$ were used to stimulate evoked responses of impulse response $h(t)$, the output would be

$$y(k) = \sum_{i=0}^{N-1} h(i)\, a_{k-i} \tag{6.30a}$$

Note when $k < i$, negative subscripts will appear; a_{-1} is defined as a_{N-1}, a_{-2} as a_{N-2}, and so on.

Equation (6.30a) can be written in matrix form

$$\begin{bmatrix} y(0) \\ \cdot \\ \cdot \\ \cdot \\ y(N-1) \end{bmatrix} = M_N \begin{bmatrix} h(0) \\ \cdot \\ \cdot \\ \cdot \\ h(N-1) \end{bmatrix} \tag{6.30b}$$

where M_N is an $N \times N$ matrix composed of the shifted M-sequences.

$$M_N = \begin{bmatrix} a_0 & a_{N-1} & a_{N-2} \cdots\cdots & a_1 \\ a_1 & a_0 & a_{N-1} & a_2 \\ \cdot & & & \\ \cdot & & & \\ a_{N-1} & a_{N-2} & a_{N-3} \cdots\cdots & a_0 \end{bmatrix}$$

Equation (6.30) could be solved by matrix multiplying both sides by the inverse of M_N. An apparently simpler alternative used by Eysholdt and Schreiner has been to use M'_N instead of M_N as the stimulating matrix, defined as

$$M'_N = 1/2(M_N - U_N) \tag{6.31}$$

Where U_N is an $N \times N$ matrix of which all terms are $+1$. It can be shown that M'_N is the inverse of M_N by the following argument:

$$M_N\, U_{N|ij} = \sum_{k=0}^{N-1} m_{ik}\, u_{kj} = \sum_{k=0}^{N-1} m_{ik} \tag{6.32}$$

where the m_{ik} are the terms of the ith row of the matrix M_N.

However, it is known from the properties of M-sequences that

$$\sum_{k=0}^{N-1} m_{ik} = 1$$

Hence from Equation (6.32),

$$M_N\, U_N = -\, U_N \tag{6.33}$$

From Equations (6.31) and (6.33),

$$\begin{aligned} M_N M'_N &= M_N[1/2(M_N - U_N)] \\ &= 1/2[M_N M_N + U_N] \end{aligned} \tag{6.34}$$

But

$$M_N M_N \big|_{ij} = \sum_{k=0}^{N-1} m_{ik}\, m_{kj}$$

$\sum_{k=0}^{N-1} m_{ik}\, m_{kj}$ is nothing more than a product of one row of the M-sequence with one column of the same M-sequence, ie, with a shifted version of itself. Thus from Equation (6.14),

$$\begin{aligned} M_N M_{N|ij} &= -1 \text{ for } i \neq j \\ &= N \text{ for i } = j \end{aligned}$$

Substituting this result into Equation (6.34) gives

$$M_N M'_N = \frac{1}{2}(N + 1)\, I_N$$

where I_N is the unit matrix. Thus if we excite the system, with the matrix M'_N giving rise to the outputs

$$\begin{bmatrix} y(0) \\ \cdot \\ \cdot \\ \cdot \\ y(N-1) \end{bmatrix} = M'_N \begin{bmatrix} h(0) \\ \cdot \\ \cdot \\ \cdot \\ h(N-1) \end{bmatrix}$$

and multiply both sides by M_N, we obtain the required solution for the impulse response in the form

$$\begin{bmatrix} h(0) \\ \cdot \\ \cdot \\ \cdot \\ h(N-1) \end{bmatrix} = \frac{2}{N+1} M_N \begin{bmatrix} y(0) \\ \cdot \\ \cdot \\ \cdot \\ y(N-1) \end{bmatrix}$$

In the classical method of estimating $h(t)$, the stimulating impulses must be separated by the decay time of this function. The above derivation shows that if the stimulating impulses can be made narrow enough so that an M-sequence of N terms can be used during the decay time of $h(t)$, then the improvement in the signal-to-noise ratio of the measurement will be of order N also.

The techniques described above for enhancing signal-to-noise ratio by exciting the system to be characterized at intervals shorter than the decay time of its impulse response are applicable to a wide variety of linear systems (eg, nuclear magnetic resonance) and have also been used in acoustics. It should be noted, finally, that the computational procedure described above, of multiplying the output signal matrix $[y]$ by the inverse of the stimulating signal matrix $[M']$, is analogous to the hardware technique used in the correlation receiver of multiplying the echo signal by variously delayed segments of the transmitted signal.

References

Bilgutay NM, Furgason ES, Newhouse VL: Evaluation of a random signal correlation system for ultrasonic flaw detection. IEEE Trans Son Ultrason, SU-23: 329–333 (1976).

Chapelon JY, Cathignol D, Fourcade C: Improved ultrasonic sensitivity using pseudo-random binary code phase-modulated signals. Ultrasonic Imaging 1: 255–264 (1979).

Elias CM: An ultrasonic pseudo-random signal-correlation system. IEEE Trans Son Ultrason SU-27: 1–7 (1980).

Eysholdt U, Schreiner C: Maximum length sequences—A fast method for measuring brain stem auditory evoked responses. Proc IEEE Conf Frontiers of Engineering in Health Care. Houston, Texas (1981) pp 306–309.

Furgason ES, Newhouse VL, Bilgutay NM, et al: Application of random signal correlation techniques to ultrasonic flaw detection. Ultrasonics 13: 11–17 (1975).

Gerson A, Perraz J, Hahnech B: Obtention d'échogrammes à structures fines grace au traitement du signal. Ann Phys Biol Med 5: 115–122 (1971).

Golay JE: Complementary series. IRE Trans Inf Theory IT-9: 118–121 (1963).

Golomb SW: Shift register sequences. San Francisco, Holden-Day Inc (1976).

Lee BB, Furgason ES: High speed digital Golay Code flaw detection system. Proc IEEE Ultrasound Conf, Chicago (1981) pp 888–891.

Lee BB, Furgason ES: An evaluation of ultrasound NDE correlation flaw detection systems. IEEE Trans Son Ultrason SU-29: 359–369 (1982).

Nahamoo D, Kak AC: Ultrasonic echo imaging with pseudo-random and pulsed sources: A comparative study. Ultrasonic Imaging 3: 1–36 (1981).

Newhouse VL, Furgason ES, Bilgutay NH: Random signal flaw detection. Proc 19th IEEE Ultrasonics Symposium (1974) pp 711–715.

Takeuchi Y: An investigation of a spread energy method for medical ultrasound systems. Part one: Theory and investigation. Ultrasonics 17 (4): 125–182 (1979).

7
Pseudo-Random Correlation Flow Measurement

D. CATHIGNOL

Introduction

Problems Encountered with Pulsed Velocimeters

Two important problems limit the possibilities of pulsed velocimeters: (1) The impossibility of augmenting the signal-to-noise ratio without degrading other parameters of clinical interest, and (2) The impossibility of augmenting the product of the maximum detectable range with that of the maximum detectable velocity. This forces the user to choose a compromise solution that is often deleterious to precision of measurement. There are essentially two limitations that have led certain authors to search for types of transmitted signals that make it possible to overcome, partially or totally, these two problems. To better understand the choices made by these authors, it appears to us necessary to recall briefly why the signal-to-noise ratios of pulsed velocimeters are weak, and why they are subject to problems of range velocity ambiguity.

The signal-to-noise ratio of any type of system provided with its matched filtering depends on the energy received during the measurement time. In the case of a pulsed Doppler velocimeter, it is clear that improvement of the signal-to-noise ratio can only be obtained through (1) an augmentation of the peak power, (2) an increase in the duration of the transmission, or (3) a decrease of the repetition period. The first possibility presents two difficulties: First, it is not possible to indefinitely augment the peak transmitted power because of the risk of transducer breakdown and because of unpredictable biological effects. Second, the biological transmission medium, as well as the water of which it is composed, is not linear. Thus the attenuation becomes more and more important, the greater the peak transmitted power.

The second possibility consists of augmenting the duration of transmission. The resolution in distance is inversely proportional to the transmission duration, and one can not increase the transmission duration without degrading the longitudinal resolution.

$$\Delta R = \frac{1}{2} c\tau_e \tag{7.1}$$

The third possibility consists of reducing the repetition period, ie, reducing the duration T_R. We recall that in that case we reduce the maximum distance explorable without ambiguity proportionally to T_R.

$$R_{max} = \frac{1}{2} c\, T_R \tag{7.2}$$

As can be seen, the augmentation of the energy of a pulsed velocimeter is difficult because of mathematical relations that exist with other parameters and that are fundamental for the clinician. Nevertheless, one sees that except for the first technique in which the transmission medium and the ultrasonic transducer are extrinsic to the transmitted signal, the two other techniques converge on the same objective, namely that of diminishing the peak to average power ratio. Thus one can understand why research teams have explored the possibilities of systems for which the ratio of peak power p_c to average power p_m equal unity or close to unity.

The second important point is the range velocity ambiguity. We recall that the maximum detectable frequency is given by

$$f_{max} = \frac{1}{2\, T_e} \tag{7.3}$$

where T_e is the sampling interval of the Doppler signal, and the maximum range explorable without ambiguity is given by

$$R_{max} = \frac{1}{2} c\, T_R \tag{7.4}$$

where T_R is the period of the emitted signal.

In the case of the pulsed velocimeter, the period separating two samples of the Doppler signal can only be equal or less than the period T_R of the transmitted signal, and we have the implicit relationship $T_e = T_R$. One should note that this equation, which is rarely mentioned, is very important because it reminds us that these two quantities can be distinct. This independence leads us to propose types of emission in which T_e and T_R either are totally independent or are linked by a relation $T_e = k\, T_R$, where k is larger than unity.

The types of proposed transmitted signal must therefore satisfy two essential criteria: (1) augmentation of the emitted energy by diminution of the ratio p_c/p_m while not affecting the parameter ΔR and (2) reduction or suppression of the dependence of T_e on T_R.

Constraints on Proposed Solution

The solution proposed for resolving the problems corresponding either to increase of the transmitted power, or to diminution of the range frequency ambiguity, must not contribute to the detriment of other parameters that are of primary importance in the biomedical domain, which is of interest here. These are as follow:

1. The signal-to-noise ratio. This parameter, already discussed above, characterizes the possibility of detecting deep lying vessels in the strongly absorbing biological medium.
2. The longitudinal range resolution ΔR. This parameter determines the minimum distance between two neighboring range cells in which we measure velocity. We will attempt to minimize this parameter.
3. The signal-to-clutter ratio for stationary clutter target. The veins and arteries are contained in a medium with numerous fixed scatterers that reflect ultrasound. The signal resulting from these must not perturb the Doppler signal. Thus, it will be important to know the degree of perturbation contributed by these stationary targets. The ratio between the Doppler signal and that of the fixed clutter targets will be defined as the signal-to-clutter ratio $SCR_{stationary}$.
4. The signal-to-clutter ratio for moving clutter. The velocity measurement in a particular volume element may be perturbed by all the other reflecting targets that are moving in the ultrasonic field. The ratio between the Doppler signal issuing from the volume element considered and the other moving targets will be defined as the signal-to-mobile-clutter ratio SCR_{moving}.
5. The range ambiguity. This is defined as the maximum distance over which it is possible to take measurements without ambiguity of the measured distance.
6. The ambiguity in frequency or velocity. This is the maximum velocity detectable without ambiguity.

One could also define other interesting parameters, notably the frequency resolution; however, the six parameters cited above appear to us to be the most fundamental and will be examined the most carefully.

As we have mentioned, it is essentially the search for the augmentation of the emitted energy that has led to the development of velocimeters in which the transmitted signal differs from an impulse. Two forms of signal have received the most attention: Gaussian noise and pseudo-random codes. These two signals are studied below. In addition, some recent results of systems based on these principles, as well as of systems transmitting a chirp, are given.

The Ambiguity Function and Its Relation to the Doppler Measurement

The basic components of a random, pseudo-random, or pulsed Doppler system that uses separate transducers are shown in Figure 7.1.

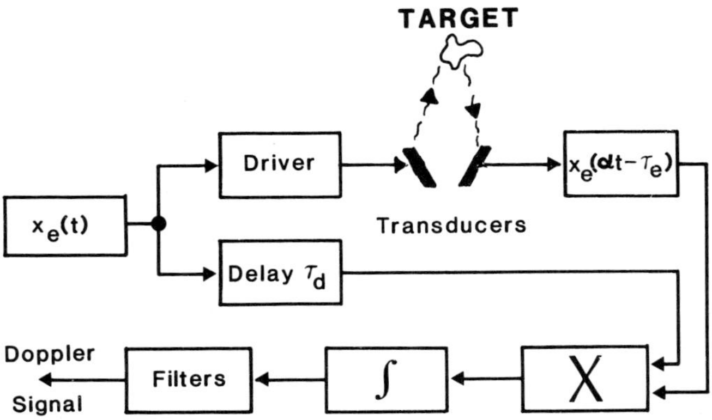

FIGURE 7.1. System arrangement common to all correlation-type velocimeters. For pseudo-random code systems the integrator includes a sample and hold circuit.

The Signal at the Output of the Correlator

Assume that the transducer emits a signal $x_e(t)$. After reflection from a moving particle of radial speed v and situated at a distance r_o, the received signal is

$$x_r(t) = k\, x_e(\alpha t - \tau_e) \tag{7.5}$$

$$\tau_e \text{ is the time of flight } = \frac{2r_o}{c}$$

$$\alpha = \frac{c - v}{c + v} \simeq 1 - \frac{2v}{c}$$

c = ultrasound velocity
v = target velocity
k = constant

The receiver correlates the received signal with the transmitted signal delayed by τ_d, giving

$$z(\tau) = k \int_0^{T_e} x_e(t - \tau_d)\, x_e(\alpha t - \tau_e)\, dt \tag{7.6}$$

In the case where the modulating signal has a small bandwidth, ie, where the spectrum of the modulated signal and the spectrum of the modulating signal do not overlap, it is always possible to write the signal $x(t)$ in the form

$$x_e(t) = R_e\left[X_e(t)\right] \tag{7.7}$$

with

$$X_e(t) = A(t)\, \exp 2i\pi f_o t \tag{7.8}$$

That is to say, in the form of a carrier of frequency f_o modulated by a complex amplitude $A(t)$.

In the same manner we define an analytic echo signal

$$X_r(t) = kA(\alpha t - \tau_e)\exp 2\pi i f_o (\alpha t - \tau_e) \tag{7.9}$$

To the degree to which the amplitude varies slowly with respect to the term $\exp 2i\pi f_o(\alpha t - \tau_e)$, which itself varies at a frequency f_o, it is possible to approximate α by 1 in the term corresponding to the amplitude, which gives

$$X_r(t) = kA\,(t - \tau_e)\exp 2\pi i f_o(\alpha t - \tau_e) \tag{7.10}$$

The analytic signal at the output of the multiplier is written as

$$u(t) = KA(t)\exp 2\pi i f_o t \cdot A^*\,(t - \tau_o)\exp - 2\pi i(f_o + f_d)\,(t - \tau_o) \tag{7.11}$$

where $f_d = \dfrac{2v}{c} f_o$

Here τ_o is the initial value of $\tau_e - \tau_d$ at $t = 0$.

At the output of the correlator the analytic form of z is

$$z\,(f_d,\tau) = \int_0^{T_e} u(t)\,dt \tag{7.12}$$

The value of $z(f_d,\tau)$ determined in this way is stored during every period T_e by a sample and hold circuit.* At the end of the time nT_e the target has covered the distance nvT_e. The delay corresponding to the time of flight between the target and the transducer is now $\dfrac{2nvT_e}{c} + \tau_o$. At that instant the value of z corresponding to the position of the target is

$$z(f_d,\tau_o + \frac{2nvT_e}{c}) = K\int_0^{T_e} A(t)\exp 2\pi i f_o t \cdot A^*(t - \tau_o - n\frac{2vT_e}{c})\exp - 2\pi i(f_o + f_d)\,[t - (\tau_o + \frac{2nvT_e}{c})]\,dt \tag{7.13}$$

$$z(f_d,\tau_o + \frac{2nvT_e}{c}) = \exp 2\pi i(f_o + f_d)\,(\tau_o + \frac{2nvT_e}{c})\,K\int_0^{T_e} A(t)A^*(t)(t - \tau_o - \frac{2nvT_e}{c})\exp - 2\pi i f_d t\,dt \tag{7.14}$$

with

$$-2\pi f_o \tau_o = \phi_o$$

$$z(f_d,\tau_o + \frac{2nvT_e}{c}) = \exp i(2\pi f_o \frac{2nvT_e}{c} - \Phi_o)\,K\int_0^{T_e} A(t)\,A^*(t)(t - \tau_o - \frac{2nvT_e}{c})\exp - 2\pi i f_d t\,dt \tag{7.15}$$

*Not shown in Figure 7.1.

The time-dependent portion of z is the ambiguity function $\mathcal{A}(f,t)$ in which the time has suffered a compression equal to $2v/c$ (Doppler compression). The integral represents the value of the ambiguity function of the modulating signal $A(t)$ for

$$\tau = \tau_o + \frac{2nvT_e}{c} \text{ and } f = f_d$$

The term in front of the integral represents a signal of the frequency f_d. Thus, a target situated initially at the distance (f_d, τ_o) gives at the correlator output a low-frequency signal the amplitude of which is the ambiguity function in the plane $f = f_d$ and for the values

$$\tau = \tau_o + \frac{2nv}{c} - T_e$$

Because the values of $z(t)$ are calculated at intervals T_e, the value obtained is a sampling of $\mathcal{A}(f, \tau)$ in the plane $f = f_d$. Figure 7.2 could represent one of these cases.

In conclusion, the evolution of $z(t)$ with time gives us information on (a) the target velocity from the carrier frequency f_d and on (b) the target position to the degree to which the envelope of $\mathcal{A}(f_d, \tau)$ presents a maximum at $\tau = 0$ that is sufficiently distinguishable for each f_d.

The performance of the system is a function of the characteristics of $\mathcal{A}(f, \tau)$ and consequently of the transmitted signal.

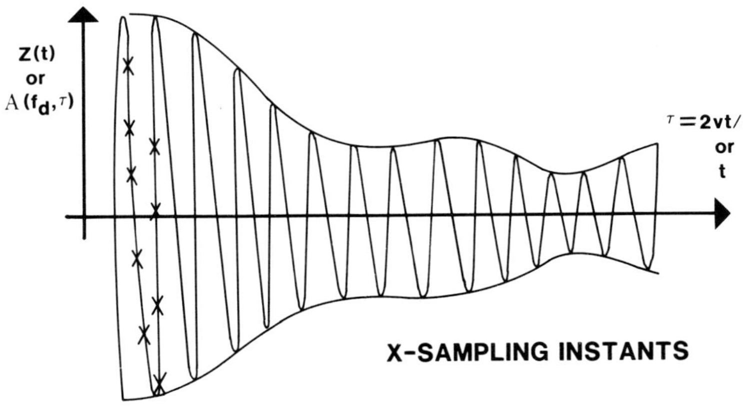

FIGURE 7.2. Correlator output signal. Its shape represents the ambiguity function in the plane $f = f_d$ and for different values $\tau = (2nvT_e/c)$. The carrier frequency represents the low-frequency signal $2f_0v/c$, and the crosses represent the sampling times at intervals of T_e.

Criteria for Choice of Transmitted Signal

Some properties of the ambiguity function permit us to give the essential characteristics of the signal to be transmitted.

1. In the plane $(0, \tau)$ the ambiguity function is the correlation function $R(\tau)$. The correlation function is the Fourier transform of the spectral density of the amplitude $A(t)$. The ability to separate two targets of known speed at different ranges thus depends on the spectral width of the emitted signal. The length of the range cell is given by

$$\Delta R = \frac{c}{2B} \tag{7.16}$$

2. In the plane $(f_d,0)$ and when the signal is narrow band, the ambiguity function is the Fourier transform of $\mid A(t) \mid^2$. The ability to separate two targets with different speeds at known ranges thus varies as the duration of the signal:

$$\Delta f_d = \frac{1}{T} \text{ or } \Delta v = \frac{c}{2Tf_o} \tag{7.17}$$

   Taking into account Equations (7.16) and (7.17), our choice must be oriented toward a signal in which the product BT is large.
3. In the case of stationary targets the value of z is identical for two successive emissions of $A(t)$. Hence these echoes give a DC level that is easy to filter. On the other hand, if the value of z varies for two distinct realizations of $A(t)$, then the signal $z(t)$ at the correlator output will be taken for a Doppler signal.
4. The Doppler signal is sampled at the frequency $F_r = 1/T_e$. The value of T_e cannot exceed $1/2f_{max}$, where f_{max} is the maximum Doppler frequency.
5. The signal-to-noise ratio of all matched systems depends uniquely on the energy received during the measurement time. Thus we will search for a signal in which the peak-to-average power ratio is as close as possible to unity.

Applications to Various Velocimeters

Pulsed Doppler System

Figure 7.3 shows a typical transmitted waveform for this type of system consisting of a sine pulse trains of duration τ_e, the period of which is T_R, and the measurement of the Doppler shift is integrated over a period T. The ambiguity function is known and is also shown in Figure 7.3. In the plane parallel to $f_d = 0$, the ambiguity function shows peaks equal to unity whenever $\tau = T$, where T_R is the repetition period of the transmitted

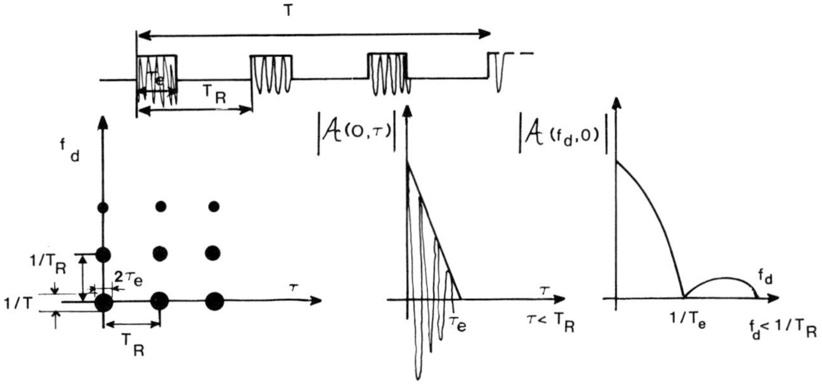

FIGURE 7.3. Ambiguity function for a burst of impulses of duration τ_e and period T_R during an integration time T. Maximum detectable range: $R_{max} = cT_R/2$; maximum Doppler frequency: $f_{dmax} = 1/2T_R$; range resolution: $\Delta R = c\,\tau_e/2$; frequency resolution: $\Delta f_d = 1/T$.

pulses. For a target with the velocity v, the signal $z(t)$ is a triangle modulated by cos $2\pi f_d t$. All the values of the parameters given earlier may be determined from the ambiguity function.

Range Resolution

The range resolution is given by

$$\Delta R = \frac{c}{2}\tau_e \simeq \frac{c}{2B} \tag{7.18}$$

Any improvement of the range resolution is obtained at the expense of a decrease in the duration of the emitted signal, which is in contradiction with the requirement of a peak-to-average power ratio close to unity.

Range Ambiguity

The maximum range without ambiguity is given by

$$R_{max} = \frac{1}{2}cT_R$$

Frequency Ambiguity

Because the signal at the output of the correlator is sampled at the frequency $1/T_e$, the maximum frequency detectable is $f_{max} = 1/2\ T_e$. In a pulsed system where the sampling frequency equals that of the transmitted signal $T_R = T_e$, combining the two Equations (7.1) and (7.2) with

$$f_{max} = \frac{2f_o v_{max}}{c} \tag{7.19}$$

we obtain

$$v_{max} R_{max} = \frac{c^2}{8f_o} \tag{7.20}$$

Influence of Fixed Clutter

Since the transmitted signal is repeated at intervals T_R, the value $z(t)$ is constant from one emission to another. After filtering, this continuous level of the signal, due to fixed targets, is eliminated.

Influence of Moving Clutter

Outside of the measurement volume, or range cell, determined by the peaks of the ambiguity function, the correlator output signal is zero, and the signal-to-clutter ratio SCR_{mobile} is infinite. This is an important property of pulsed velocimeters.

Noise Doppler System

Ambiguity Function for Noise

A transmitted signal of Gaussian noise of bandwidth B and center frequency f_o has been considered by Newhouse et al [1973]. The signal is written

$$X(t) = A(t)\exp 2\pi i f_o t \tag{7.21}$$

where $A(t)$ is a random function of time. It is no longer possible in this case to speak of an ambiguity function because $A(t)$ is not stationary. One can define an expectation value of the ambiguity $E[\mathcal{A}(f,\tau)]$, which is the average of $\mathcal{A}(f,\tau)$ and a variance that represents the nonreproducibility of $A(t)$. This parameter is essential in our case where we are interested in variations of $z(t)$ for two successive transmissions. $E[\mathcal{A}(f,\tau)]$ has been shown to have a central peak of width $1/T_e$ along the frequency axis and $1/B$ along the time axis. Outside this peak $E[(\mathcal{A}(f,\tau)]$ tends toward $1/BT_e$ (Figure 7.4) with a variance ε^2 equal to $1/BT_e$.

The essential properties that follow as far as the range frequency resolution and the influence of fixed or moving targets are studied in more detail below in connection with a practical system.

Typical System

The basic components of a random-signal Doppler system that uses separate transducers for transmitting and receiving are shown in Figure 7.1. The generator is a continuous or intermittent noise source. A corresponding continuous or intermittent noise signal is reflected by scatterers in the fluid and correlated with a delayed version of the transmitted signal. The correlation is composed of a multiplier and a low-pass filter. The delay

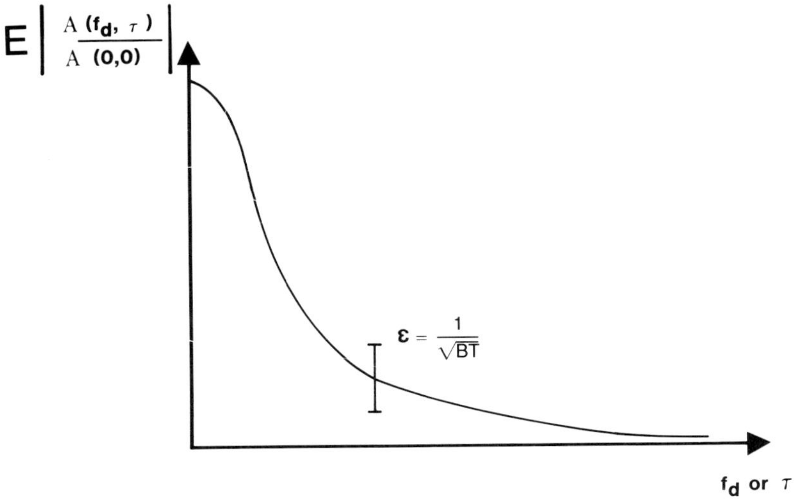

FIGURE 7.4. Expectation of $A\ (0,\ \tau)$ or $A\ (f,\ 0)$.

is obtained either by an ultrasound delay line or, better, by the storage in memory of the emitted signal followed by a delayed readout.

We have seen that the correlator output signal is a low-frequency signal whose amplitude is given by the value of the ambiguity function for the pair of values (f,τ). For slow speeds the ambiguity function $\mathcal{A}(f,\tau)$ is similar to the correlation function $R(\tau)$ of the transmitted signal.

In the case of a random signal, the correlation function is constant if the integration is made over infinite time. In that case it is equal to the Fourier transform of the spectral density; for integration times less than infinite, it differs from the Fourier transform by a quantity ε which depends on the nature of the transmitted signal, its duration, the integration time, and the sample chosen.

Newhouse et al [1973] studied the case where the emitted signal has a spectral density

$$S(f) = \frac{(B/2)^2}{(B/2)^2 + (f - f_o)^2} \tag{7.22}$$

The correlator output signal at each integration is given by

$$z(t) = E\left\{R(\alpha t - \tau)\right\} \cos 2\pi f t + \varepsilon \tag{7.23}$$

If the variations of $R(\tau)$, which exist between two consecutive samples, are negligible,

$$z(t) = R\ (\alpha t - \tau) \cos 2\ \pi f t \tag{7.24}$$

and because $R(\tau)$ is the Fourier transform of the spectral density,

$$z(t) = \frac{B}{2} \exp - \pi B(\alpha t - \tau) \cos 2\pi f_d t \tag{7.25}$$

SYSTEM ANALYSIS

Range Resolution

We recall that the range resolution is determined by the ambiguity function in the plane $(0,\tau)$ and is expressed by $\Delta R \simeq c/2B$. Thus for a given spectral bandwidth, the range resolution is independent of the form of the transmitted signal.

In the case of pulsed velocimeters, the width of the spectral density is related to $B = 1/\tau_e$, where τ_e represents the width of the transmitted pulse. Therefore, one can only augment the range resolution by reducing the duration of the transmitted signal, that is to say, by reducing the emitted energy and thus the signal-to-noise ratio. On the other hand, in the case of a continuous random noise signal, the spectral density is independent of the duration, thus permitting a reduction of the peak transmitted energy, while conserving the range resolution.

Ambiguity in Range and Velocity

There is no range ambiguity because the ambiguity function has only one central peak since the noise signal is never repeated.

If one chooses the time constant of the integrator in a manner such that

$$T_e \leqslant \frac{1}{2 f_{\max}} \tag{7.26}$$

there will be no velocity ambiguity. It appears, therefore, that the emission of a random signal permits the elimination of range-velocity product ambiguity problems.

Influence of Fixed Target Echoes

The signal $z(t)$ in the presence of fixed target echoes is written as

$$z(t) = E\{\mathscr{A}(0,\tau)\} + \varepsilon = E\{R(\tau)\} + \varepsilon \tag{7.27}$$

$z(t)$ varies from one realization to another. This variation versus time is similar to that of a Doppler signal. Thus, it is necessary to evaluate this effect.

In the case of a Gaussian random signal it was shown earlier that the error ε obtained in estimating the correlation function is a Gaussian signal of zero mean and of expectation $R(0)/\sqrt{BT_e}$. Thus the ratio between the signal due to a moving target of reflecting power a^2 situated at a distance

τ_e and the signal due to a fixed echo of reflecting power b^2 is given by the signal-to-clutter ratio

$$\mathrm{SCR}_{\text{stationary}} = \frac{a}{b}\sqrt{BT_e} \tag{7.28}$$

In this equation T_e cannot be increased indefinitely since it is related to the maximum detectable Doppler frequency. Thus, to see if a signal is adequate or not, it is necessary to consider a concrete example, for instance, that of a velocimeter at 5 MHz used in current clinical practice. For this apparatus, the bandwidth of the emitted spectral density is limited by the transducer to $B = 2$ MHz, and the maximum Doppler frequency encountered is 5 KHz.

Since T_e cannot exceed $1/2\, f_{\max}$, the product BT_e is of order 200. It is therefore necessary that the power ratio between fixed targets and moving targets does not exceed 200. Unfortunately, in medical application this value is often higher.

Influence of Moving Clutter

In contrast to the pulsed Doppler velocimeter, the ambiguity function $A(f,\tau)$ is different from zero outside the peak. The ratio between the signal amplitude due to a target situated in the range volume of a power a^2 and that of a moving clutter target situated at a distance τ_e of reflecting power b^2 is given by:

$$\mathrm{SCR}_{\text{moving}} = \frac{a}{b}\sqrt{BT_e} \tag{7.29}$$

For n particles this ratio is approximately

$$\mathrm{SCR}_{\text{moving}} \simeq \frac{a}{\left(\sum_{i=1}^{n} b_i^2\right)^{1/2}}\sqrt{BT_e} \tag{7.30}$$

Signal-to-Noise Ratio

The improvement in power signal to thermal noise ratio produced by a correlation receiver is

$$\frac{\mathrm{SNR}_{\text{out}}}{\mathrm{SNR}_{\text{in}}} = \eta BT_e \tag{7.31}$$

with η equal to the on-off ratio of the transmitted signal. Amir and Newhouse [1984] showed that the power signal to thermal noise ratio improvement for a pulsed velocimeter equals B_dT_e, where B_d is the bandwidth of the Doppler output signal and T_e is the repetition period. Thus, for a given density of the transmitted signal, a given integration time or repetition period and the same peak power, the improvement of the SNR of a noise system over that of pulsed system is $\eta B/B_d$. The ratio B/B_d can be shown to equal the larger value c/v.

In conclusion, this signal—which shows, on the one hand, a noticeable improvement of the signal-to-noise ratio and, on the other hand, a total absence of range velocity ambiguity in the domain considered—is of interest in the particular case where the ratio of the reflecting powers between the target to be detected and the clutter signal is above unity.

M-Code Doppler System

Ambiguity Function in M-Sequences

Cathignol et al [1980] have proposed the use of a circular M-sequence as a transmitted signal. The practical advantages of such a signal have been dealt with in a previous section. The ambiguity function of such a signal is given in Figure 7.5.

In the plane $f = 0$, we find the correlation function studied in an earlier section, with a constant level of lateral time lobes equal to $1/N \simeq 1/BT_R$. In this plane the ambiguity function is periodic for $\tau = T_R$. Outside this plane it can be shown that the ambiguity function presents side lobes whose maximum values reach $1/\sqrt{N}$.

Typical System

Basic Principle

This is identical to that of Figure 7.1, except that the noise source is replaced by an M-sequence, which phase modulates a carrier frequence f_o. This solution already presents all the inherent advantages of the signals that have been discussed in Chapter 6. The sequence is cyclic and the integration time is equal to the length of the complete M-sequence.

System Analysis

Range Resolution

In the plane $(0,\tau)$ the ambiguity function is a triangle of base width 2θ, where θ is the smallest period of the pseudo-random code. Two targets will be resolved if

$$\Delta R = c\theta/2 \simeq c/2B \tag{7.32}$$

One finds again the known fact that the range resolution depends only on the bandwidth B of the transmitted signal.

In the plane $f = f_d$, $\mathcal{A}(f_d,\tau)$ is no longer a triangle, but always exhibits a peak at $\tau = 0$, whose breadth is of the order of $2\ \theta$. For this reason, we may consider that the resolution is of the order of $c/2B$ for all frequencies.

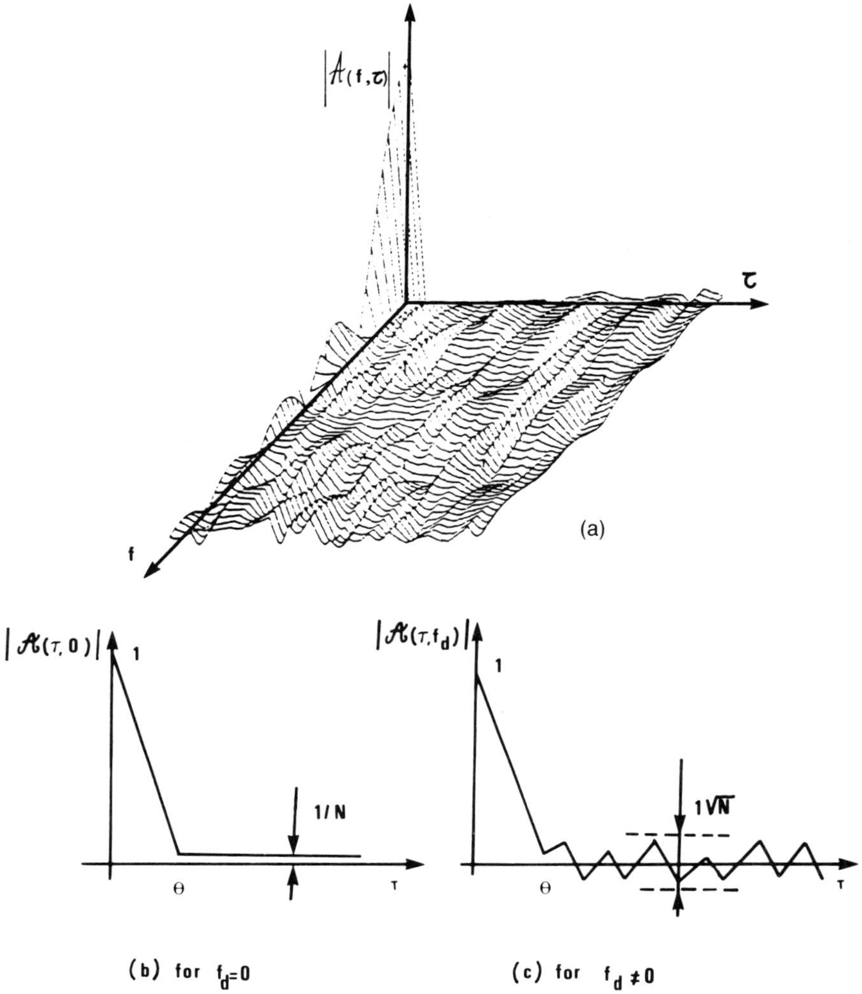

FIGURE 7.5. **(a)** Example of the ambiguity function of an M-sequence of 127 states after Masson [1979]. **(b)** for $f_d = 0$; **(c)** for $f_d \neq 0$.

Range Velocity Ambiguity

The signal is repetitive with period T_R. The problems encountered with respect to range velocity ambiguity are identical to those posed by pulsed emission velocimeters as summed up in Equation (7.20).

Influence of Fixed Clutter Targets

The M-sequence Doppler systems proposed by Waag et al [1972] and Ohtsuki and Okujuma [1976] use a correlator formed of a multiplier followed by a low-pass filter. Under these conditions the signal at instant t_i differs from that at the instant t_j. It follows that at the output of the low-pass filter, just as in the case of noise velocimeters, the signal will vary with time and will be interpreted like a Doppler signal. If the signal is sampled at period T_R, this problem disappears, since the values $z(t)$ are constant from one realization to another. It is therefore possible, just as for monochromatic emission in pulsed emission velocimeters, to eliminate fixed echoes totally by filtering out the DC signals.

Influence of Moving Clutter Targets

In the case of a pulsed velocimeter, the energy of the correlation output signal due to a target not situated at $\tau = \tau_d$ is null, and the signal-to-clutter ratio is infinite. This characteristic is a considerable advantage for this type of emission.

When the speed of the moving clutter is small, the ambiguity function may be approximated by the auto-correlation function and the ratio between the signal received for a target of strength a^2 at $\tau = \tau_d$, and a mobile clutter target of strength b^2 situated at $\tau \neq \tau_d$ becomes aN/b, where N equals the number of states of the code ($N \simeq BT_R$). Comparison with Equation (7.29) shows that for slowly varying clutter targets, the pseudo-random transmitted signal provides a greater reduction of clutter noise than is provided by noise velocimeters.

In the case of more rapidly moving clutter targets, it is not possible to approximate the ambiguity function with the auto-correlation function. In this case the amplitude of the clutter signal is not reduced by a factor $N \sim BT_R$ but by a weaker compression factor defined as $1/\sigma$. The value of the correlator output clutter power σ^2 depends on the number of states of the M-sequence, on the velocity of the clutter target, and on its range. Cathignol [1986] calculated the root-mean-square (RMS), (σ_m), of σ^2 averaged for all the positions of the clutter target. Figure 7.6 gives the value σ_m as a function of the Doppler frequency for the three values of the number of states 255, 127, and 63.

Beyond 500 Hz the value of σ_m increases very rapidly and approaches the value $1/\sqrt{N}$, corresponding to that of the noise velocimeter. Cathignol [1986] calculated the SCR_{moving} for the case of a vessel of diameter D. Let us consider a weakly divergent ultrasonic beam that has a small diameter compared with the diameter D of the vessel (see Figure 7.7). If ΔR is the length of the system range cell, the Doppler signal power corresponding to the targets located between R, and $R + \Delta R$ is proportional to the range cell volume, ie, to ΔR. Hence, the average correlator output voltage is

$$s_d = K\,(\Delta R)^{1/2} \tag{7.33}$$

Similarly, the Doppler signal corresponding to clutter targets is

$$s_{\text{clutter}} = K \left\{ \int_{0}^{R} \sigma^2\,(r)\,dr + \int_{R\,+\,\Delta R}^{D} + \sigma^2\,(r)\,dr \right\}^{1/2} = s_c \tag{7.34}$$

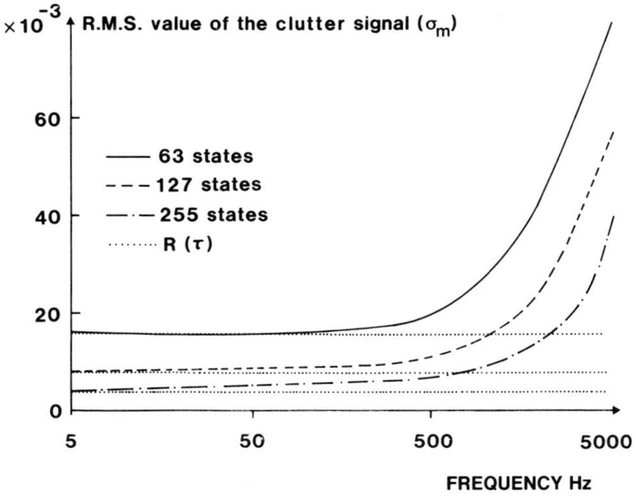

FIGURE 7.6. These curves give the RMS value of the time compression $1/\sigma_m$ averaged for all the positions of the target as a function of Doppler frequency and the states numbers of the M-sequences. Dotted lines correspond to the result obtained with the approximation $A(\alpha, \tau) = R(\tau)$.

$\sigma^2(r)$ is the clutter-to-signal-power ratio for the range r.

If the number of reflecting targets is sufficiently great, s_c behaves like noise, ie,

$$s_c = K\sigma_m(D - \Delta R)^{1/2} \tag{7.35}$$

and the voltage signal-to-clutter ratio is

$$\frac{s}{s_c} = \frac{1}{\sigma_m}\left(\frac{\Delta R}{D - \Delta R}\right)^{1/2} \tag{7.36}$$

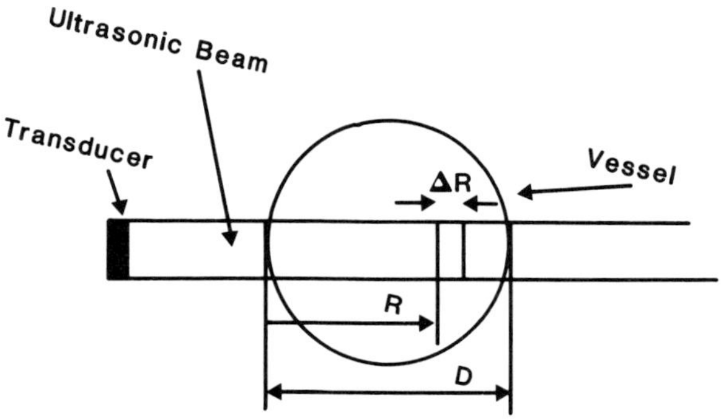

FIGURE 7.7. Ultrasonic beam crossing blood vessel.

Figures 7.8 a and b give the signal-to-clutter ratio as a function of the diameter for three numbers of states 255, 127, 63, and two values of $(\alpha - 1) = 10^{-3}$ and 10^{-5}.

In the case of a frequency f_o equal to 5 MHz, the range resolution is, respectively, ± 0.3 mm, ± 0.6, and ± 1.2 mm for the number of states equal to 255, 127, and 63. For the same frequency, the Doppler shifts corresponding to $\alpha - 1 = 10^{-3}$ and $\alpha - 1 = 10^{-5}$ are 5 KHz and 50 Hz. To obtain an acceptable error (< 10%) in measuring the Doppler frequency, it is necessary that the RMS value of the Doppler signal be at least three times the noise level. This is the case for the three numbers of states (255, 127, 63) and for $(\alpha - 1) = 10^{-3}$ up to vessel diameters of 2 cm. However, for larger vessels or for cardiologic examinations the signal-to-clutter ratio is too low to allow for a precise velocity measurement for any number of

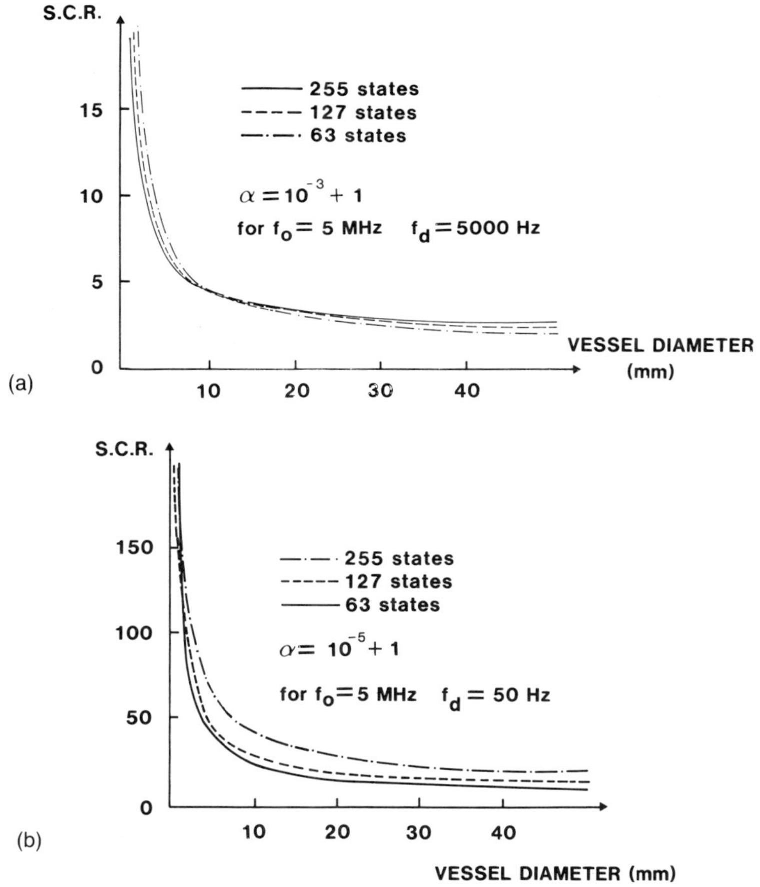

FIGURE 7.8. Signal-to-clutter ratio as a function of the vessel diameter for the two values of $(\alpha - 1) = 10^{-3}$ **(a)** and $(\alpha - 1) = 10^{-5}$ **(b)**.

states. The above calculation confirms the major interest of pseudo-random velocimeters, namely, the possibility of transmitting a substantial amount energy for the detection of small, deep-lying vessels. In the frequently encountered case where it is only necessary to detect the presence of a vessel, the operator has to choose the smallest possible number of states for a given T_R, thus reducing the transmitted bandwidth B. In this case the received signal and the signal-to-noise ratio increase because the length of the system range cell increases. The signal-to-clutter ratio is little affected because the reduction of the compression factor BT_R is compensated by the fact that the number of moving clutter targets decreases when the range cell increases.

Double M-Code Doppler System

Since an M-sequence Doppler system emits periodic signals, we should expect to encounter ambiguity problems analogous to those found in conventional pulsed radio-frequency (RF) Doppler systems. The maximum distance to the target without ambiguity is given by

$$R_{\max} = \frac{cT_R}{2} \tag{7.37}$$

where T_R is the period of the signal. The maximum frequency of the Doppler output is

$$f_{\max} = \frac{1}{2} \Big/ T_e \tag{7.38}$$

where T_e is the time separating two samples of the Doppler signal at the correlator output. In a pulse Doppler system, as in an M-sequence Doppler system, $T_e = T_R$ and we find the known relation (7.20).

In a multi-M-code Doppler system T_e and T_R are dissociated. It is possible to choose $T_R = \beta T_e$, with $\beta > 1$. The relation (7.20) becomes

$$R_{\max} v_{\max} = \frac{\beta c^2}{8 f_o} \tag{7.39}$$

and we have reduced the range velocity product by a factor β.

System Analysis

Range Frequency Ambiguity

For technical reasons, we consider only the case $\beta = 2$. However, the demonstration that follows holds true whatever the value of β. Consider a binary signal composed of two segments T_1 and $T_2 = T_e$ equal and independent. As the total signal repeats itself, each $T_R = T_1 + T_2$ (Figure 7.9a), and the maximum range without ambiguity is given by

$$R_{\max} = T_R \frac{c}{2} \tag{7.40}$$

The integration is made during the times $(0, T_R/2)$ then $(T_R/2, T_R) = T_e$. If we choose $T_R = 2T_e$, Equation (7.39) becomes

$$R_{max} V_{max} = \frac{2c^2}{8f_o} \tag{7.41}$$

and we have increased the range-velocity product by a factor 2.

The system properties concerning the signal-to-noise ratio as well as the range resolution are conserved. Only the influence of fixed and mobile clutter targets is modified.

Influence of Fixed Targets

We have seen that the fixed target echoes are totally eliminated if the signal at the output of the correlator is stationary, that is to say, if the correlation product does not vary from one integration to another. Since the code sequences T_1 and T_2 are different, there is no reason why the correlation product during the period $(0, T_R/2)$ should be the same as that obtained during the time $(T_R/2, T_R)$. Thus, the output of the sample and hold circuit will be in the form of a square wave (Figure 7.9b). This signal could be taken for a Doppler signal. However, this inconvenience is elim-

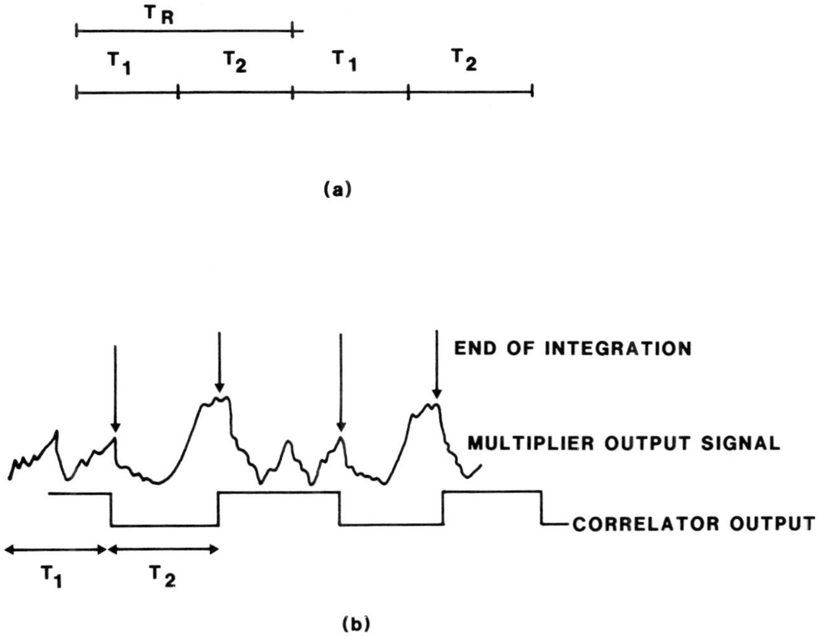

FIGURE 7.9. Waveforms of the double M-code Doppler system. **(a)** The emitted signal is composed of two pseudo-random sequences T_1 and T_2. **(b)** As the sequences T_1 and T_2 are different, the correlation product during $0 - T_R/2$ is different from that obtained during time $T_R/2 - T_R$, and the sample-and-hold outputs look like a binary signal.

inated by the scheme shown in Figure 7.10. The integration and the reset are made at intervals $T_R/2$; the value found at the end of the period $T_R/2$ is fed to one channel, and the value found at the end of the period T_R is fed to a second channel. In the presence of fixed target echoes, the signals at the output of two channels are continuous and are filtered out.

In the case of a moving target situated at $t = \tau_d$, sending the signal first on one channel then on another does not modify in any way the result obtained. In the case of simultaneous fixed targets and mobile targets, the output of the hybrid correlator is the sum of the above operations.

Influence of Moving Targets

As in the case of a single M-sequence, we can define a signal-to-clutter ratio due to moving targets. If one considers that two sequences T_1 and

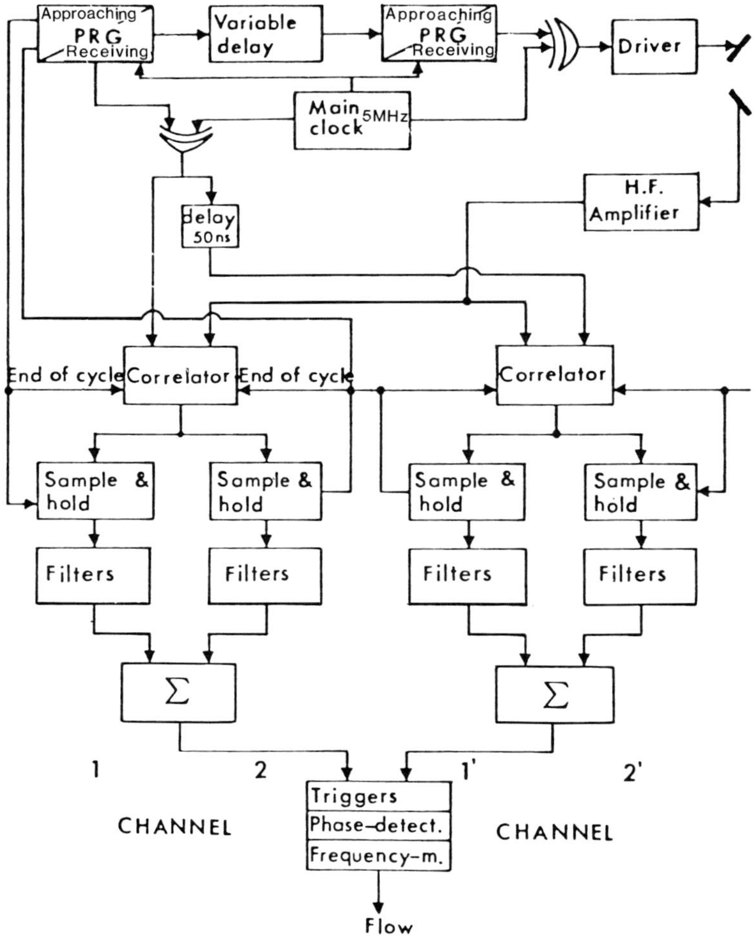

FIGURE 7.10. The double M-code Doppler system.

T_2 have the same properties as two sequences of Gaussian noise, then in this case one again finds the ratio

$$\mathrm{SCR} = a/b\ \sqrt{\mathrm{BT}_e} \tag{7.42}$$

Realization of a Practical System [Cathignol et al, 1980]

The choice of parameters depends on the specific application, taking into account the priority given to each of the following constraints: maximum detectable Doppler frequency, maximum exploration depth without ambiguity, and optimized signal-to-clutter ratio.

The block diagram of Figure 7.10 corresponds to the case of a directional velocimeter, which can detect a maximum frequency $f_{\max}$ for a theoretical maximum depth $R_{\max}$ at f_o. The maximum detectable frequency $f_{\max}$ determines the integration time T_e by Equation (7.38) and the maximum depth of exploration $R_{\max}$ in Equation (7.37).

The number of states N of the two M-sequences must be maximal for obtaining the best signal-to-clutter ratio. Taking into account that the passband B of the transducers is estimated at 50% of the transmitted frequency f_o, the number of states N of each code is determined by $T_R f_o/2$.

For all the M-sequences of which the number of states is N, two are chosen for which the function of interambiguity is minimal. As with the

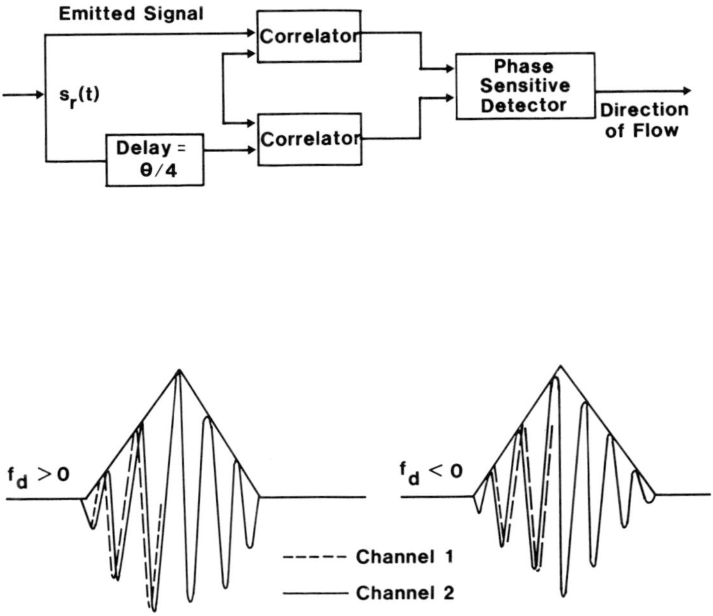

Figure 7.11. Method for obtaining flow polarity. A part of the transmitted signal is delayed by $\theta/4$; this delay corresponds to the phase-shift of $\pi/2$ introduced in conventional Doppler systems.

code echo detection systems of Chapter 6, the M-sequences are generated by shift registers and phase modulate the carrier frequency f_o. In the same way the delay τ_d of the reference channel is obtained from two other shift registers identical to the first ones and delayed by this period.

The principle for determining the sense of the flow is given in Figure 7.11. The delay $\theta/4$ corresponds to the phase shift or $\pi/2$ introduced in the continuous Doppler system. This method is preferred over that obtained by frequency offset for which the practical realization is technically more difficult.

The measure of the frequency of the Doppler signal is made taking into account the various precautions formulated by different authors, by means of a zero crossing detector followed by a frequency meter.

On this same principle Cathignol et al [1980] developed a system that makes it possible to measure the velocity simultaneously at ten distinct points. This improvement allows the visualization in real time of a blood vessel velocity profile. This particular requirement calls for the realization of ten channels of demodulation identical to that shown in Figure 7.10.

Clinical Applications

The examples described below illustrate the possibilities of this instrument (Figures 7.12 and 7.13).

Shape of Carotid Bifurcation

We show through these three profiles the shape of the carotid bifurcation. In the top photograph the two carotid arteries are not yet separated. One can see two distinct zones that correspond to the beginning of two separate flows, one in the internal carotid artery and the other in the external carotid artery. At the end of diastole, a continuous flow remains in the two zones, as is the case for the common carotid artery. The absence of flow at the end of diastole appears only in the middle photograph where the two carotid arteries are actually distinct. In the bottom picture the carotid arteries are separated. The inversion of the position of the two carotid arteries arises from the fact that the transducer looks from bottom to top.

Profile of the Region of the Abdominal Aorta

This figure has the goal of demonstrating the superiority of the pseudorandom noise velocimeter compared with a pulsed Doppler system. It shows a recording made in the region of the abdominal aorta. The artery is situated at a depth of 40 mm. Flow measurement in this region of the artery with a pulsed Doppler system and with the same peak power at the same frequency has been found to be impossible because the signal-to-noise ratio is too feeble.

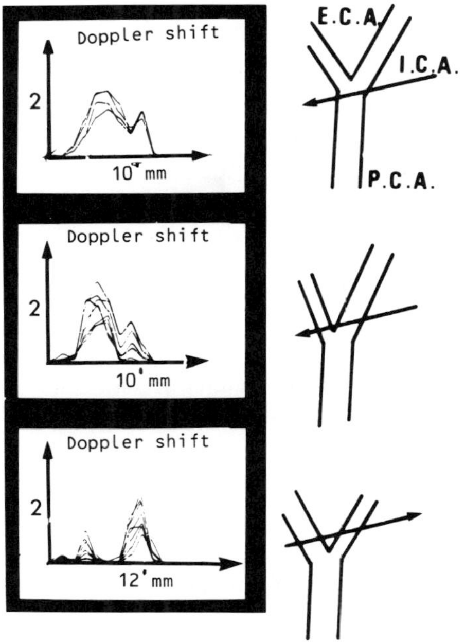

FIGURE 7.12. Doppler frequency profiles near the carotid birufcation. (*Top*) Beam crossing slightly below bifurcation; (*Middle*) beam crossing slightly above bifurcation; (*Bottom*) beam crossing well above bifurcation. E.C.A. = external carotid artery, I.C.A. = internal carotid artery, P.C.A. = common carotid artery.

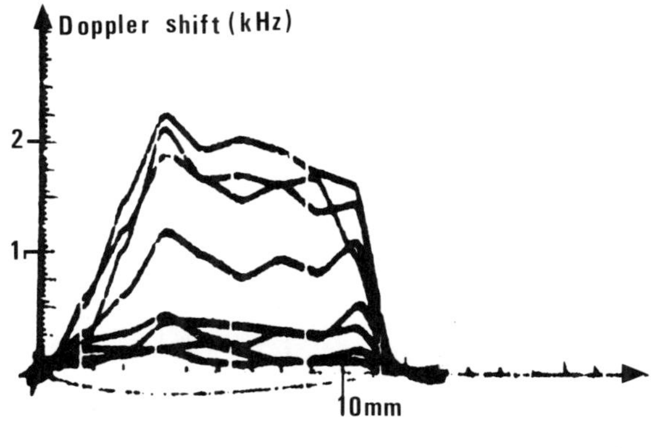

FIGURE 7.13. Velocity profiles made by the pseudo-random code Doppler of Cathigonol et al. [1980] in the abdominal aorta. The profiles shown are taken at ½-second intervals.

Other Systems

We describe here other systems proposed in the literature. These different systems have not always been studied in depth, but have been researched to improve one or another parameter. The repercussions of this on the other parameters have not always been taken into account.

Doppler System Using Cross-Correlation Techniques

In the case of pseudo-random velocimeters, it has been shown above that moving clutter is not totally eliminated because $\mathcal{A}(f_d,\tau)$ differs from zero away from the region $\tau = \tau_o$. Shiozaki et al [1979] proposed a system where transmitted and received signals are modulated and demodulated by two different sequences, $F_1(t)$ and $F_2(t)$ (Figure 7.14). The two sequences are chosen in such a way that their cross-correlation is zero away from the peak (Figure 7.15). Let $F_1(t)$ be a pseudo-random signal as in Figure 7.14, with period T_R generated by a clock pulse of period θ_c. Then $F_2(t) = F_1(t)$ if $H(t) = 0$ and unity if $H(t) = 1$.

In the case where the speed of clutter targets is low, we have seen that the ambiguity function may be approximated by the cross-correlation function in the same way as the interambiguity function may be approximated by that of cross-correlation. Since this is zero outside the region

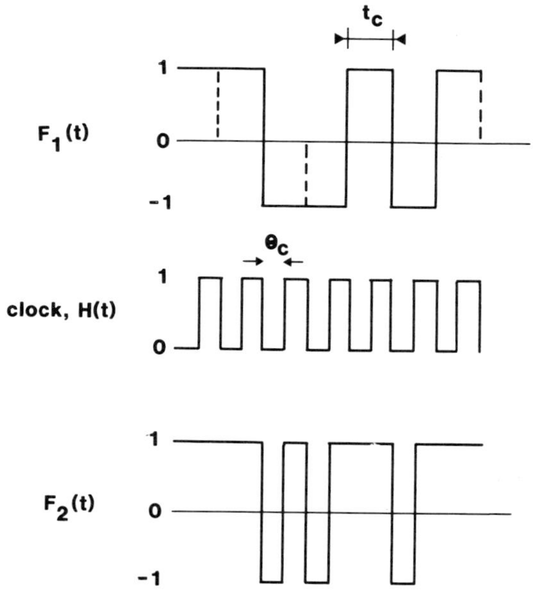

FIGURE 7.14. Code suggested by Shiozaki et al [1979]. $F_1(t)$ is an M-sequence. $F_2(t)$ is defined as a function the value of which equals the value of $F_1(t)$ when a clock pulse of $t_c/2$ is 0 and which equals unity when the clock pulse is 1.

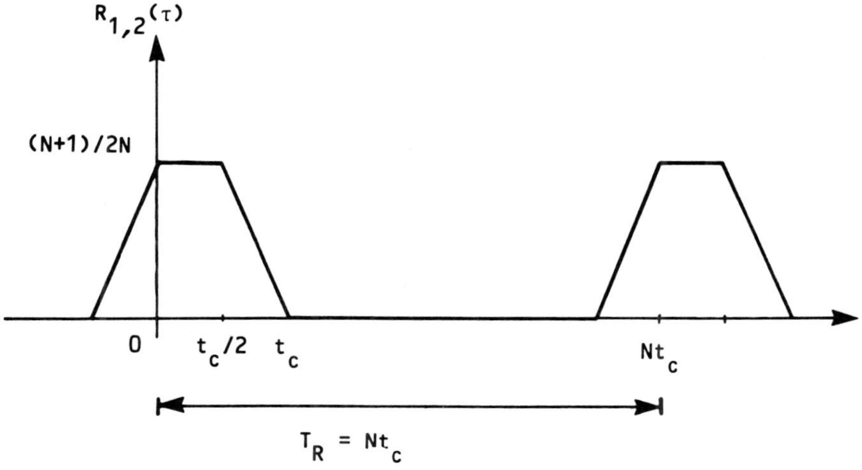

FIGURE 7.15. The intercorrelation function $R_{1,2}(\tau)$ of the codes of Figure 7.14. The intercorrelation function $R_{12}(\tau) = 1/T\int_0^T F_1(t)F_2(t)d\tau$ is periodic in T_R. This function is zero in the interval $(t_c, NT_c - T_c/2)$ modulo Nt_c and equals $(N + 1)/2N$ in the interval $(O, t_c/2)$ modulo Nt_c.

$\tau = \tau_o$, all the moving clutter targets not situated at $\tau = \tau_o$ are eliminated, so that the signal-to-clutter ratio becomes infinite (Figure 7.15).

Figure 7.16 shows a block diagram for this type of system, and Figure 7.17 compares its signal-to-clutter ratio with that of a classical pseudo-random velocimeter.

It is to be feared, however, that an increase of target velocity will lead to a rapid decline of the signal-to-clutter ratio and, thus, that the results obtained under physiological conditions will be similar to those obtained with classic pseudo-random velocimeters.

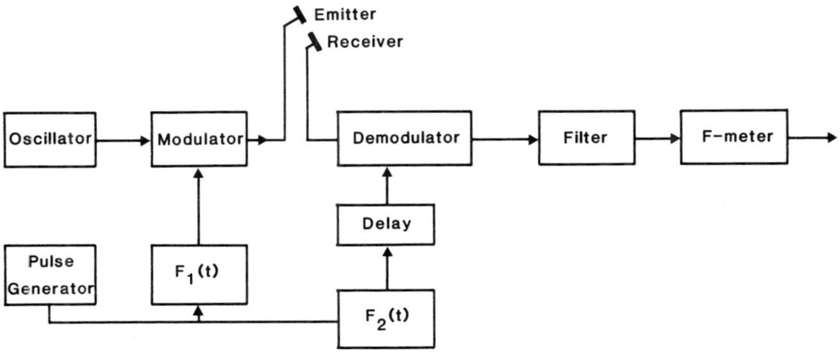

FIGURE 7.16. Doppler velocimeter using cross-correlation techniques, as proposed by Shiozaki et al [1979].

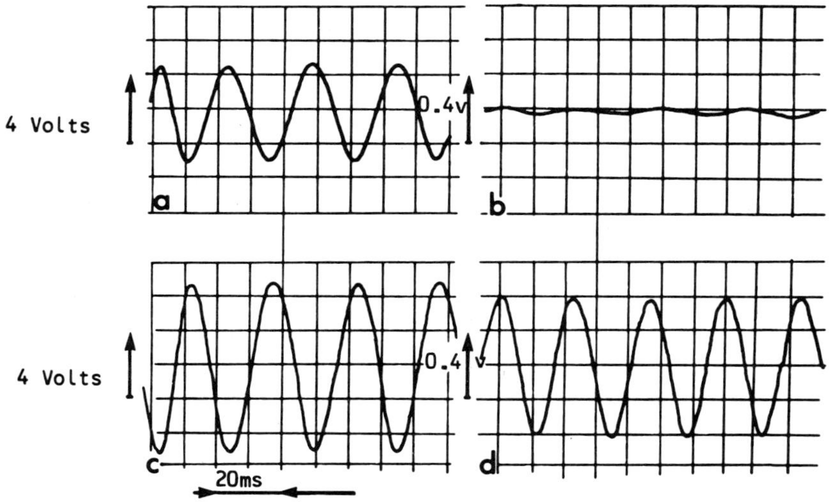

FIGURE 7.17. Output signal waveforms. **(a)** Doppler signal for targets situated at $\tau = \tau_0$ in the cross-correlation method. **(b)** Doppler signals obtained for the same targets now situated at $\tau \neq \tau_0$ (these are now clutter), in the cross-correlation-method. **(c)** Doppler signals obtained for targets situated at $\tau = \tau_0$ in the M-sequence method. **(d)** Doppler signals obtained for the same targets, now situated at $\tau \neq \tau_0$ in the M-sequence method.

DOUBLE M-SEQUENCE METHOD

We have seen in a preceding section that the problems of range-velocity ambiguity are the same whether we use a velocimeter with pulsed or pseudo-random emission.

To augment the value of the range-velocity product, Ohtsuki and Okujuma [1976] proposed replacing one pseudo-random sequence by the product of two sequences. We recall that in the case of modulation by one M-sequence only, the signal after demodulation is as follows: if $\tau = \tau_d$, a signal of frequency f_d; if $\tau \neq \tau_d$, a signal whose spectrum is shown by Figure 7.18. Note that for such a target, the amplitude of the spectral line at the frequency f_d is N times weaker than for the case where $\tau = \tau_d$.

In a system using a single M-sequence, a high-pass filter is used to eliminate the spectral lines at multiples of the frequency $1/Nt_c$. Thus any Doppler frequencies higher than the cut-off frequency of this filter will not be detected. Enlarging the passband of this filter would allow passage of the spectral components shown in Figure 7.18 (for example, the lines shown as R_1 and R_2 in the figure). The aim of the double M-sequence system is to reduce or eliminate these spectral lines. The product of a first by a second sequence is called a double M-sequence, as shown in Figure 7.19(a). The period of the first sequence is short. The second sequence has a much longer period, but is obtained from a clock having the

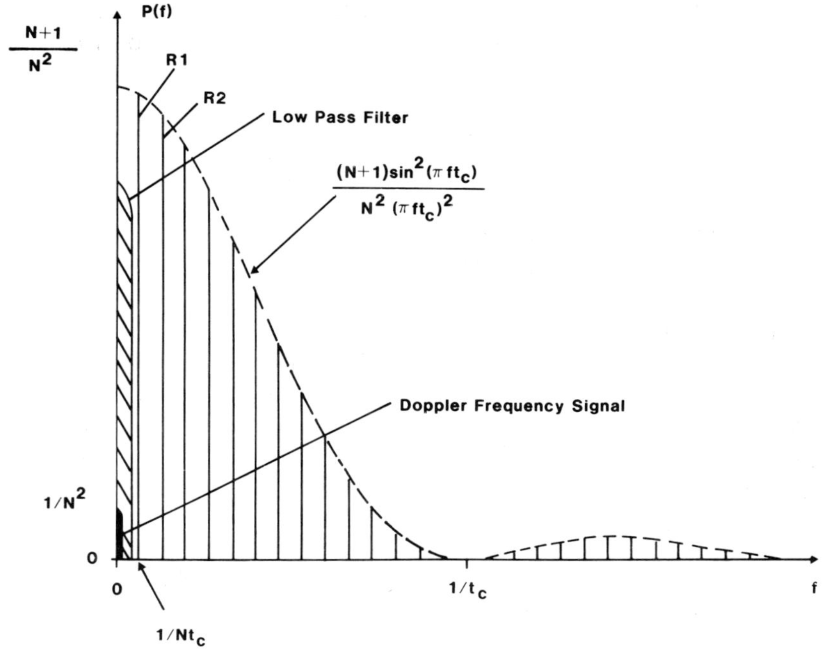

FIGURE 7.18. Power spectrum of M-sequence signal. It is not possible to expand the low-pass filter bandwidth because in this case the undesired spectral lines R_1 and R_2 would no longer be rejected. This constraint limits the maximum frequency detectable without ambiguity.

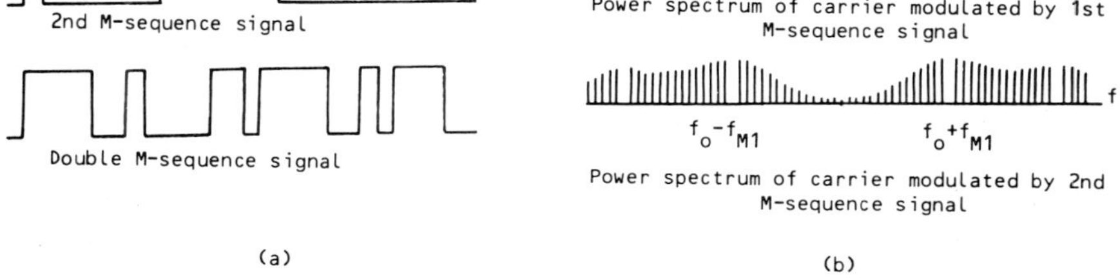

FIGURE 7.19. **(a)** Double M-sequence signal and **(b)** power spectra of the double M-sequence signal proposed by Ohtsuki et al [1976].

same frequency as that of the first M-sequence; hence, the spectrum of the emitted signal now appears like that of a carrier modulated by the first sequence convolved with the spectrum of the second sequence. Since the clock frequencies of the two sequences are the same, the amplitudes of the spectral components situated in the neighborhood of the frequency f_o are weakened.

The signal received after demodulation is the product of the double M-sequence multiplied by itself, delayed by τ, and multiplied by the carrier, which has undergone the Doppler shift. The spectrum of this signal has been studied experimentally, using a double M-sequence that is the product of a sequence of seven states and a sequence of 15 states generated by a clock at 125 kHz. The carrier is at 2 MHz. The spectra after demodulation have been studied for different values of τ. As expected, for $\tau = 0$ there is solely a component at $f = 2$ MHz. For other values of τ, the "parasitic" spectral lines are removed from the vicinity of 2 MHz. However, for $\tau = 7t_c$, important spectral lines appear in the neighborhood of the 2 MHz frequency (Figure 7.20).

The above results seem to show that it should be possible to detect relatively high velocities without being troubled by lines situated at $1/Nt_c$. In the case of very high velocities, the approximation of supposing that the M-sequences are simply delayed is not totally justified, and a more complete study of the echo spectrum is necessary. As in the case of velocimeters with only one single M-sequence, it is probable that the echoes of stationary targets are not eliminated by the simple low-pass filter.

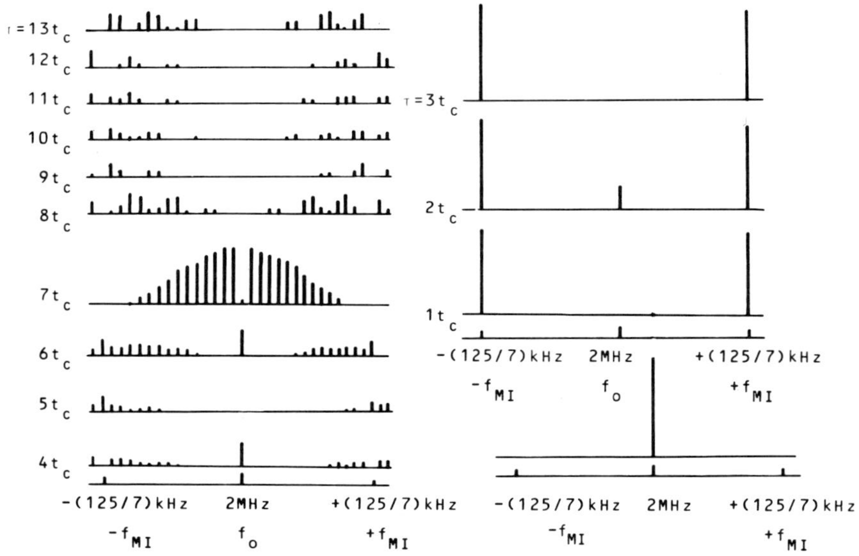

FIGURE 7.20. Spectra of demodulated signal in double M-sequence method.

Pulsed Noise Doppler Signal

In the case where one is not particularly interested in peak-to-average power ratio close to unity, it is possible to reduce the influence of fixed and moving clutter targets while conserving the principal advantage of noise velocimeters, namely, that of the absence of range velocity ambiguity by transmitting noise in the form of pulses rather than continuously [Bendick and Newhouse, 1974]. The repetition frequency of the noise pulses need only satisfy the sampling condition, since there is no range ambiguity. The duration of the noise pulses has no influence on the range resolution, since $\Delta R = c/2B$.

The limitations of this type of velocimeter are related to the problems of clutter because, as discussed above, successive signals from stationary targets even outside the range cell are not totally decorrelated with one another. For these same reasons, when the duration of the transmitted signal is larger than the longitudinal resolution of the system, the echo reflected by scatterers outside the range cell leads to a clutter noise signal that degrades the signal-to-clutter ratio.

Random-Pulsed Noise Signal

A variation of the noise system described above has been proposed by Jethawa et al [1975]. In this system the transmitted pulses are randomly staggered. This essentially relaxes the requirement that the pulse repetition rate be greater than twice the maximum expected Doppler frequency. The authors demonstrate that the peaks in the ambiguity function, $\mathcal{A}(f,\tau)$, which occur at intervals of f_r for periodic pulses, will occur at Lf_r, where L is the total possible number of different intervals that are possible in the pulse repetition code (ie, the probability of any one position is $1/L$). It must be remembered that decreasing the velocity ambiguities must result in a resolution trade-off. The total volume under $\mathcal{A}(f,\tau)$ is constant. Thus, diminishing the number of peaks associated with the Doppler ambiguity will result in increasing the function elsewhere, such as by broadening of the widths of the remaining peaks (close-target separability) or increasing the height of the pedestal (wide-target interaction or self-clutter).

Frequency-Modulated Ultrasonic Doppler Flowmeter

McCarty and Woodcock [1975] proposed to utilize a signal modulated linearly in frequency for simultaneously measuring the position of fixed targets and the speed of mobile targets. The system is similar to that of Doppler velocimeters (Figure 7.21), except that the local oscillator is replaced by a generator modulated in frequency. The signal received after reflection from the different interfaces is multiplied with a portion of the emitted signal. Since the frequency of the emitted signal varies linearly

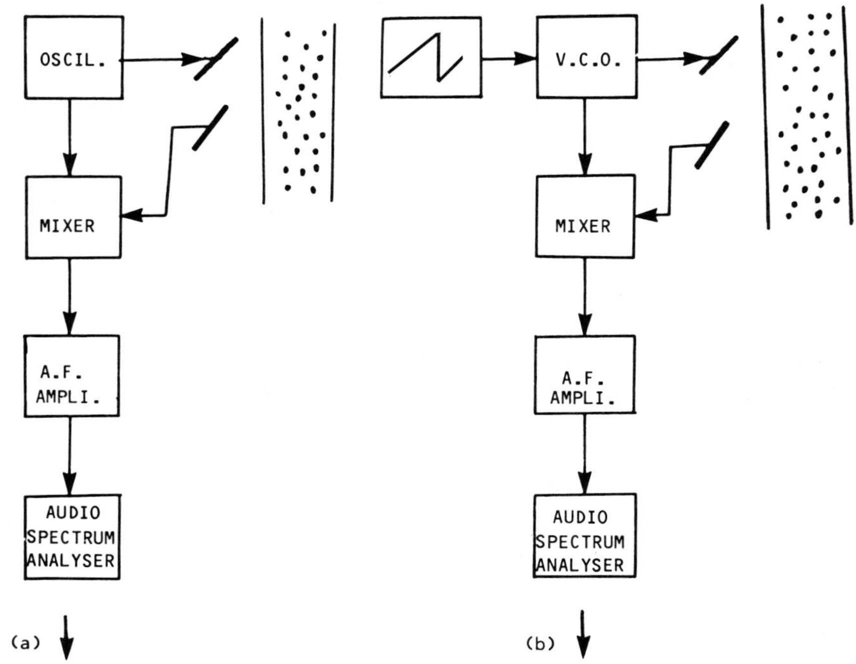

FIGURE 7.21. Circuit diagrams of systems incorporating **(a)** the conventional Doppler velocimeter and **(b)** the frequency-modulated flowmeter similar to that proposed by McCarty and Woodcock [1975].

with time, the frequency at the output of the multiplier is proportional to the distance between the target and the transducer (Figure 7.22)

$$\Delta f = \frac{2R}{c}\frac{df}{dt}$$

where Δf is the shift in the frequency at the output of the multiplier, R is the range of the target, and df/dt is the rate of variation of the emitted frequency. The range resolution depends directly on the spectral resolution of the multiplier output signal. In the case of moving targets, the echo frequency at time t_i is no longer f_i but $f_i \pm f_d$. At this instant, the frequency coming from the local oscillator is

$$f_i + \frac{2R}{c}\frac{df}{dt}$$

and the frequency at the output of the multiplier is

$$(f_i + \frac{2R}{c}\frac{df}{dt}) - (f_i \pm f_d)$$

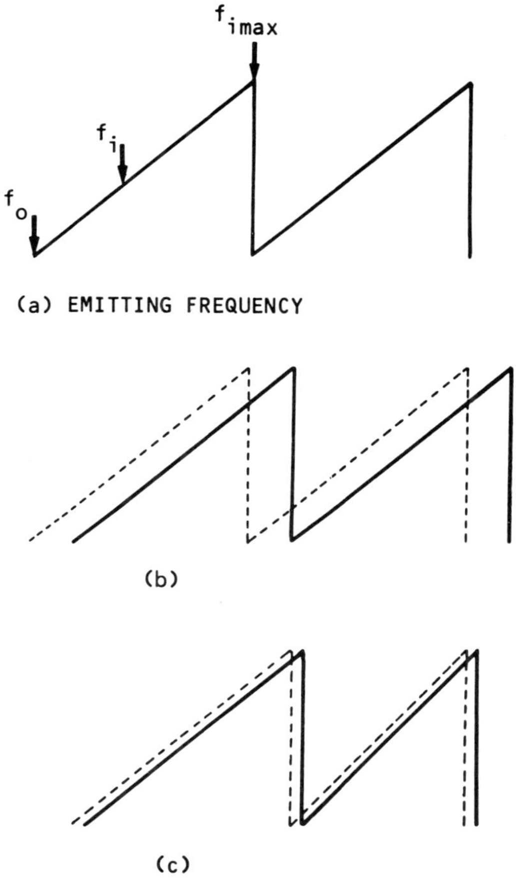

FIGURE 7.22. Principle of range measurement using a frequency-modulated system. **(a)** Frequency variation with time of the emitted signal. **(b)** Frequency variation of echo signal corresponding to a fixed target at $\tau = 2R/c$. The shift in frequency between the received signal and that transmitted from the oscillator is:

$$\Delta f = \frac{2R}{c}\frac{df}{dt}.$$

(c) Time variation of the frequency of the echo signal corresponding to a moving target located at $\tau = 2R/c$. The frequency difference between the echo signal and the local oscillator signal is

$$\Delta f = \frac{2R}{c}\frac{df}{dt} + f_d.$$

It is necessary to make sure that $2R/c\ df/dt$ is always larger than f_{max}. The frequency of the multiplier output will be higher or lower than $2R/c\ df/dt$, depending on whether the moving target is approaching or receding from the transducer. Positive results have been obtained in the separate determination of the range of fixed targets and the velocity measure of moving particles. However, the simultaneous measurement of both the speed and the position of targets seems extremely delicate because of problems of range velocity ambiguity associated with this type of modulation. Even if, as the authors state, it is possible to distinguish wall echoes simply because of their strong reflecting power, one cannot count solely on this criterion for distinguishing with confidence between stationary and mobile targets.

A frequency-modulated system consisting of both up and down sweeps as proposed by Rihaczek [1969] should ameliorate certain of the above problems. However, the work in this area is too specific for us to affirm that the problems posed by the multitude of reflecting targets, and the dynamic range requirements posed by the large strength difference between the fixed and moving targets echoes, can be resolved.

References

Amir I, Newhouse VL: On the signal to noise Ratio: Enhancement of the Doppler process. Ultrasonics pp 231–239 (1984).

Bendick PJ, Newhouse VL: Ultrasonic random signal flow measurement system. J Acoust Soc Am 56(3): 860–865 (1974).

Cathignol D: Signal to clutter ratio in pseudo random Doppler flowmeter. Ultrasonic Imaging 8: 272–284 (1986).

Cathignol D, Fourcade C, Chapelon JY: Transcutaneous blood flow measurement using pseudo-random noise Doppler flowmeter. IEEE Trans Biomed Eng 27(I): 30–36 (1980).

Jethawa CP, Kaveh M, Cooper GR, et al: Blood flow measurements using ultrasonic pulsed random signal Doppler system. IEEE Trans Biomed Eng SU22(1): 1–11 (1975).

Masson J: Tracés de fonctions d'ambiguité. Synthèse de signaux à partir de la fonction d'ambiguité. Rapport du Centre d'Etude des Phénomènes Aléatoires et Géophysiques. Institut National Polytechnique (1979).

McCarty K, Woodcock JP: Frequency modulated ultrasonic Doppler flowmeter. Med Biol Eng 13(1): 59–64 (1975).

Newhouse VL, Cooper GR, Feigenbaum H: Ultrasonic blood velocity measurement using random signal correlation techniques. Digest of the Xth International Congress on Medical and Biological Engineering, Dresden, 1973.

Ohtsuki S, Okujuma M: Double M-sequence method and M-sequence pulsed method for ultrasonic Doppler velocity meter. Bull Res Lab Precis Mach Electronics (Tokyo Inst Technol) March 1976.

Rihaczek AW: Principles of high resolution radar. New York, McGraw-Hill (1969) chap 2, pp 10–39.

Shiozaki A, Senda S, Kitabatake A, et al: A new modulation method with range resolution for ultrasonic Doppler flow sensing. Ultrasonics (November 1979).

Waag RC, Myklebust JB, Rhoads WL: Instrumentation for non invasive cardiac chamber flow rate measurement. Ultrason Sympos Proc IEEE 72: 74–77 (1972).

8
Pseudo-Random Characterization of Time-Varying Media

VERNON L. NEWHOUSE

Introduction

Studies of the propagation characteristics of moving media usually have as goals a visibility improvement of objects being viewed through the medium, the gathering of information about the medium itself, or the improvement of communication through the medium. The first two of these objectives are or may prove to be important in ultrasound in connection with imaging through fluid-borne scatterers as well as fluid-parameter characterization, and all three objectives are important in the field of underwater sound. Ultrasound materials characterization and imaging through random but stationary media constitute other possible applications of the techniques described here. Other applications are possible in nondestructive testing, where the movement of a transducer over a stationary medium often produces signals that are similar to those produced by a medium moving past a transducer.

Modeling the Medium

The techniques described below have been developed in connection with underwater sound propagation and take into account sound absorption, reflection, refraction, and scattering from spatially distributed moving targets, such as waves, as well as transmitter and receiver motion. When a medium is illuminated with narrow band radiation, it is found that its propagation characteristics can be represented by a time-varying filter whose output is $H(t,\xi)$ at time t in response to a unit impulse at time $t - \xi$. The output $y(t)$ of such a filter for an input signal $x(t)$ is given by

$$y(t) = \int x(t - \xi)\, H(t,\xi)\, d\xi \tag{8.1}$$

$H(t,\xi)$ is known as the bitemporal response of a variable parameter filter [Middleton, 1972] and is a real, time-varying random variable.

Suppose now that $H(t,\xi)$ has a Fourier transform defined by

$$U(\nu,\xi) = \int e^{-i2\pi\nu t} H(t,\xi)\, dt$$

Then Equation (8.1) can be written

$$y(t) = \int \int x(t - \xi)\, U(\nu,\xi)\, e^{i2\pi\nu t}\, d\nu\, d\xi \tag{8.2}$$

This equation is useful in characterizing an underwater medium filled with moving scatterers, provided that we now consider $x(t)$ to be a narrow band signal written in complex form. With this definition of $x(t)$, Equation (8.2) implies that the echo at time t of a signal $x(t - \xi)$ emitted ξ seconds earlier is multiplied by $U(\nu,\xi)$ and Doppler shifted in frequency by ν Hz. In addition, $U(\nu,\xi)$ can be regarded as the density of the scatterers giving Doppler shift ν to echoes reflected at time $t - \xi$.

The formalism discussed heretofore, in which scatterers at range delay ξ multiply a received signal by $U(\nu,\xi)$ and frequency shift it by ν, is only appropriate for narrow band signals. For broad band signals the Doppler effect is better represented by a time scaling of the reflected signal. This takes into account that a signal $x(t)$ reflected from a scatterer moving with velocity v produces an echo given by

$$y(t) = x(t - \frac{2v}{c}t - \phi_o)$$

where ϕ_o is a function of the range of the scatterer.

To represent this time scaling effect, we may, following Jourdain [1973], use the "compression operator" $G(\lambda,\xi)$, defined so that an incident signal $x(t)$ produces an output signal

$$y(t) = \int \int G(\lambda,\xi)\, x(\lambda t - \xi)\, d\lambda\, d\xi \tag{8.3}$$

$G(\lambda,\xi)$ can be regarded as the density of scatterers with velocity $(c/2)(1 - \lambda)$ at range delay ξ and has been used in this way in calculating the Doppler spectra of moving scatterers illuminated with broadband signals [Laval, 1954; Newhouse et al, 1976].

Measurement of the Medium Parameters

Classical Sampling Method

The most straightforward method of estimating the function $H(t,\xi)$ as a function of range and time [Thielè, 1972] is to transmit impulses or short bursts of radiofrequency (RF) at times $t_1 t_2 \cdots t_i$. The echo of the impulse transmitted at time t_i corresponds to a sample function of $H(t_i + \xi,\xi)$. The time interval between the transmission times must be larger than Δ, the reverberation time of the medium. Thus the sampling frequency of the process $H(t,\xi)$ can be no greater than $1/\Delta$, and hence the maximum measurable bandwidth B over which ν in $U(\nu,\xi)$ can be measured, is limited by the sampling theorem to

$$B\Delta < \frac{1}{2}$$

This relation defines a so-called sub-dispersive channel. Note also that this limit is related to the range-velocity product limit of pulsed Doppler systems.

Interambiguity Function Technique

The limitation of the classic or pulse (echo) method of estimating $H(t,\xi)$ is that it must estimate the time variation of this function using samples separated by a time equal or greater than the reverberation time of the system. This limitation can be overcome, though at a price in frequency resolution, by a technique in which the target's rate of change of phase at each range is estimated from a single echo, instead of being estimated from a series of samples, as in the classic technique.

The success of noise Doppler correlation systems in measuring fluid velocity with no range-velocity product limitation suggests the use of input-output signal cross-correlation for estimating medium transmission characteristics, possibly with random or pseudo-random noise as the transmitted signal. This concept is analyzed below and is shown to be feasible. Its accuracy is found to depend on the sharpness of the peak of the ambiguity function of the transmitted signal.

Noise as the Transmitted Signal

To bring out some of the limitations of correlation techniques for medium characterization, we will first analyze the case of noise as the transmitted signal, even though noise is usually not the optimum type of signal for this application.

If the medium output signal $y(t)$ due to noise is multiplied by the input noise existing at a time τ earlier, we will have from Equation (8.1)

$$y(t)\,\mathrm{x}(t - \tau) = \int x(t - \tau)\, x(t - \xi)\, H(t,\xi)\, d\xi \tag{8.4}$$

If the noise signal has bandwidth B, then samples of it taken at intervals larger than $1/B$ are uncorrelated with one another. Hence, to obtain a reliable mean $R_x^T(\tau - \xi)$, for the product $x(t - \tau)\, x(t - \xi)$, one must integrate for a time $T >> 1/B$. If $H(t,\xi)$ varies sufficiently slowly, we can integrate Equation (8.4) for a time T during which $H(t,\xi)$ can be considered to remain effectively constant. Applying this procedure to Equation (8.4), we find

$$E^T\{y(t)\, x(t) - \tau)\} = \int R_x^T\,(\tau - \xi)\, H(t,\xi)\, d\xi \tag{8.5}$$

where $E^T\{\ \}$ stands for the time average over period T. If the noise autocorrelation function $R_x^T(\tau)$ can be approximated by $\delta(t)$, then Equation (8.5) becomes

$$E^{T}\{y(t)\,x(t-\tau)\} = H(t,\tau)$$

Even in the case where $H(t,\xi)$ varies so rapidly with time that it can not be estimated by the above method, which produces samples of it averaged over time T, its Fourier transform averaged over the period T can nevertheless be estimated by calculating the so-called interambiguity function of the input and output signals.

For two signals $x_1(t)$ and $x_2(t)$ the interambiguity function will be defined as

$$\varphi_{x_1x_2}(f,\tau) = \int_{-\infty}^{\infty} x_1(t)\,x_2^*(t-\tau)\,e^{-i2\pi ft}\,dt \tag{8.6}$$

which for $x_1(t) = x_2(t)$ is the well-known Woodward ambiguity function whose shape for a given radar signal is a measure of its ability to discriminate between different target ranges and velocities.

Calculating an approximate interambiguity function φ_{yx}^T between the output $y(t)$ and the input $x(t)$ of a medium by integrating over a period T, we have from Equation (8.2), following Jourdain [1973],

$$\varphi_{yx}^T(f,\tau) = \int_0^T x^*(t-\tau)\,e^{-i2\pi ft} \left[\int\int U^T(\nu,\xi)e^{i2\pi\nu t}\,x(t-\xi)\,d\nu\,d\xi\right] dt \tag{8.7}$$

where T is the length of the signal $x(t)$ if $x(t)$ is nonrandom, or where T is long enough to provide an adequate estimate of φ_{yx}, if $x(t)$ is random. U^T is defined as the average of U taken over the period T. Rewriting Equation (8.7) gives

$$\varphi_{yx}^T(f,\tau) = \int\int U^T(\nu,\xi)\,\varphi_x^T(f-\nu,\tau-\xi)\,d\nu\,d\xi \tag{8.8}$$

where φ_x^T is defined as the value of $\varphi_{xx}(f,\tau)$ obtained by integrating over a period T. Equation (8.8) is equivalent to the two-dimensional convolution of $U^T(\nu,\xi)$ and $\varphi(f,\tau)$.

If we can make the assumption that φ_x^T is zero for all values of τ and f other than zero, then from Equation (8.8) the long-time expectation of φ_{yx} becomes

$$\varphi_{yx}^T(\tau,f) = \varphi_{xx}^T(0,0)\,\mathrm{U}^T(f,\tau)$$

The two methods of intercorrelating the input and output signals that have just been described only provide us with an average over time T of the function $H^T(t,\xi)$ and $U^T(\nu,\xi)$. To obtain information on the velocity at various ranges, say ξ_o, it is necessary to compare samples $H(t_o + nT, \xi_o)$, $H(t_o + [n+1]T, \xi_o)$ etc, which, as shown above, limits our measurements to targets whose position does not change rapidly within the time T. To overcome this limitation we need to be able to directly measure the function $G(\lambda,\xi)$. Jourdain [1973] has shown that this can be done by time scaling one of the two signals being correlated. Time scaling of a signal can be accomplished by sampling it, storing the samples, and then reading them out at a different rate.

To adapt the method just described—ie, of forming the interambiguity function for estimating $U^T(\nu,\xi)$—to estimating the compression operator $G(\lambda,\xi)$, we use the time-scaled ambiguity function defined by Swick [1969],

$$\chi_{x_1x_2}(k,\tau) = \int_0^\infty x_1^*(kt - \tau)\, x_2(t)\, \mathrm{dt}$$

$\chi(f,\tau)$ is an ambiguity function suitable for broadband signals whose peak lies at $k = 1$, $\tau = 0$.

Thus, if we multiply the medium output $y(t)$ by a scaled and time-shifted version of the input signal $x(t)$, then from Equation (8.3) we obtain

$$y(t)\, x^*(kt - \tau) = \int \int G(\lambda,\xi)\, x^*(kt - \tau)\, x(\lambda t - \xi)\, d\lambda\, d\xi$$

If we now integrate over a period T, which is chosen to be long enough to produce a satisfactory average for the product $x^*(kt - \tau)\, x(\lambda t - \zeta)$ but sufficiently short that $G(\lambda, \zeta)$ remains effectively constant, we obtain

$$\chi_{yx}^T(k,\tau) = \int \int \frac{1}{\lambda}\, G^T(\lambda,\xi)\, \chi_{xx}^T(k/\lambda, \tau - k\xi/\lambda)\, d\lambda\, d\xi \qquad (8.9)$$

where G^T is defined as the average of G taken over time T.

Under the assumption that $\chi(k,\tau)$ is zero unless $k = 1$ and $\tau = 0$, Equation (8.9) leads to

$$\chi_{yx}^T(k,\tau) = \frac{\chi_{xx}^T(1,0)}{k}\, G^T(k,\tau) \qquad (8.10)$$

The method of estimating $G^T(\lambda,\xi)$ described above and shown schematically in Figure 8.1 can be replaced with either of two other mathematically equivalent procedures, shown in Figures 8.2 and 8.3, which may be more convenient in practice [Jourdain, 1973; Jourdain and Jourdain, 1977]. In the technique shown in Figure 8.2 the medium output signal $y(t)$

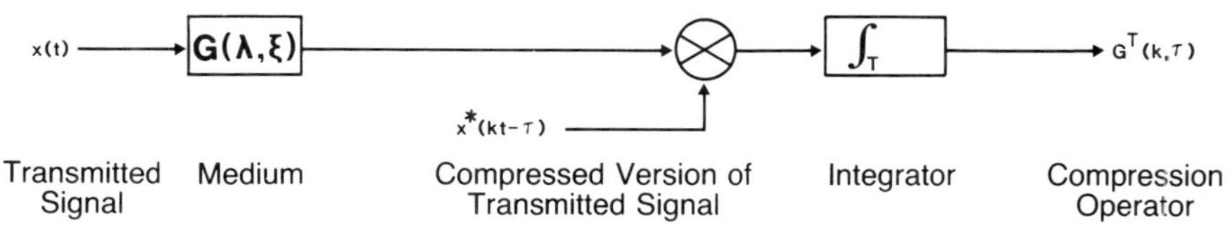

FIGURE 8.1. Technique for estimating the compression operator $G(\lambda, \xi)$ [After Jourdain, 1973.]

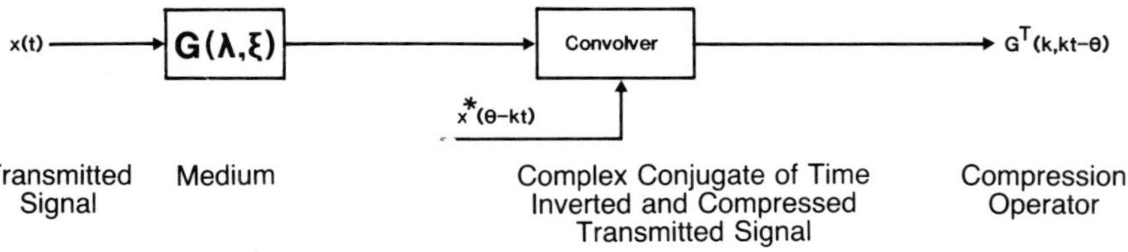

FIGURE 8.2. Technique for estimating the compression generator using a bank of matched filters [after Jourdain, 1978].

is passed through a filter of form $x^*(\theta - kt)$ matched to the time compressed input signal. The output of this filter can be shown to be

$$y(t) * x(\theta - kt) =$$

$$- \int \int \frac{1}{\lambda} G^T(\lambda,\zeta)\, \chi^T_{xx}\, (\frac{k}{\lambda}, -\theta + k[t - \frac{\xi}{\lambda}])\; d\lambda\, d\xi \qquad (8.11)$$

where $G^T(\lambda,\xi)$ is defined as the average of $G(\lambda,\xi)$ taken over time T and $*$ stands for convolution.

To the extent to which we can neglect the width of the peak of the interambiguity function χ_{xx}, Equation (8.11) can be written

$$y(t) * x^*(\theta - kt) = -\chi^T_{xx}(1,0)\, G^T(k, kt - \theta) \qquad (8.12)$$

This equation indicates that the output of the filter in Figure 8.2 provides us with values of $\frac{-\chi^T_{xx}(1,0)}{k} G^T(k,\tau)$ at time $t = (\tau + \theta)/k$.

Yet another alternative is provided by the system shown in Figure 8.3, which time scales the output signal of the medium, but only requires a single matched filter. In this case, the medium output signal $y(t)$ after time and amplitude scaling followed by convolution with the filter shown in Figure 8.2 having impulse response

$$h(t) = x^*(\theta - t)$$

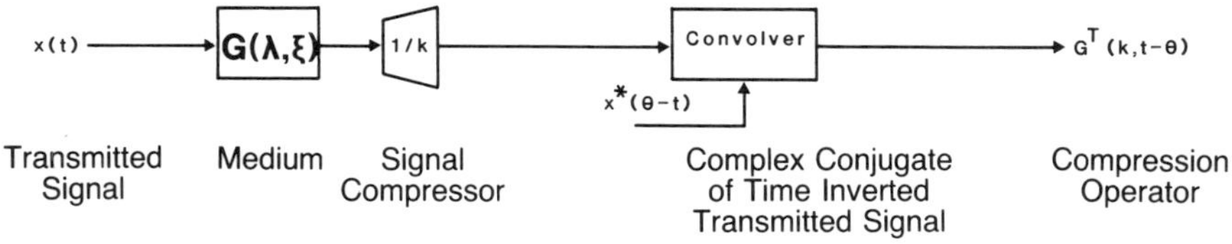

FIGURE 8.3. Technique for estimating the compression generator using time scaling and a single matched filter [after Jourdain and Jourdain, 1977].

becomes

$$\frac{1}{k} y(\frac{t}{k}) * x^*(\theta - t) = \frac{1}{k} \int \int G^T(\lambda,\xi)\, \varphi^T_{xx}(\frac{k}{\lambda}, t - \theta - \frac{k}{\lambda} \xi)\, d\lambda\, d\xi$$

which, on approximating χ^T_{xx} by a delta function, becomes

$$y(\frac{t}{k}) * x^*(\theta - t) = \chi^T_{xx}(1,0)\, k\, G^T(k,t - \theta) \qquad (8.13)$$

which indicates that the arrangement of Figure 8.3 delivers an output signal equal to $\chi^T_{xx}(1,0)\, k\, G^T(k,\tau)$ at time $t = \theta + \tau$.

Choice of Transmitted Signal

The resolution in range provided by the techniques previously described depends on the width of the peaks of the ambiguity function φ and χ along the range axis, and the resolution in velocity depends on the width of φ along the frequency axis or the width of χ along the k axis, where $k = 1 - 2v/c$. The width of both of these functions along the τ axis is $1/B$, which leads to the range resolution of $c/2B$, where B is the bandwidth of the transmitted signal. In the same way it can be shown that for both φ^T and χ^T, the Doppler frequency resolution equals $1/T$, where T is the time over which the transmitted signal is integrated. Hence to obtain the best product of range and velocity resolution one must, as always, maximize the time bandwidth product TB of the transmitted signal.

Thus, if we want to maximize the time resolution of $U^T(\nu,\xi)$, or of $G^T(\lambda,\xi)$ by keeping T small, frequency resolution in measuring the parameter ν or λ must be sacrificed. Note that if we wish to simultaneously maximize both T and B we must choose a random or pseudo-random signal for which B and T are independent. (The reader will recall that for a deterministic signal such as a burst of RF of length T, the bandwidth $B = 1/T$ is determined once T is chosen.)

Another factor that limits the maximum length of signal T is the minimum difference in range delay between moving targets. If this is less than T, echoes from different moving targets arrive at the same instant, thus worsening the input signal-to-noise ratio of the system. It has been shown above that the noise due to a stationary target automatically becomes DC and can be filtered out if a periodic transmitted signal is used. However, for two moving targets the input signal-to-noise ratio may be inacceptable unless T is reduced sufficiently to separate them in time. The degree of interference between different targets depends, of course, on the time side lobes of the signal used. This indicates that m-sequence codes will be better than noise, and Golay codes would be better than either. At this time, however, experiments have not been made with this type of code.

Experimental Results

Systems using intercorrelation techniques have been described for measuring both the range delay-velocity space density $U(\lambda,\xi)$ [Jourdain and Faure, 1978] and the diffusion constant $D(\lambda,\xi)$* [Jourdain and Jourdain, 1977]. Measurements of $D(\lambda,\xi)$ with a moving receiver were made in an underwater channel using a 512-step code with a duration of 2.56 seconds and a carrier frequency of 800 Hz. Estimates of the interambiguity function of the code employed and of the measured diffusion constant are shown in Figures 8.4 and 8.5, respectively. The resolution along the λ axis has the expected value of $1/T\nu_o$, where ν_o is the carrier frequency and T is the length of the transmitted code. The existence of waves with a frequency of approximately 0.14 Hz is probably responsible for the series of peaks along the λ axis at nearly zero range in Figure 8.5.

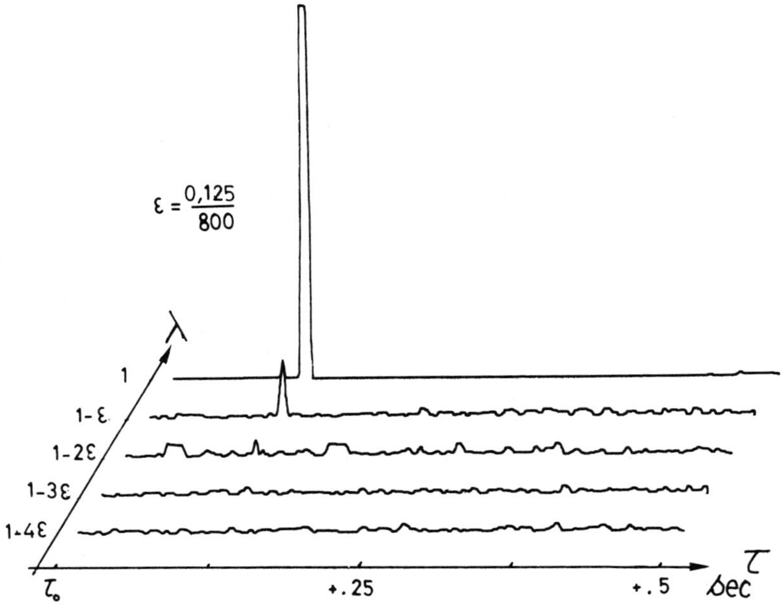

FIGURE 8.4. Estimates of the interambiguity function of a 512-step code with a duration of 2.56 seconds and a carrier frequency of 800 Hz. [After Jourdain and Jourdain, 1977.]

*$D(\lambda,\xi)$ is the variance of the time-varying portion of $G(\lambda,\xi)$.

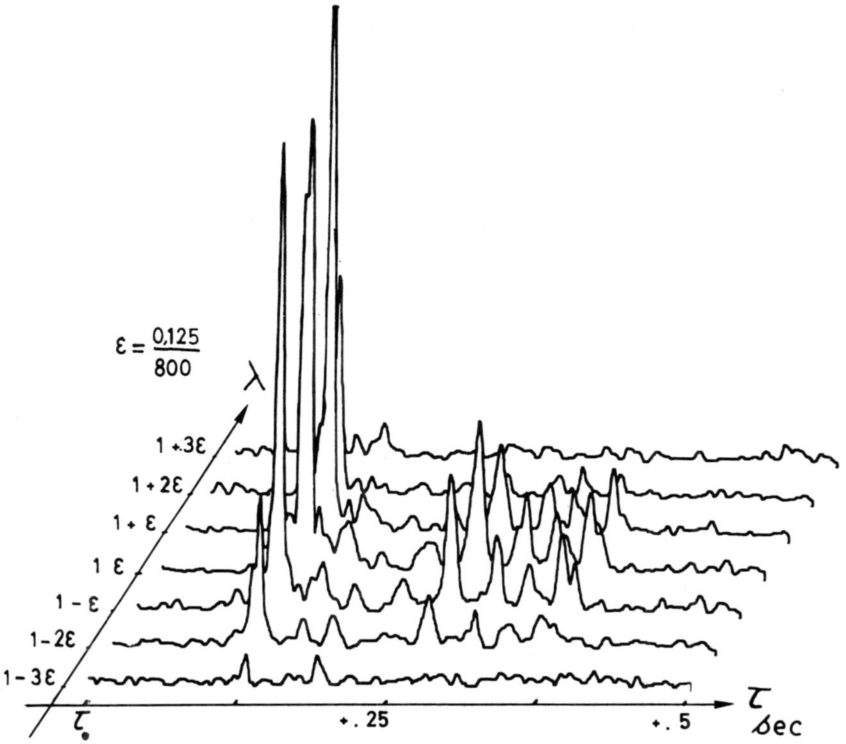

FIGURE 8.5. Estimates of the diffusion constant $D(\lambda, \tau)$ made in an underwater channel using the code of Figure 8.4 and a moving receiver. [After Jourdain and Jourdain, 1977.]

Comparison Between Sampling and Interambiguity Techniques

It has been shown that the so-called interambiguity Doppler systems can estimate target velocity with only a single echo if the signal-to-noise ratio is good enough. Thus, they do not suffer from the range-velocity product limit, an advantage that is shared by noise and pseudo-random code Doppler systems. The interambiguity Dopplers are, however, inferior in frequency resolution to the sampled Dopplers.

As shown above, the frequency resolution of the interambiguity technique is

$$\Delta f_{\text{ia}} \sim \frac{1}{T}$$

where T is the integration time of the emitted signal. T is also equal to the length of the transmitted signal if this is periodic. If a code of n terms is used, then the above equation is written

$$\Delta f_{\mathrm{ia}} \sim \frac{1}{T} = \frac{B_{\mathrm{o}}}{n}$$

where B_{o} is the bandwidth of the transmitted signal.

For any Doppler system that uses sampling (whatever its transmitted signal), it can be shown [Newhouse et al, 1976] that for a single velocity target, the bandwidth of the Doppler output signal is

$$B_d \sim \frac{2v}{c} B_{\mathrm{o}}$$

This is therefore the frequency resolution Δf_{s} of sampling Dopplers. Hence the ratio of the frequency resolution of sampling Dopplers to that of intercorrelation Dopplers is

$$\frac{\Delta f_{\mathrm{s}}}{\Delta f_{\mathrm{ia}}} \sim n \frac{2v}{c}$$

which for most cases of interest is smaller than unity. Hence the intercorrelation technique can measure more rapidly changing velocities than the sampling systems, but has a correspondingly worse velocity resolution.

References

Altes AR: Supression of radar clutter and multipath effects for wideband signals. IEEE Trans Inf Theor (1973) pp 344–345.

Jourdain G: Caracterisation d'un milieu de transmission variable par un modele de FAPV. Ann Telecom 28: 413–422 (1973).

Jourdain G, Faure B: Identification active de milieux constitues par des reflecteurs mobiles et immobiles. CETHEDEC 56: (1978).

Jourdain G, Jourdain JY: Methods et techniques de estimation de fonction de diffusion. Proc Colloque National Dur Le Traitement Du Signal et Ses Applications, Nice, April 1977.

Laval R: Transformation aleatoire des signaux par la propagation Nata Colloquium, Grenoble, 1954.

Middleton D: A statistical theory of reverberation and similar point order scattered fields. Part I, Waveforms and the general process. Part II. Moments, spectra and special distributions. IEEE Trans Inf Theor 1t 18, 1: (1972).

Newhouse VL, Bendick PJ, Varner W: Analysis of transit time effects on doppler flow measurement IEEE Trans Biomed Eng BME-23: 381-386 (1976).

Swick DA: An ambiguity function independent of assumption about bandwidth and carrier frequency. NRL Rep 6471 USA (1969).

Thiele R: Anwendung der Theorie zeitveranderlicher Filter zur Beschreibung des Flachwasserschallkanals (Application of the time-variant filter theory to the description of the shallow water sound). Thesis T.U., Hanover (1972).

9 Pseudo-Random Multimode Operation

Vernon L. Newhouse

Basic System

In this concluding chapter of our review, we discuss the use of pseudo-random codes for labeling the various possible transmission modes of a phased array so that their echoes can be distinguished even when several modes are transmitted simultaneously. The advantages of this procedure were discussed in Chapter 5 and include lowered side lobes and improved field coverage when viewing moving targets.

The basic multimode system concept and its advantages over conventional phased arrays are brought out by the following analysis.

The ultrasound amplitude produced at point r by a transducer array when fed by a single frequency wave train through a set of delay lines as shown in Figure 9.1 may be described by the function $D_e(r)$, which is called the emission directivity. The form of $D_e(r)$ depends on the relative phases or time delays imposed on the signals supplied to the individual (transducer) elements.

The echoes reaching the n elements of a transducer array can be individually delayed by independent delay lines as shown in the figure and then summed, giving rise to a reception directivity $D_r(r)$. This may be defined as the output of the transducer summing circuit when the array is illuminated by a single frequency point source at r. The overall directivity $D_{er}(r)$ of the array, which may be defined as the amplitude of the summing circuit output when the array is fed by its source signal and receiving corresponding echoes from a point reflector at position r, may be written as

$$D_{er}(r) = D_e(r)D_r(r) \tag{9.1}$$

This equation illustrates the fact that the overall directivity of an array can be modified by phase delaying the emitted and the received signals and that these two sets of phase delays need not necessarily be the same.

Many phased array systems are in common use, eg, for two-dimensional imaging, in which ultrasound beams are sent successively in different di-

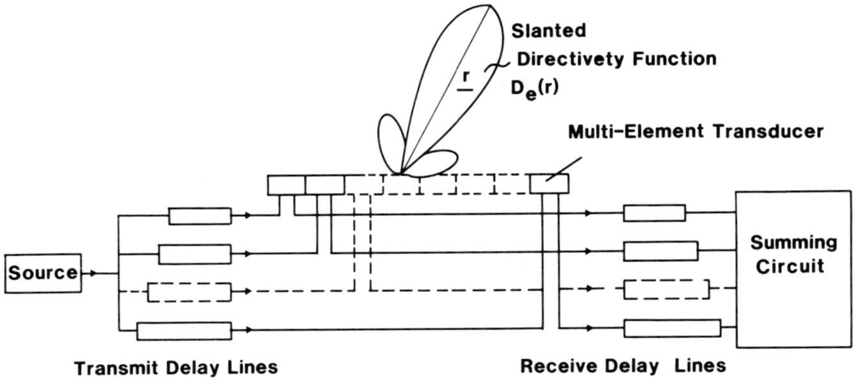

FIGURE 9.1. Operating principle of a "phased array" using time delays for transmission and reception.

rections. Other applications exist, and are discussed later, in which different emission and reception modes are used to improve the overall directivity function $D_{er}(r)$. In both emission and reception, time could be saved if it were possible to operate an array in several different modes at the same time. A system using noise or coded signals, which makes simultaneous multimode operation possible, is shown in Figure 9.2 [Newhouse and Furgason, 1978].

The system illustrated consists of a three-element array of which each element is fed by two incoherent noise sources through two separate delay lines. The lines fed by the two noise sources are phased to send beams in different directions as shown. The echoes received by the array are passed through two other sets of delay lines feeding two correlators, each of which is supplied with its own reference signal. The output of correlator 1 suppresses all echoes due to noise source 2 and can thus be said to produce an output corresponding to an overall directivity

$$D_{er1}(r) = D_{e1}(r)D_{r1}(r)$$

where $D_{e1}(r)$ is determined by the delay lines fed by noise source 1 and $D_{r1}(r)$ by the delay lines that feed correlator 1. It is clear that the system can be extended to arrays with more elements than the three shown in the figure and that the number of simultaneous independent modes used can be increased by increasing the number of noise sources, correlators, and delay lines.

An interesting, but as yet untested, variation of the system of Figure 9.2, suggested by Tournois [1980], is shown in Figure 9.3. Here n codes are transmitted through n transducer elements, and the corresponding echoes are received by a single nondirectional receiver that feeds n filters matched to the transmitted codes. The n outputs of these matched filters are processed by a beam-forming network whose outputs conform to m specific beam directions. This system thus performs beam forming even

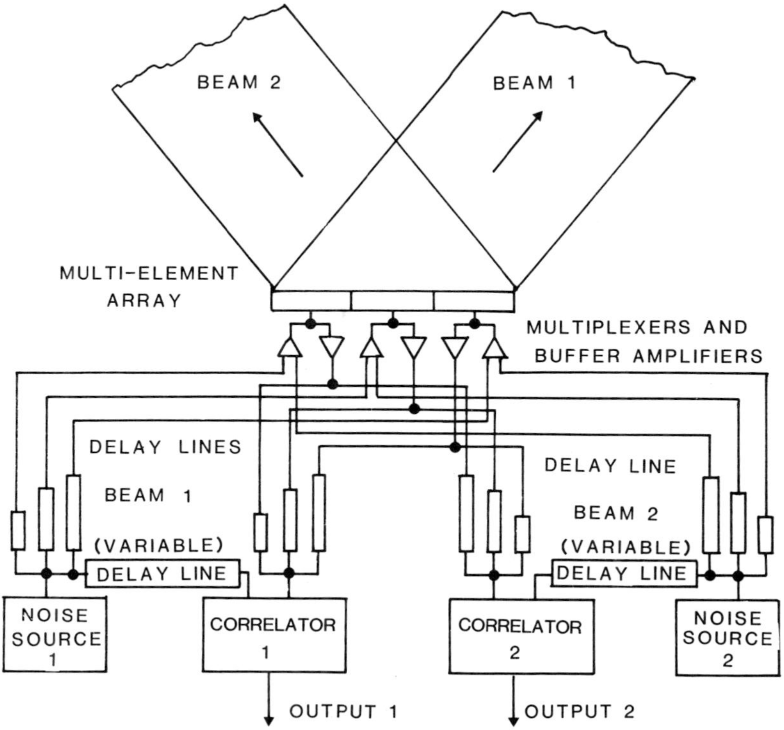

FIGURE 9.2. Principle of multimode phased array using noise as the transmitted signals [Newhouse and Furgason, 1978].

though the beam-forming network is associated with the nondirectional receiver rather than the n transmitting transducers. For more elaborate techniques using multi-element arrays for reception as well as transmission, reference should be made to the original paper.

Potential Applications

The most obvious application of simultaneous mode operation is in two-dimensional imaging, particularly of moving targets. As shown in Chapter 5, if the m beams used to form an image are emitted successively at intervals T, determined by the maximum range R, then the time F needed to illuminate a frame is related to the maximum range R by

$$F = m \frac{2R}{c}$$

Hence, by simultaneous beam emission, either the range R or the number of beams m can be increased (thus improving the lateral resolution), or the frame time F can be decreased.

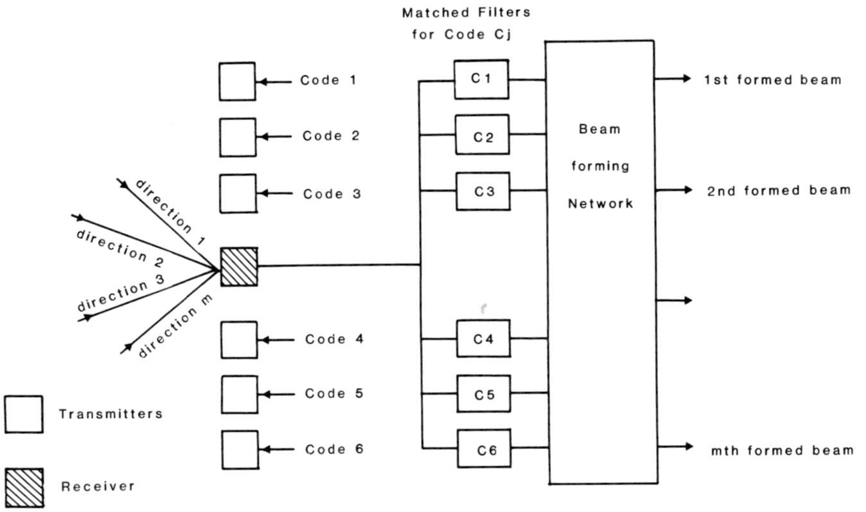

FIGURE 9.3. A system proposed by Tournois [1980] employing a beam-forming network following a nondirectional receiver.

Inspection of Figure 9.1 shows that it is possible to emit several beams simultaneously from a phased array, using a single source waveform that is subjected to different delays. However, it is demonstrated below that emitting many beams of the same signal simultaneously in different directions results in the loss of the transmission directivity, thus changing the effective overall directivity from the value given in Equation (9.1) to

$$D_{er}(r) \simeq D_e(r)$$

This reduces the lateral resolution and greatly increases the side lobes of any single beam. Hence the emission of many beams in many directions at the same time requires some type of orthogonal coded signal set, if it is not to be accompanied by serious degradation to the lateral resolution.

Another application of multimode operation is for side-lobe suppression. Given an array with an overall directivity function shown in Figure 9.4a say, one may, after emitting this mode, emit the mode shown in Figure 9.4b, and then subtract the resulting echoes from the ones obtained from mode a. The result of this procedure is equivalent to the overall directivity pattern shown in Figure 9.4c, which is seen to lack one side lobe. Mathematically, this is equivalent to summing a series of the overall directivity functions of the type shown in Equation (9.1) to produce a new function

$$D_{\mathrm{ern}}(r) = \Sigma a_n D_{en}(r) D_{rn}(r) \tag{9.2}$$

It would seem that any number of other side lobes can, in principle, be removed successively with this technique. In ultrasound imaging this pro-

FIGURE 9.4. Proposed technique for side-lobe suppression using subtraction that could be used with multimode operation. (**a**) Unmodified directivity function transmitted with code 1. (**b**) Independently emitted mode emitted with code 2 whose main lobe is parallel to the first side lobe of the pattern emitted with code 1. (**c**) Effective directivity showing suppression of side lobe produced by subtracting echoes of pattern *b* from those of pattern *a*.

cedure seems only to have been used by Burkhardt et al [1975] and then only with $n = 2$. Presumably the limited number of terms used was due to the time required to emit the different modes. With an orthogonal set of emitted signals, however, all the modes shown in Equation (9.2) could be emitted simultaneously.

An alternative procedure for improving lateral resolution, which is too well-known to need description here, is that of spatial deconvolution. In this procedure, also, beams are emitted over a range of angles to acquire the echoes needed for the procedure. Here again, simultaneous emission of the many beams required will save time.

Another possible application of multimode operation is to use a phased array so as to focus a beam simultaneously at several points along its length. In the already existing technique of dynamic focusing, a pulse is transmitted along an unfocused or weakly focused beam, and during reception the delays imposed on the echo are changed with respect to time so as to optimize the focus of the reception directivity $D_r(r)$ for all ranges. Assuming that the transmitted pulse train was unfocused so that $D_e(r) \simeq 1$, this procedure gives an overall directivity of

$$D_{er}(r) = 1 \times D_r(r)$$

Using an orthogonal signal set, with each signal focused with directivity $D_e(r)$ at a different range along a single line, the overall directivity function would approach

$$D_{er}(r) = D_e^2(r)$$

For a Gaussian beam for which

$$D_e(r) = \text{const} \exp\left[-\frac{z^2}{\sigma^2}\right]$$

where z is taken at right angles to the beam and σ is the effective focal point radius, the improvement in the effective radius in going from $D_e(r)$ to $D_e^2(r)$ is only $1/\sqrt{2}$; however, the improvement in the side lobe level, which gives a corresponding improvement in dynamic range, may be important.

Experimental Results

An experiment using three-dimensional transducers and two-dimensional noise sources, which was used by Lee and Furgason to test the feasibility of the simultaneous mode transmission scheme, is shown in Figure 9.5.

A linear array composed of two outer 1/2-in.-disk transducers and one central 1/4-in.-disk transducer was used. The central transducer had a more divergent beam than the beams of the two outer transducers and received echoes from both of the targets: two brass rods that were 0.156 and 0.125 in. in diameter, respectively, and whose axes were oriented normal to the plane of the array (the plane of the page in Figure 9.5). The two brass targets were located so that sound from only one of the outer transducers was incident upon each target. This setup avoided confusion by making it possible to associate each mode with only one of the targets.

Two simultaneous modes were produced by transmitting noise signals from two noise sources A and B through the outside elements of the array as shown, and then receiving both these signals with the central element of the array. The two modes were then isolated, upon reception, by using

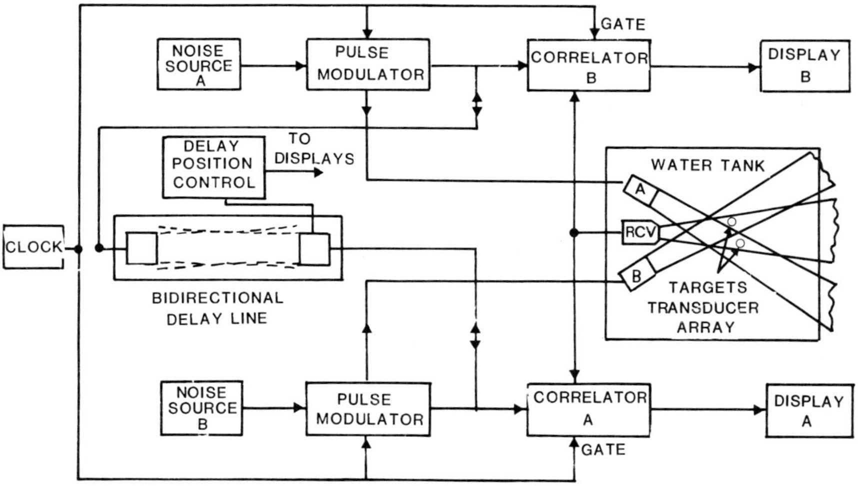

FIGURE 9.5. Simultaneous mode transmission experiment [Lee and Furgason, 1980].

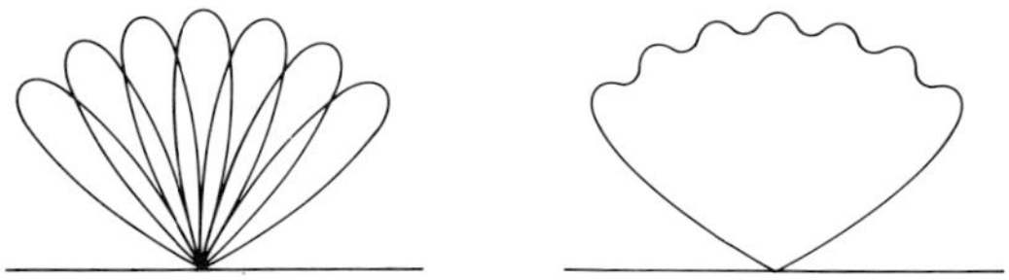

FIGURE 9.6. (**a**) Directivity functions of a set of modes. (**b**) Net directivity function produced by simultaneous emission of the modes shown in Figure 9.6a.

correlation receivers. In this scheme an analog bidirectional water delay line is used to produce identical delays for the reference signals of both modes at the same time. The reference signals of the two modes travel from opposite ends of the analog delay line, pass through each other within the water medium, and are received at the corresponding ends of the delay line. The correlators of both modes are gated off by an external synchronizing signal when the undelayed reference signals are present at the input of each correlator. This prevents a possible direct correlation of the undelayed reference signals with electronic crosstalk, which was present owing to the relatively large transmitted signals A and B.

The outputs of both correlators A and B were recorded simultaneously and showed no visible interference between the two modes.

The above results prove that incoherent noise signals can be simultaneously processed by a piezoelectric transducer array without interfering with one another.

Comparison with Conventional Systems

Although simultaneous multinoise mode systems have not yet been built, it is possible to make some estimates of how their performance would compare with multimode systems using conventional wave forms.

Use of Multiple Frequencies

Different array modes can be transmitted (and received) on different frequencies if enough bandwidth is available throughout the system. In sonar and radar the targets are often much larger than the wavelength. In that case the required fractional bandwidths $\Delta f/f_o$ for the signals used are much less than unity (Δf is taken to be the bandwidth of the signal and f_o its center frequency.) Consequently, the array bandwidths may then be much larger than the signal bandwidth, which should make simultaneous multimode operation using different frequencies possible. In ultrasound imaging, however, the targets are usually of the order of one wavelength or less and thus the fractional bandwidth of the signals is of order one. There is then not enough bandwidth available in the transducer array for multifrequency operation to be possible.

Simultaneous Radiofrequency (RF)

Another alternative to the simultaneous transmission of coded signals is to simultaneously transmit short trains of a single frequency. The feasibility of this alternative depends on the degree of spatial overlap of the different modes. A case that is particularly simple to analyze is that of an array of beams equally distributed over a sector, as shown in Figure 9.6a. In that case the ultrasound amplitude in the target region is given by 9.6b so that the emission directivity $D_e(r)$ may be said to approach unity in the sector of interest. Hence the overall directivity $D_{er}(r)$ [see Equation (9.1)] approaches $D_r(r)$. In other words, for simultaneous RF, $D_{er}(r) \simeq D_r(r)$. For simultaneous noise or sequential RF, $D_{er}(r) = D_e(r)D_r(r) \simeq D_r^2(r)$.

Thus when simultaneous RF is used instead of simultaneous noise, the overall directivity is degraded from $D_r^2(r)$ to $D_r(r)$, with a corresponding degradation of the main beam's lateral resolution and of the side lobes. Side-lobe degradation in turn degrades the dynamic range of the system.

Thus when high resolution, dynamic range, and high speed are required, the extra complexity of a simultaneous noise/code system should be justified.

Sequential RF Bursts

It has been pointed out that if we are presented with a phased array system in which N modes are used in succession, its operation can be speeded up by a factor N, by transmitting all the N modes simultaneously using N independent noise or code signals. Because of overlap of the various modes, transmitting them simultaneously worsens the instantaneous signal-to-noise ratio with which each mode is received at the input of its respective correlator. However, the ability to integrate over the relatively long noise or code signals used improves this signal-to-noise ratio, although it can never be as good as it is with each mode being emitted in succession. The effect of simultaneous mode operation on system signal-to-noise ratio is calculated below. It is found that the degradation in signal-to-noise ratio is negligible if the original system signal-to-noise ratio is dominated by thermal receiver noise. If this is not the case, it depends on the degree of overlap of the beams corresponding to the various modes.

Consider a system of the type shown in Figure 9.2 emitting N beams, each one of which intercepts a target that gives rise to a signal $E_i(t)$ at the input of correlator i. In addition, each correlator also receives a clutter signal whose instantaneous power $\tau C_i^2(t)$ is proportional to the length τ of the noise bursts used. $\sqrt{\tau}\,\overline{C}_i$, the time average amplitude of the clutter signal, is a function of the clutter targets viewed by beam i. Thus, if the signal power reaching correlator i is defined as $E_i^2(t)$, and the remainder of the power reaching the correlator is referred to as "noise," then the instantaneous signal-to-noise power ratio at the input of correlator i may be written as SNR (before correlation)

$$= \frac{E_i^2(t)}{n^2(t) + \tau C_i^2(t) + \sum_{\substack{j=1 \\ j \neq i}}^{N} \mu_{ij}^2 \left[E_j^2(t) + \tau C_j^2(t)\right]} \tag{9.3}$$

Here $n(t)$ is the instantaneous thermal noise voltage at the input of each correlator and μ_{ij} is the cross-coupling coefficient, ie, the amplitude at the input of correlator i associated with unit amplitude signal arriving at correlator j,

$$\mu_{ij} = \frac{\int D_{\text{eri}}(r)\, D_{\text{erj}}(r)\, dr^3}{\int D_{\text{eri}}^2(r)\, dr^3}$$

where the volume integrals are taken over all space.

After correlation with a reference signal of length τ and bandwidth B over m repetition periods, all the signals in the denominator are reduced by the factor $1/B\tau m$ because they are incoherent with this reference signal. Thus the signal-to-noise ratio after correlation becomes

$$\text{SNR}\big|_{\text{after correl}} = \text{SNR}\big|_{\text{before correl}} \times B\tau m \tag{9.4}$$

If mode i were transmitted by itself, using an RF signal of length $1/B$, the signal-to-noise ratio would have been

$$\text{SNR}\big|_{\text{single mode}} = \frac{E_i^2(t)}{n^2(t)\ ;\ + C_i^2(t)/B} \tag{9.5}$$

Thus the change in signal-to-noise ratio associated with going from successive RF modes to simultaneous noise of equal bandwidth and therefore equal longitudinal resolution is

$$\frac{\text{SNR}\big|_{\text{after correl.}}}{\text{SNR}\big|_{\text{single mode}}}$$

$$= \frac{n^2(t) + C_i^2(t)/B}{n^2(t) + \tau C_i^2(t) + \sum_{i \neq j}^{N} \mu_{ij}^2\left[E_j^2(t) + \tau C_j^2(t)\right]} \times B\tau m \tag{9.6}$$

If we ignore subscripts and average with respect to time

$$\frac{\text{SNR}\big|_{\text{after correl.}}}{\text{SNR}\big|_{\text{single mode}}} = \frac{(n^2\,\text{B} + C^2)\tau m}{n^2 + \tau C^2 + (N-1)\mu^2\left[E^2 + \tau C^2\right]} \tag{9.7}$$

Equation (9.7) shows that if thermal noise dominates, the signal-to-noise ratio of the simultaneous noise mode system is always better than that of the single mode RF system even for $m = 1$, owing to the suppression of the thermal noise by the correlation process. If thermal noise is negligible, then Equation (9.7) shows that the signal-to-clutter ratio is worsened by multiple operation for $m = 1$. If the signal-to-(noise plus clutter) ratio of the multimode system is to equal that of the single mode system, then the right hand side (r.h.s.) of Equation (9.7) must be set equal to one. If the multimode system is to be faster than the single-mode system using N modes sequentially, we must have

$$m < N$$

which gives

$$\mu^2 < \frac{N(n^2B\tau + C^2\tau) - n^2 - \tau C^2}{(N - 1)(E^2 + \tau C^2)}$$

assuming that $N >> 1$, $B\tau >> 1$ we obtain finally

$$\mu^2 < \frac{\tau(n^2B + C^2)}{E^2 + \tau C^2}$$

Equation (9.7) shows that for a multicode system transmitting independent noise signals, the signal-to-clutter ratio is optimized by keeping the cross-coupling coefficient μ between the various beams as small as possible.

Choice of Optimum Transmitted Signal

The experimental work on simultaneous multimode operation reviewed above, as well as the estimates of system performance, are for systems using noise as a transmitted signal. If pairs of Golay codes that have zero time side lobes could be transmitted instead of noise, the self-clutter terms $C_i^2(t)$ of Equation (9.3) would vanish, greatly improving the signal-to-noise ratio. Additional improvement could be obtained if the various Golay code pairs had zero cross-correlation with one another. In that case correlation of the echoes from one beam with its specific Golay code pair would suppress the cross-coupling echoes from all the other beams that carry different Golay codes. These cross-coupling echoes correspond to the terms under the summation sign in Equation (9.3).

Sarwate and Pursley [1980] showed that eight Golay code pairs exist that have zero cross-correlation with each other. However, these codes are only two bits long, which is too short for practical application. If these eight codes are lengthened, using the procedures described by Golay [1961], the zero cross-correlation property is no longer retained between all the pairs. Nonetheless, Lee and Furgason [1985] showed that a multimode system using these Golay codes is superior to systems using noise or m-sequence codes under conditions where the input signal-to-noise ratio is below a certain maximum that is greater than unity. For input signal to receiver noise ratios less than unity, the performance of the Golay code system and the m-sequence or random signal system is the same and is again better than that of the conventional pulse-echo system transmitting sequential signals of the same bandwidth. For the clutter-limited case it was determined that there is a maximum clutter-to-receiver noise ratio below which the simultaneous transmission system will perform better than sequential pulse-echo systems. Also, the Golay code system provides improved performance over a greater range of signal-to-noise ratios than does the m-sequence or random signal system, except in the case of integration over only one repetition period for which the performance of all three systems is again equivalent.

Lee and Furgason also investigated the performance of Golay codes for moving targets and found that as for m-sequence codes, their desirable properties deteriorate at high velocities. However, it was shown that "the Golay code system will still produce self-noise cancellation of greater than 20 dB for velocities up to 30% of the maximum velocity that a pulse-echo system can track." In general, these authors determined that there are many situations (defined more fully in the original reference) for which simultaneous transmission systems using Golay codes, m-sequence codes, or random signals will provide performance superior to that of conventional sequential pulse-echo systems.

References

Burckhardt CB, Grandchamp PA, Hoffman H: Focussing ultrasound over a large depth with an annular transducer—An alternative method IEEE Trans Sound Ultrason Su-22: 11–15 (1975).

Golay MJ: Complementary Series. Proc IRE Trans on Inf Theory IT7: 82–87 (1961). Lee BB, Furgason ES: Use of noise signals for multi-mode operation of phased arrays JASA 68: 320–328 (1980).

Lee BB, Furgason ES: Pseudo-random codes for single-mode and simultaneous multi-mode operation in ultrasonic imaging systems. Tech Rep TR-EE 85-10, School of Electrical Engineering, Purdue University, West Lafayette, Indiana (1985).

Newhouse VL, Furgason ES: The use of noise signals for multi-mode operation of phased arrays Proc 3rd Symp Ultrasonic Imaging and Tissue Characterization. N.B.S., Gaithersburg, Maryland (1978).

Sarwate DV, Pursley MB: Crosscorrelation properties of pseudorandom and related sequences. Proc IEEE 68: 593–610 (1980).

Tournois P: Acoustical imaging via coherent reception of spatially balanced transmissions. Proc 1980 Ultrason Symp IEEE (1980) pp 747–750.